Container Terminals and Cargo Systems

Kap Hwan Kim
Hans-Otto Günther
Editors

Container Terminals and Cargo Systems

Design, Operations Management, and Logistics Control Issues

With 138 Figures and 72 Tables

Springer

Professor Kap Hwan Kim
Pusan National University
Department of Industrial Engineering
Jangjeon-dong, Kumjeong-ku
Busan 609-735
Korea
kapkim@pusan.ac.kr

Professor Dr. Hans-Otto Günther
TU Berlin
Department of Production Management
Wilmersdorfer Straße 148
10585 Berlin
Germany
hans-otto.guenther@tu-berlin.de

Papers of this volume have been published in the journal OR Spectrum.

Library of Congress Control Number: 2006939791

ISBN 978-3-540-49549-9 Springer Berlin Heidelberg New York

Springer is part of Springer Science+Business Media

springer.com

Production: LE-TEX Jelonek, Schmidt & Vöckler GbR, Leipzig
Cover-design: WMX Design GmbH, Heidelberg

SPIN 11935964 42/3100YL - 5 4 3 2 1 0 Printed on acid-free paper

Table of Contents

Part 1:

Introduction

Kap Hwan Kim · Hans-Otto Günther

Container terminals and terminal operations

1 Container traffic

Over the recent years, the use of containers for intercontinental maritime transport has dramatically increased. Figure 1 exhibits the growth of world container turnover. Starting with 50 million TEU (twenty feet equivalent unit) in 1985 world container turnover has reached more than 350 million TEU in 2004. A further continuous increase is expected in the upcoming years, especially between Asia and Europe.

Since their introduction in the 1960s containers represent the standard unit-load concept for international freight. Transhipment of containers between different parties in a supply chain involves manufacturers producing goods for global use, freight forwarders, shipping lines, transfer facilities, and customers. Container terminals primarily serve as an interface between different modes of transportation, e.g. domestic rail or truck transportation and deep sea maritime transport. As globally acting industrial companies have considerably increased their production capacities in Asian countries, the container traffic between Asia and the rest of the world has steadily increased (cf. Wang (2005)). For instance, from 1990 to 1996 total container traffic volume between Europe and Asia doubled, whereas in the same period total container flow between Europe and the Americas went up by only 10%.

A few facts highlight the ever increasing importance of maritime container transportation (cf. Brinkmann (2005), Lee and Cullinane (2005), and Steenken et al. (2004)).

Kap Hwan Kim
Dept. of Industrial Engineering, Pusan National University, Jangjeon-dong, Kumjeong-ku, Busan 609-735, Korea, E-mail: kapkim@pusan.ac.kr

Hans-Otto Günther
Dept. of Production Management, Technical University Berlin, Wilmersdorfer Str. 148, D-10585 Berlin, Germany, E-mail: hans-otto.guenther@tu-berlin.de

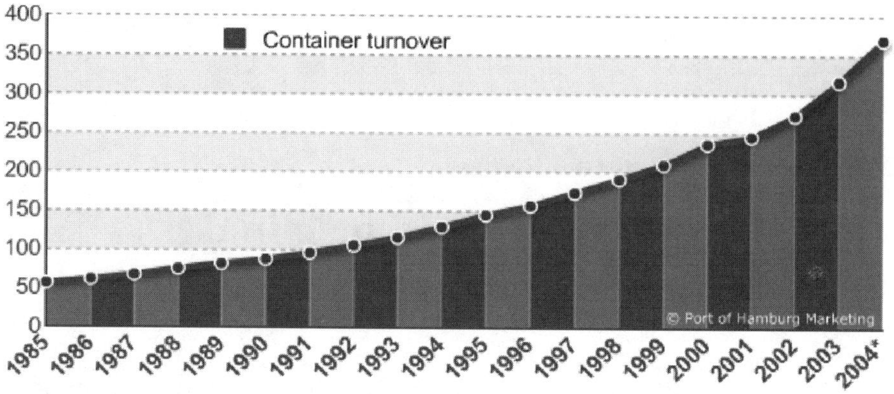

Fig. 1 Development of world container turnover (Unit: million TEU)
(Source: www.hafen-hamburg.de/en/index.php?option=com_content&task=view&id=96&Itemid=126; visited on June 2, 2006)

- Since regular sea container services began 1961 with routes between the East Coast of the United States and ports in Central and South America, the fraction of container transportation in the world's deep-sea cargo rose to more than 60%. Some major maritime freight routes are even containerized up to 100%.
- The transportation capacity of the worldwide container fleet has almost doubled during the past 10 years. At the same time, the transportation capacity of a single vessel rose steeply, culminating in the recent generation of 10,000 TEU container vessels.
- While the worldwide gross national product increased from 1990 to 2003 by about 50%, world container turnover tripled in the same period.
- In 1997 as much as 93.7% of the piece goods handled in the port of Hamburg were packaged in containers.

As a consequence, the number and capacity of seaport container terminals increased considerably, although investments for deep-sea terminals and the related infrastructure expansions almost reach one billion EURO, as it is reported from the latest deep-sea container terminal project at Wilhelmshafen, Germany. At the same time, there is an ongoing trend in the development of seaport container terminal configurations to use automated container handling and transportation technology, particularly, in countries with high labour costs. Hence, manually driven cranes are going to be replaced by automated ones and often automated guided vehicles (AGVs) are used instead of manually operated carts.

Driven by huge growth rates on major maritime container routes, competition between container ports has considerably increased. Not only handling capacities of container terminals worldwide got larger and larger. Moreover, significant gains in productivity were achieved through advanced terminal layouts, more efficient IT-support and improved logistics control software systems, as well as automated transportation and handling equipment. For instance, in the port of Singapore, container turnover per employee quintupled from 1987 to 2001.

In the scientific literature container terminal logistics have received increasing interest. Many papers have been published dealing with individual strategic, operational and control issues of seaport container terminals. Recent overviews can be found in Vis and de Koster (2003), Steenken et al. (2004), Murty et al. (2005), Kim (2005) as well as Günther and Kim (2005).

2 Container terminal operations

Although seaport container terminals considerably differ in size, function, and geometrical layout, they principally consist of the same sub-systems (see Figure 2). The ship operation or berthing area is equipped with quay cranes for the loading and unloading of vessels. Import as well as export containers are stocked in a yard which is divided into a number of blocks. Special stack areas are reserved for reefer containers, which need electrical supply for cooling, or to store hazardous goods. Separate areas are used for empty containers. Some terminals employ sheds for stuffing and stripping containers or for additional logistics services. The truck and train operation area links the terminal to outside transportation systems.

The chain of operations for export containers can be described as follows (see Figure 3). After arrival at the terminal by truck or train the container is identified and registered with its major data (e.g. content, destination, outbound vessel, shipping line), picked up by internal transportation equipment and distributed to one of the storage blocks in the yard. The respective storage location is given by row, bay, and tier within the block and is assigned in real time upon arrival of the container in the terminal. To store a container at the yard block, specific cranes or lifting vehicles are used. Finally, after arrival of the designated vessel, the container is unloaded from the yard block and transported to the berth where quay cranes load the container onto the vessel at a pre-defined stacking position. The operations necessary to handle an import container are performed in the reverse order.

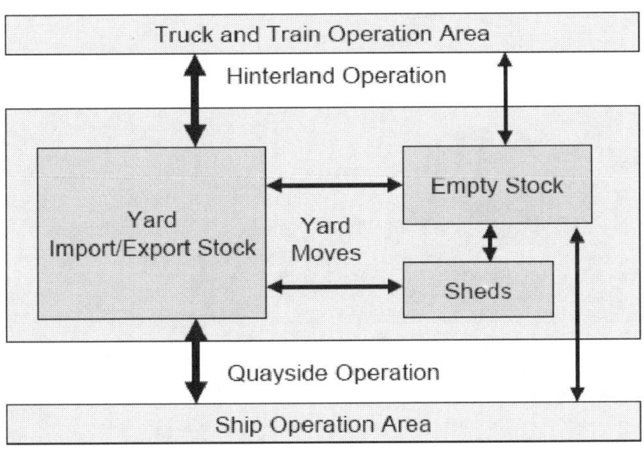

Fig. 2 Operation areas of a seaport container terminal and flow of transports (Source: Steenken et al. (2004), p. 6)

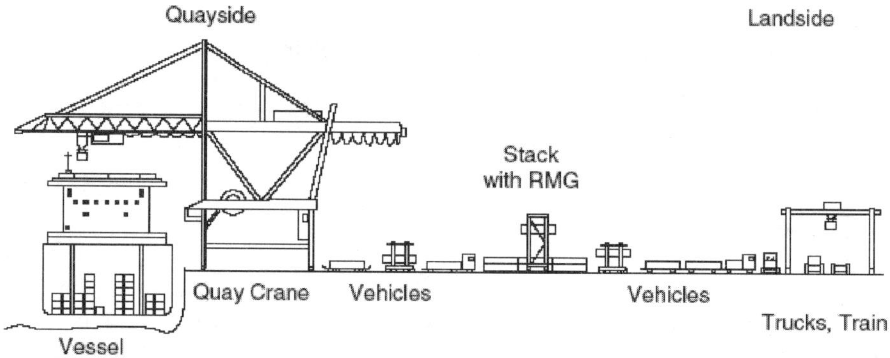

Fig. 3 Transportation and handling chain of a container
(Source: Steenken et al. (2004), p. 13)

Scheduling the huge number of concurrent operations with all the different types of transportation and handling equipment involved is an extremely complex task. In view of the ever changing terminal conditions and the limited predictability of future events and their timing, this control task has to be solved in real time.

Seaport container terminals greatly differ by the type of transportation and handling equipment used. Regarding quay cranes, single or dual-trolley cranes can be found. The latter employ an intermediate platform for buffering the loaded or unloaded container. The most common types of yard cranes are rail-mounted gantry (RMG) cranes, rubber-tired gantry (RTG) cranes, straddle carriers, reach stackers, and chassis-based transporters. Of these types of cranes only RMG cranes are suited for fully automated container handling. Figure 4 exhibits the working principle of the different types of handling equipment and their comparative performance figures with respect to the number of TEUs, which can be stored per hectare.

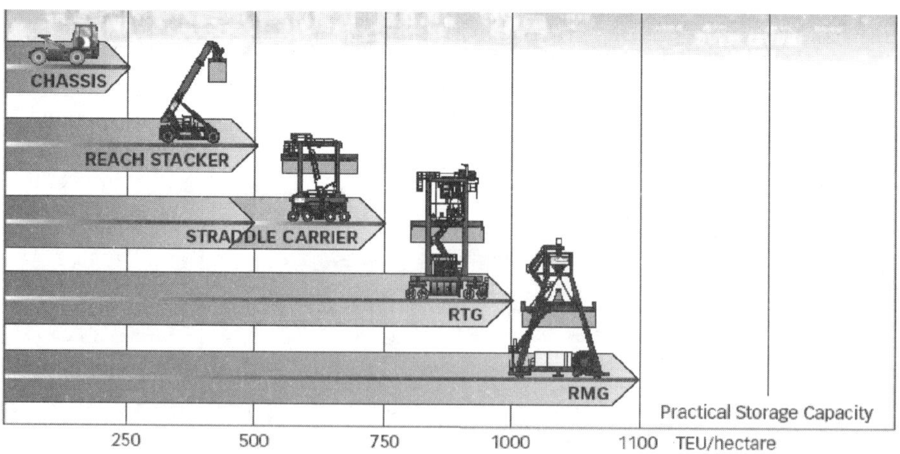

Fig. 4 Different types of handling equipment
(Source: www.kalmarind.com; visited on January 2, 2006)

Different types of vehicles can be used both for the ship-to-yard transportation and the interface between the yard and the hinterland. The most common types are multi-trailer systems (MTS) with manned trucks, automated guided vehicles (AGVs), and automated lifting vehicles (ALVs). The latter ones, in contrast to AGVs, are capable of lifting a container from the ground by themselves (cf. Vis and Harika, 2004; Yang et al., 2004). However, despite their superior handling capabilities ALVs have not yet gained widespread use in container terminals.

3 Planning and logistics control issues of container terminals

A container terminal represents a complex system with highly dynamic interactions between the various handling, transportation and storage units, and incomplete knowledge about future events. There are many decision problems related to logistics planning and control issues of seaport container terminals. These problems can be assigned to three different levels as shown in Figure 5: terminal design, operative planning, and real-time control. In the following a brief overview of these planning and control levels and their relationship to the various kinds of terminal equipment is given.

Terminal design problems have to be solved by facility planners in the initial planning stage of the terminal. These problems have to be analyzed both from an economic as well as a technical feasibility and performance point of view. In particular, construction of a completely new terminal site and the use of automated equipment require huge investments. From the various design problems, only the most important ones shall be highlighted. For a more detailed overview see Steenken et al. (2004).

- *Multi-modal interfaces*: In contrast to their Asian counterparts, most European container terminals are laid out as multi-modal facilities, i.e. they are directly linked to railway, truck and inland navigation systems. The integration of these different modes of transportation has a major impact on the design of the entire terminal.

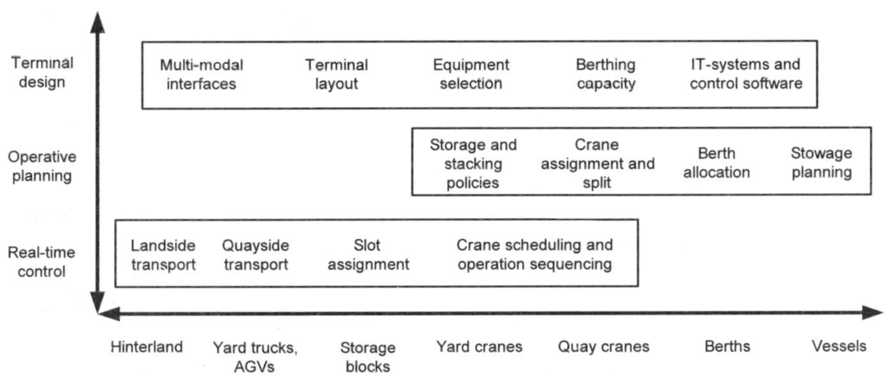

Fig. 5 Logistics planning and control issues in seaport container terminals

- *Terminal layout*: The storage yard, transportation guide paths, and quays represent the major entities of each container terminal. Their capacity and spatial arrangement heavily determine the performance of the terminal configuration. Terminal layout also includes the reservation of certain areas for reefer or hazardous goods containers, empty containers or non-standard-size containers.
- *Equipment selection*: Different types of equipment can be used for handling and transportation within the terminal. They primarily differ by their degree of automation and their performance figures. Currently, there is an ongoing trend to make increased use of automated storage cranes und driverless vehicles, although these types of equipment raise complex logistics control problems.
- *Berthing capacity*: The global performance factor of a container terminal is given by its seaside dispatching capacity. The berthing capacity not only determines the number and size of the vessels that can be served, but also the requirements for storage yard space and the fleet size of vehicles etc.
- *IT-systems and control software*: Finally, logistics control in large-sized container terminals is a tremendously complex task, which requires real-time decisions on matching handling tasks with the corresponding equipment units and the provision of detailed information about each individual container. Different modes of software and IT support as well as use of sophisticated optimization tools are issues of considerable importance.

The level of *operative planning* (cf. Steenken et al. (2004)) comprises guidelines and basic planning procedures for performing the various logistic processes at the terminal. Since decentralized planning is the only realistic mode to govern logistics control of automated container terminals, the entire logistics control system is subdivided into various modules for the different types or groups of resources. Hence, specific issues arise in planning and scheduling the use of key resources for a short-term planning horizon of several days or weeks.

- *Berth allocation:* Before arrival of a ship, the required berthing space has to be allocated taking the prospective time the ship spends in the terminal into account. Additional constraints arise from the availability of cranes and the berthing and crane requirements of other vessels which already moor at the quay or are expected to arrive shortly.
- *Crane assignment and split:* To load and unload a large container vessel, several quay cranes are used. First it has to be decided which individual cranes are to be assigned to the various ships considering the accessibility of cranes at the berth and the impossibility to exchange cranes between different berths at the terminal. Second the cranes operating at one ship have to be assigned to different sections or hatches of the ship.
- *Stowage planning and sequencing:* Shipping lines have to decide which positions within the ship are assigned to specific categories of containers considering container attributes such as destination, weight or type of the container. Based on this given assignment, the terminal operator decides which

individual container has to be stored at the specific slots within the vessel. This final slot-assignment heavily affects the loading and unloading sequence of containers. Based on the stowage plan, planners in container terminals determine the sequence of unloading inbound containers and of loading outbound containers. For the outbound containers, in addition to the loading sequence for individual containers, the slot in the vessel into which each outbound container will be stacked must be determined at the same time. The unloading and loading sequences represent a major input for determining the yard crane's and vehicle's schedules.

- *Storage and stacking policies:* Large container terminals in Europe store a total of several 10,000 containers with average dwell times of 3-5 days and daily turnover of 10-20,000 containers. The storage area is separated into blocks, which are organized into bays, rows and tiers. Policies for assigning individual storage locations and stacking of containers are ruled by the objective to expedite the necessary storage and retrieval operations as far as possible and to avoid reshuffling of containers within the block. Specific issues include the reservation of dedicated storage areas for import and export containers and the planning of remarshalling operations for stacked containers.
- *Workforce scheduling:* Workforce is another important resource in container terminals. Rosters and schedules for workers to operate equipment must be generated in advance.

Container terminals represent highly dynamic and highly stochastic logistics systems, which do not allow pre-planning of detailed transportation and handling activities for a look-ahead horizon of more than 5-10 minutes. Hence, *real-time control* of logistics activities is of utmost importance. Real-time control (or real-time planning) is usually triggered by certain events or conditions and requires that the underlying decision problem is solved within a very short time span, in practice usually within less than a second. Real-time decisions include the assignment of transportation orders to vehicles and routing and scheduling the vehicle trips for landside transportation as well as for transportation between the berth and the storage yard, the assignment of storage slots to individual containers, and the determination of detailed schedules and operation sequences for quay and stacking cranes.

4 Overview of the book

Apart from this introductory section, this book is divided into two further Parts 2 and 3. The subsequent Part 2 focuses on seaport container terminals while the final Part 3 considers other types of cargo systems, e.g. vehicle distribution, air and maritime cargo systems as well as issues of revenue management and collaboration between forwarding enterprises.

Part 2 comprises eleven papers on seaport container terminals. Due to the complexity of automated container terminals, highly sophisticated control strategies are needed for the operation and control of the equipment. In addition, the design and the performance analysis of terminal configurations are issues of major practical importance.

The first paper by *J.A. Ottjes, H.P.M. Veeke, M.B. Duinkerken, J.C. Rijsenbrij* and *G. Lodewijks* presents a generic simulation model structure for the design and evaluation of multi-terminal systems. The authors apply their modelling approach to the existing and the future terminals in the Rotterdam port area. Experimental results show the requirements for deep-sea quay lengths, storage capacities, and equipment for inter-terminal transport.

A simulation study to compare three different transportation systems for the overland transport of containers between container terminals is presented in the paper by *M.B. Duinkerken, R. Dekker, S.T.G.L. Kurstjens, J.A. Ottjes* and *N.P. Dellaert*. The simulation model is applied to a realistic scenario taken from the Rotterdam port area. The numerical results give insight into the different characteristics of the transport systems and their interaction with the handling equipment.

In the subsequent paper, *R. Moorthy* and *C.-P. Teo* analyse the home berth problem, i.e. the preferred berthing location for a set of vessels scheduled to call at the container terminal on a weekly basis. They model this problem as a rectangular packing problem on a cylinder and use a sequence pair based simulated annealing algorithm to solve the problem. Extensive computational studies show the efficiency of the proposed modelling approach.

In their paper, *E. Kozan* and *P. Preston* model the seaport terminal system with the objective of determining the optimal storage strategy and container-handling schedule. They present an iterative search algorithm that integrates a container-transfer with a container location model in a cyclic fashion to determine both optimal locations and corresponding handling schedules. Results are analysed and compared with current practise at an Australian port.

A mixed-integer linear programming model for storage yard management in transhipment hubs is presented by *L.H. Lee, E.-P. Chew, K.C. Tan* and *Y. Han*. To solve large-sized problem instances, two heuristic solution procedures are developed. The first is a sequential method while the second is based on column generation. Finally, it is shown that the heuristics find near-optimal solutions in a reasonable amount of time.

Stacking policies for containers at an automated container terminal are addressed by *R. Dekker, P. Voogd* and *E. van Asperen*. They provide a comprehensive overview of stacking policies used in practise. Specifically, they consider several variants of category stacking, where containers can be exchanged during the loading process. In a numerical study, different stacking policies are compared.

The next paper by *E.K. Bish, F.Y. Chen, Y.T. Leong, B.L. Nelson, J.W.C. Ng* and *D. Simchi-Levi* analyses discharging and uploading operations of containers to and from ships. Specifically, the authors address the dispatching of vehicles to containers so as to minimize the service time (makespan) of a ship. To solve this problem they develop heuristic dispatching algorithms that generate optimal or near-optimal solutions.

In the paper by *M. Grunow, H.-O. Günther* and *M. Lehmann* strategies for dispatching Automated Guided Vehicles (AGVs) at automated seaport container terminals are analysed and evaluated using a scalable simulation model. The authors develop a so-called pattern-based heuristic which utilizes the dual-load capability of AGVs. Results of the simulation study reveal that this heuristic outperforms conventional dispatching heuristics known from flexible manufacturing systems.

Another type of dispatching strategies for AGVs is proposed by *D. Briskorn, A. Drexl* and *S. Hartmann*. They present an alternative formulation of the job-vehicle assignment problem which is based on a rough analogy to inventory management. In a simulation study, it is shown that the inventory-based model leads to better productivity of the terminal than the due-time-based model formulation.

In automated container terminals, situations occur where different equipment units directly or indirectly request each other to start a specific process. Hence, all of the affected resources are involved in a deadlock. *M. Lehmann, M. Grunow* and *H.-O. Günther* develop different methods for the detection and resolution of deadlocks occurring in the resource-assignment phase. The suitability of these methods is shown in a comprehensive simulation study.

Another type of deadlocks arising in the traffic control of AGVs in seaport container terminals is investigated by *K.-H. Kim, S.M. Jeon* and *K.R. Ryu*. They develop an efficient deadlock prediction and prevention algorithm Their approach guarantees deadlock-free reservation schedules of grid blocks in the guide path for AGVs to cross the same area at the same time. The proposed method was tested in a simulation study.

Part 3 comprises six papers on different types of cargo systems. The first paper by *D.C. Mattfeld* and *H. Orth* addresses the planning of transportation and storage capacity over time. They consider intermodal transhipment terminals used for the import and export of large volumes of automotives and develop an evolutionary algorithm for determining a period-oriented capacity utilization strategy. As a result, a balanced distribution of vehicle movements over the periods of the planning horizon is achieved.

The next paper by *H. C. Huang, C. Lee* and *Z. Xu* considers a large air cargo handling facility composed of two identical cargo terminals. To balance the workload between the two terminals, a stochastic mixed-integer linear programming model is developed and efficiently solved. The simulation results based on data from a large international air port show that the proposed algorithms effectively balance the workload and the cargo service time is considerably reduced.

In their paper, *D. Li, H.-C. Huang, A.D. Morton* and *E.-P. Chew* develop an integrated model for incorporating the cargo routing problem into fleet assignment. Their solution approach is based on Benders decomposition and simultaneously determines the optimal assignment of fleets (types of airplanes) to flight-legs and the routing of cargo over the network within reasonable amount of computational time.

A mathematical programming based approach for revenue management in cargo airlines is the topic of the paper by *P. Bartodziej, U. Derigs* and *M. Zils*. Their approach deals with making capacity reservations for expected cargo demand over a certain period of time, e.g. a year, so as to maximize the expected profit contribution. As the number of booking request per week for a major cargo airline is extremely large, an issue of considerable practical importance is to answer customer enquiries in near real-time.

The paper by *L.H. Lee, E.-P. Chew* and *M.S. Sin* also deals with issues of revenue management. They show that, in a sea cargo application, the optimal policy to allocate the capacity of a ship is a threshold policy, i.e. to base decisions on the acceptance of customer orders on the remaining capacity of the ship. An efficient heuristic procedure is proposed to solve this problem.

The final paper by *M.A. Krajewska* and *H. Kopfer* presents a model for collaboration among independent freight forwarding enterprises. Their model is based on theoretical foundations of combinatorial auctions and game theory. They show that through their collaboration model additional profit for a coalition of freight forwarders and for each participant can be gained. Therefore, the proposed model provides a useful basis for developing application-specific profit sharing mechanisms in the freight forwarding business.

5 Final remarks

The primary objective of this book is to reflect recent developments in design, operations management and logistics control of automated container terminals and cargo systems and to examine related research issues of quantitative analysis and decision support. It comprises reports on the state of the art, applications of quantitative methods, as well as case studies and simulation results. Seventeen papers previously published in "*OR Spectrum – Quantitative Approaches in Management*" have been selected for publication in this volume. All papers have been peer-reviewed according to the standards of the journal.

This book has greatly benefited from the cooperation among the authors, reviewers, and editors. We would like to express our sincere thanks to the reviewers for their excellent and timely refereeing. Last, but not least, we thank all the authors for their contributions which made this book possible.

References

Brinkmann A (2005) Seehäfen: Planung und Entwurf, Springer, Berlin, Heidelberg (in German)

Günther HO, Kim KH (2005) Container Terminals and Automated Transport Systems. Springer, Berlin, Heidelberg

Kim KH (2005) Models and methods for operations in port container terminals. In: Langevin A, Riopel D (eds) Logistics Systems: Design and Optimization, pp 213-243. Springer, New York

Lee TW, Cullinane K (2005) World Shipping and Port Development. Palgrave, Basingstoke

Murty KG, Liu J, Wan YW, Linn R (2005) A decision support system for operations in a container terminal. Decision Support Systems 39: 309-332

Steenken D, Voß S, Stahlbock R (2004) Container terminal operation and operations research – a classification and literature review. OR Spectrum 26: 3-49

Vis IFA, Harika I (2004) Comparison of vehicle types at an automated container terminal. OR Spectrum 26: 117-143

Vis IFA, de Koster R (2003) Transhipment of containers at a container terminal: An overview. European Journal of Operational Research 147: 1-16

Wang M (2005) The rise of container transport in Asia. In: Lee TW, Cullinane K (eds) World Shipping and Port Development, pp 10-35. Palgrave, Basingstoke

Yang CH, Choi YS, Ha TY (2004) Simulation-based performance evaluation of transport vehicles at automated container terminals. OR Spectrum 26: 149-170

Part 2:

Container terminals

Jaap A. Ottjes · Hans P. M. Veeke · Mark B. Duinkerken · Joan C. Rijsenbrij · Gabriel Lodewijks

Simulation of a multiterminal system for container handling

Abstract A generic simulation model structure for the design and evaluation of multiterminal systems for container handling is proposed. A model is constructed by combining three basic functions: transport, transfer, and stacking. It can be used for further detailing of the subsystems in the terminal complex while preserving the container flow patterns in the system. The modeling approach has been applied to the complete set of existing and future terminals in the Rotterdam port area, using forecasts of containers flows, statistical data from existing terminals, expert opinions, and conceptual designs of the new port area called “second Maasvlakte”. Experimental results including the requirements for deep-sea quay lengths, storage capacities, and equipment for interterminal transport are shown. Further traffic flows on the terminal infrastructure are determined, and the consequences of applying security scanning of containers are evaluated.

Keywords Container terminal · Simulation · Process interaction method · Strategic · Conceptual design

1 Introduction

Container terminals play an important role as a node in many supply chains. A container terminal is an area for container transshipment between various transport modalities. The main modalities are deep-sea, short-sea, inland waterway, road, and rail. Container flows worldwide are growing very rapidly and it is expected that this growth will continue during the next decades. A new generation of deep-sea container vessels, with a capacity of 8,000–10,000 “20-ft container equivalent

J. A. Ottjes (✉) · H. P. M. Veeke · M. B. Duinkerken · J. C. Rijsenbrij · G. Lodewijks
Faculty of Mechanical, Maritime and Materials Engineering, Delft University of Technology, Mekelweg 2, 2628 CD Delft, The Netherlands
E-mail: j.a.ottjes@tudelft.nl, h.p.m.veeke@tudelft.nl, m.b.duinkerken@tudelft.nl, j.c.rijsenbrij@tudelft.nl, g.lodewijks@tudelft.nl

units" (TEU), is coming. Even larger vessels are under development. These developments urge container main ports to reconsider their equipment and logistics or even to expand.

Recently, the Dutch government decided to extend the Rotterdam port area with the so-called "second Maasvlakte" (MV2) to be reclaimed from the North Sea. This area will be mainly used for container handling and it is anticipated that a number of container terminals will be established on it. We will call this a "multiterminal." The question to be answered is how to arrange these terminals. One theoretical option is to have a number of autonomous deep-sea terminals, owned by different parties, each facilitating all necessary modalities. These are called "compact" terminals. Another possibility is to aim at functional specialization per modality. This would imply separate terminals for deep-sea, rail, barge, and truck handling. These are called "dedicated" terminals. Between these two extremes, there are numerous mixed multiterminal configurations possible. Each multiterminal will need transportation facilities between the individual terminals, the so-called interterminal transport (ITT).

A research project has been carried out with the main objective to evaluate conceptual multiterminal designs for the second Maasvlakte, including interterminal transport systems, in coherence with the existing terminals on the first Maasvlakte (MV1). It is assumed that both intra- and interterminal transport will take place with automated guided vehicles (AGV). An AGV autonomously drives to its destination, but needs an external device for loading and unloading a container.

This paper covers the first part of the research project, concerning a strategic logistic simulation study of the complete set of existing and planned future terminals in the Rotterdam Maasvlakte area to support the design activities for MV2. The research goal of this study is to determine, for a number of conceptual multiterminal designs, the requirements with respect to the number of AGVs, the capacity of the interterminal transport infrastructure, the sea berth length, the stacking capacity, and the influence of safety measures. It was decided to use simulation to be able to deal with stochastic effects already in the early design stage. In a later phase of the design process, detailed studies on subsystems like individual terminals and the interterminal transport of containers are anticipated. Therefore, the simulation model should be developed in such a way that it is reusable as a common basis for the parallel development of detailed submodels and can be used as a consistent input framework for the submodels.

This paper is organized as follows: In "Container terminals," we analyze the problem under investigation and formulate the research demands. "Container terminal modeling" presents the model framework, and "Model construction" discusses the model construction. "Modeling the Maasvlakte terminals" describes the application of the model and a selection of the results. "Results" contains conclusions and future research topics.

2 Container terminals

In this section, some typical container-terminal related issues will be analyzed, and the research question will be formulated.

2.1 Push and pull

In main port container terminals, all activities are concentrated on serving deep-sea vessels. These vessels are unloaded and loaded with high priority. The unloaded containers (the "import" containers) are distributed over the hinterland, and the containers to be loaded (the "export" containers) are collected from the hinterland. Consequently, the unloading and loading process sets the pace for all other logistic activities on the terminal. Logistically speaking, a vessel "pushes" its import load onto the terminal and it "pulls" its export load from the terminal. This push–pull mechanism will be the basis for the model. We call the deep-sea container flow the "originating" flow. All other flows to and from the terminal are related to this originating flow and will be called "derived" flows. A distinction is made between short-sea feeder services and autonomous short-sea shipping. Containers from feeders are counted to be derived, and containers of autonomous short-sea ships are considered originating.

The handling capacity of a main port container terminal primarily depends on the number and capacity of deep-sea quay cranes available. In general, the quay length should never be a restriction to the timely serving of deep-sea vessels. The terminal storage (stacking) capacity, the interterminal transport capacity, and landside handling capacities are to be derived from the deep-sea capacity and should not delay the deep-sea handling process.

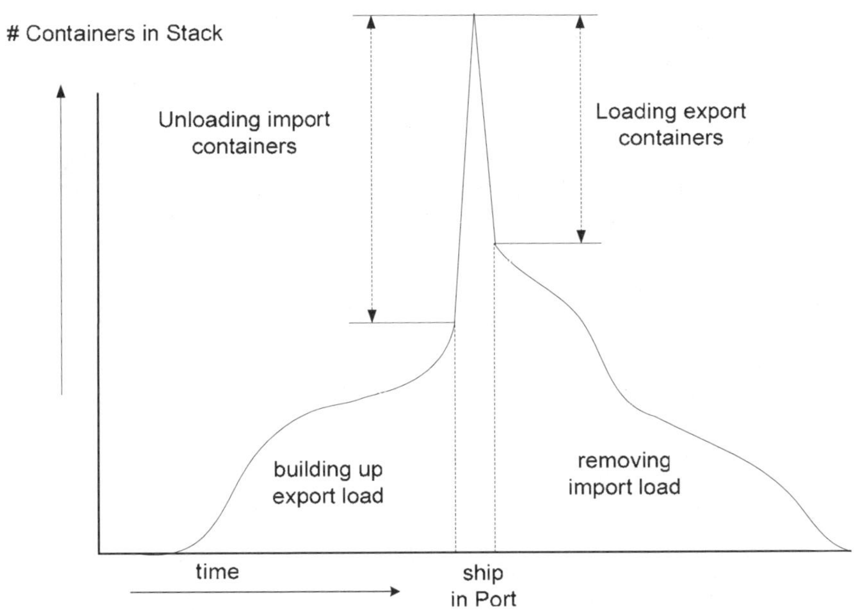

Fig. 1 The variation of the stack contents as a result of a visit of one deep-sea ship

2.2 Dwell times

Both import and export containers are usually stored temporarily in a "marine stack" close to the deep-sea quay area. On the landside, for similar reasons, smaller decoupling stacks are used, for example, in rail service centers (RSC) and barge service centers (BSC). Consequently, a container may reside in one or more successive stacks during its stay on a container terminal.

We define "dwell time" as the total time a container spends in one or more terminal stacks. Several factors may influence container dwell times, such as time tables and availability of hinterland connections, the influence of custom regulations, and typical supply-chain related influences, such as the time the container owner decides to fetch his imported containers or to supply his containers for export. In the current practice, dwell times are in the order of days. The average stacking space needed is linearly proportional to the average container dwell time [12]. The stack capacity needed, however, can be temporarily much higher than the average value. One of the main causes is the incidental overlap of deep-sea ships that are handled (so-called "clashing"). One ship gives a peak in stack contents as shown in Fig. 1.

In traditional design practice, the stack capacity needed is determined by calculating the average capacity and multiplying that with a peak factor. Rule of thumb values of the peak factor used vary from 1.2 to 1.3. The influence of the new generation of container vessels and intensive interterminal transport on peak factors is not yet known.

2.3 Modal split and interterminal transport

Import containers are distributed over the hinterland via the landside modalities or even back to deep-sea vessels. Export containers are collected from hinterland locations and arrive with various transport modalities. This phenomenon is called "modal split." Modal split demands imply that containers may have to be transported between modalities and stacks and even between different stacks during their stay on the multiterminal complex. This results in interterminal transport to be carried out by AGVs. For design purposes, the peak number of AGVs in use and the resulting traffic flows are particularly essential.

2.4 Security activities

In the Rotterdam container terminals, the so-called "container scan" procedure has been introduced. Each container terminal should be prepared to send a random part of its container flow via an X-ray scanner that is able to detect illegal or dangerous cargo. This will result in extra handling and transport activities and, consequently, more AGVs will be required.

2.5 Research question

The research question is defined as follows:

Determine for a number of conceptual multiterminal designs the averages and peak values of:

- The occupation of the deep-sea quays
- The number of AGVs for interterminal transport
- The storage capacity of the stacks
- The traffic flows on the interterminal infrastructure
- The influence of security-related extra handling and transport of a part of the total container flow

The model structure should be suitable to represent any multiterminal configuration and allow further integrated detailed studies. It should be reusable as common basis for the parallel development of detailed submodels and as a consistent input framework for the submodels.

3 Container terminal modeling

After reviewing related literature, we propose a general concept for modeling multiterminal systems and discuss the modeling technique used for constructing the model.

3.1 Literature review

Many studies concerning container terminals focus on specific parts of the terminal such as yard operations [9], berth scheduling [11, 15], or dispatching of AGVs [10]. Other works relate to the evaluation of container terminals at an aggregate level, using simulation [13, 15, 24]. In [16], some probable consequences of the introduction of large container ships on the landside infrastructure are discussed. Extensive literature reviews on container terminal characteristics and classification of container terminal research are given in [17] and [18]. Subjects discussed include terminal logistics, storing and stacking logistics and optimization methods, transport optimization, and simulation systems. The conclusion is that, until now, the focus has been on optimizing several separate parts of the logistic chain of a container terminal and that there is a need for integrated optimization [18]. Also an approach is proposed to apply simulation at various levels of detail in the entire process of terminal design [17]. The first step here is the functional design in which the required number of quay cranes, quay length, and stacking capacity is to be determined. The advantage of using simulation already in the strategic stage is the fact that stochastic aspects can be taken into account. In [19], the process interaction modeling approach has been used to support the design process at all stages. This approach allows the expansion of the model from the initial functional level to a detailed operational level. Though relevant literature can be found regarding design and control of container terminals, no work was found about the integrated design and evaluation of a complex of interacting terminals.

3.2 Modeling concept

A single container is taken as the load unit and the individual means for transport, transshipment, and stacking are taken as equipment units. We abstract from the physical view to a functional view of handling containers [8]. Each type of container handling can be modeled—both aggregated and in detail—by a composition of three elementary functions:

1. *A transport function* for transporting containers
2. *A transfer function* for transshipping containers
3. *A stacking function* for storing containers

The relationships between the three elementary functions are shown in Fig. 2.

For each container, at least one transfer function must be executed. A transfer function can be coupled to both transport and stack functions. For example, a set of quay cranes (transfer function) can be coupled to an AGV system (transport function); a set of stacking cranes (transfer function) can be coupled to a stack area (stack function). The loads of ships, trains, trucks that supply or collect containers are represented by a stack function that is temporarily coupled with a transfer function.

Any container terminal can be represented by a composition of connected elementary functions. In Fig. 3, a compact terminal is shown, modeled with the elementary functions. On the sea quayside, a transfer function resembles the quay cranes that load and unload sea ships, while on the landside trains, trucks, barges, and interterminal transport are handled. One central stack serves all modalities. In terms of production logistics there is a resemblance to dedicated cell production [23]. The theoretical benefits are a reduction of order lead time, less work in progress, lower material handling costs, and simplified planning and control procedures.

Figure 4 shows a dedicated deep-sea terminal and a dedicated rail terminal or rail service center (RSC) connected by the interterminal transport function. Each

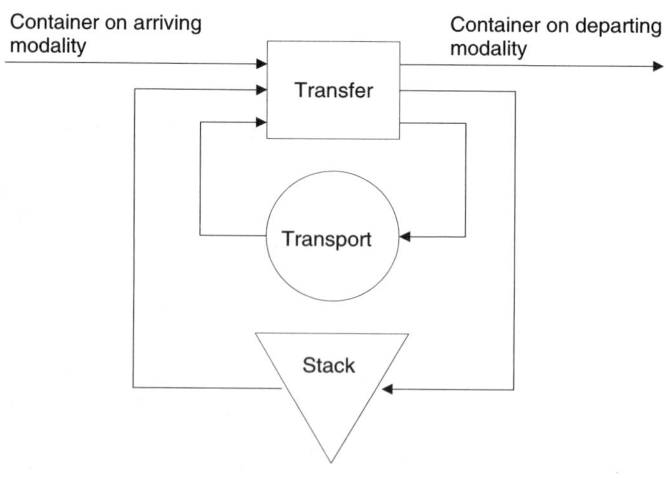

Fig. 2 Relationships between elementary functions

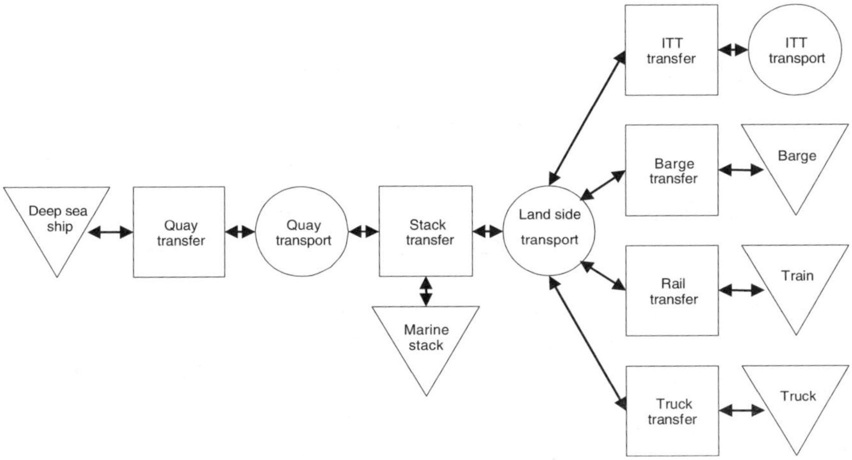

Fig. 3 A fully equipped terminal called "compact" terminal

terminal has its own operational structure. Both the deep-sea terminal and the RSC need a stack and a transfer function to (un)load ITT AGVs. Advantages are flexibility and efficient use of equipment and a higher container throughput per hectare compared to the compact configuration. A disadvantage will be increased interterminal transport. In terms of production logistics, this configuration is related to "job shop" production.

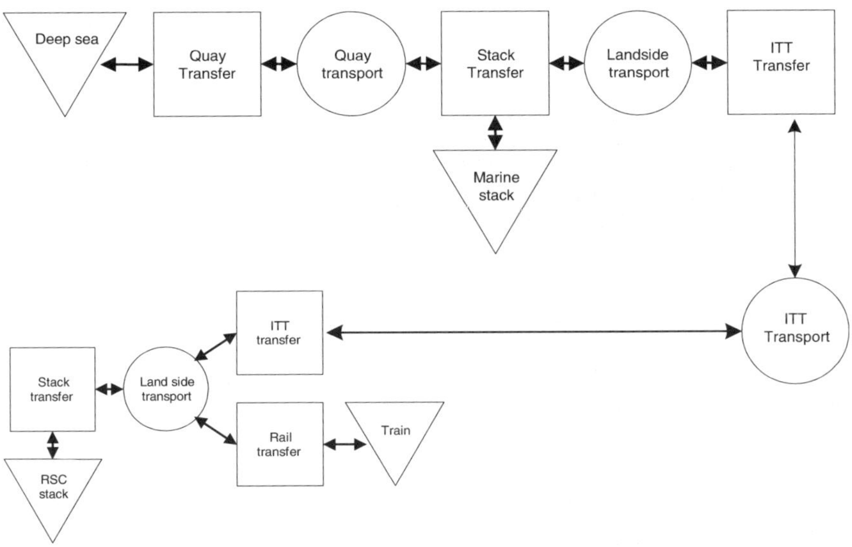

Fig. 4 A "dedicated" deep-sea terminal and a "dedicated" rail terminal (RSC) interconnected via interterminal transport

3.3 Modeling technique

In this work, the "process interaction method" is used, which is a combination of event scheduling and activity scanning [4, 25]. It consists of identifying the system elements and describing the sequence of actions for each one. The process interaction method can be summarized as follows: (1) decompose the system into relevant element classes, preferably patterned on its real world structure. An element class is characterized by its attributes. The state of each instance of a class is defined by the state or value of its attributes. (2) Determine the "living" element classes and assign a process description to these classes, making use of simulation-time consuming commands like "work, wait, drive, suspend" and "standby." A process governs the behavior of each instance of the element class. The simulation package applied has to take care of the proper sequencing of all scheduled time periods.

The process interaction approach allows defining processes initially at just the functional level and refining these processes toward both tactical and operational levels, using the same model framework [19]. Simulating at the functional level has the advantage that stochastic influences can be taken into account to obtain statistics about traffic flows, stacking volumes, and equipment capacities needed for initial design purposes. At the tactical and operational levels, optimization and control algorithms can be implemented and tested in the integral simulation environment using a consistent input. Here, choices about what type of equipment to use have to be made. Interterminal transport can be executed in many ways, for example, using straddle carriers, multitrailers, single or multiloaded automated guided vehicles and automated lifting vehicles [6, 14].

Process interaction modeling has a near-resemblance to object-oriented modeling, in which procedures and functions called "methods," besides attributes, are owned by object classes. In this sense, the process description of an element class is a method of the class. The first language applying process interaction is "Simula" [1]; two recent tools are Silk [7] and Tomas [19, 20].

4 Model construction

In this section, we will explain the construction of the model in terms of the process interaction approach. After describing the elementary functions, the modeling of the deep-sea ships with containers will be discussed.

4.1 Modeling the elementary functions

A transport function is represented by a "transport system" owning a set of "transporters." A transfer function is represented by a "transfer system" owning a set of "transfer units." The element class definitions with the main attributes and the processes are described in pseudo-code. The element classes and attributes are:

TransportSystem	
–MyTransporters	Set with all transporters
–AvailableTransporters	Set with available transporters
–MyTransferSystems	Set of Linked Transfer Systems
–ContainersToTransport	Set with containers to be transported
–TRANSPORTER_ AVAILABLE	Method: allocates or creates transporter
–PROCESS	Method: describes activities as a function of time
Transporter	
–MyTransportSystem	Referenceto Transport system
–ContainerToTransport	Assigned container to transport
–Destination	Final Destination of a transport job
–Route	Set of route points to destination Arrival time at next route point
–ArrivalTime	Arrival Time at next route point
–PROCESS	Method: describes activities as a function of time
TransferSystem	
–MyTransferUnits	Set with all transfer units
–AvailableTransferUnits	Set with available transferunits
–MyTransportSystems	Set with linked transport systems
–MyStacks	Set with linked stacks
–TRANSFERUNIT_ AVAILABLE	Method: allocates or creates Transfer unit
–PROCESS	Method: describes activities as a function of time
TransferUnit	
–MyTransferSystem	Reference to Transfer System
–ContainerToTransfer	Assigned container to be transferred
–PROCESS	Method: describes activities as a function of time

The stack function is represented by a "stack system" with a set of "stack units."

A stack system is always connected to at least one transfer system. The main task of the stack system is to keep containers. In the model, the stack function is further used to monitor and keep guard over the containers it has in store. As soon as a container has reached the end of its dwell time, a transfer request for that container is issued via the stack system to the proper transfer system, taking into account the transfer capacity of the receiving transfer system. Both transport and transfer systems have their own set of equipment units to carry out their tasks. In the MV2 case, the transport units or transporters are automated guided vehicles and transfer units are quay cranes or stacking cranes. There are two possibilities. (1) The number of equipment units of an elementary function is "restricted." Consequently, it may happen that a unit is needed and that all units are in use. In that case, containers have to wait and the simple "first in first out" assignment rule for units and containers is applied. The waiting times in this case are a measure for the performance of the subsystem. (2) The number of equipment units is "not restricted." In the model, this is realized by dynamically creating a new unit any time a unit is requested and no unit is available. The new units are added to the set of units of the corresponding system. Consequently, there will be no delay due to a lack of units. Here, the final measure is the total number of units created, indicating the upper limit of units needed so as not to delay the container flow.

In general, the number of equipment units for handling the originating flows will be restricted and equipment for the derived flows will be not restricted. Next, the processes of the classes of the elementary functions will be shown in pseudo-code.

The process of a Transport_System:

Repeat

Wait Until (ContainersToTransport >0 And TRANSPORTER_AVAILABLE)
Select container And remove it from ContainersToTransport
Select Transporter And remove it from AvailableTransporters
Assign container to Transporter
Determine route between source and destination of ContainerToTransport
Start the process of the Transporter

The process of a Transporter

Repeat Until arrived at destination

Calculate Arrival Time at next route point
Drive until Arrival_Time
Register Arrival
Add ContainerToTransport to the ContainersToTransfer of the Transfer_System of Destination
Enter Available_Transporters of MyTransportSystem
Suspend

The process of a Transfer_System:

Repeat

Wait Until (ContainersToTransfer > 0 And TRANSFERUNIT_AVAILABLE)
Select Container And remove it from ContainersToTransfer
Select TransferUnit and Remove it from AvailableTransferUnits
Assign Container to TransferUnit
Start the Process of the TransferUnit

Process of a Transfer_Unit:

If ContainerToTransfer is in a Stack Then Remove it from Stack
Determine the TransferTime
Work TransferTime
If ContainerToTransfer arrives at a Stack Then

If this Stack is ContainerToTransfer's Destination Then

Register ContainerToTransfer's throughput time
Remove ContainerToTransfer from Simulation

Else
Add ContainerToTransfer to the Stack

Else

Determine the TransportSystem according to the route of the ContainerToTransfer
Add ContainerToTransfer to ContainersToTransport of the proper TransportSystem

Enter AvailableTransferUnits of MyTransfer_System
Suspend

4.2 Modeling deep-sea ships

A number of days before a deep-sea ship arrives, export containers, destined for that ship, start arriving at the multiterminal complex. The time of arrival of a container is its dwell time before the departure of its sea ship. The dwell time distributions are considered to be input. At the arrival of a ship, all its export containers are supposed to have arrived in the marine stack. After the ship has arrived, its import containers are unloaded and at first stored in the marine stack. Within some time period, they are successively moved toward their next mode of transport directly or via another stack. The route a container follows in a multiterminal goes from stack to stack, including temporary stacks associated with the modalities of arrival and departure. Typically, an import container visits the marine stack of the arrival terminal, and an export container visits the marine stack of its departure terminal. A container may also stay in a local modality stack, for example, the rail stack. The following ship and container element classes and their attributes are defined:

Ship
–Arrival time
–Terminal
–'stack' with import containers to be unloaded
–'stack' for export containers to be loaded
Container
–import or export
–Myship (deep sea connection)
–Stacks to stay in
–Dwell time per stack
–Land side destination or origin

4.3 Input generation

In the early design phase of future large-scale systems, usually only very rough data, based on forecasts, is available. For the Port of Rotterdam, there are rough estimates of yearly throughput of containers for a period of some 20 years ahead. However, for investigation of the influences of the stochastic aspects of container flows on handling, transport, stacking capacity, and infrastructure, a much smaller timescale than 1 year is needed. Consequently, a very important step is to decompose the rough year-based data into a much smaller timescale in the order of magnitude of hours. For that purpose, arrival patterns of deep-sea ships, modal split statistics, and dwell time distributions are needed. A team of experts has advised on this matter on the basis of experience and current practice. To obtain consistency in the model input, a separate generator model was developed to generate ship arrivals and all related import and export container data [2]. The generator model creates a simple berth planning based on the proposition that deep-sea ship arrivals are scheduled on berth occupation level.

4.4 Implementation

The simulation model has been implemented at a platform that supports process interaction modeling and allows switching between stand-alone mode and distributed mode [21] http://www.tomasweb.com; Website regarding simulation software applied in this work; last check of address: June 2005. The model can accomplish distributed work by transforming all element classes to member models, thus forming a distributed model structure [3, 5].

5 Modeling the Maasvlakte terminals

In this section, the application of the model will be described and the model input is explained.

5.1 Fully developed MV2

MV2 will be developed gradually. We focus on the fully developed stage, probably reached in 2025. The basic container flow predictions for that stage are listed in Table 1. The flow is expressed in TEU per year. This number can be converted into the number of real containers (boxes) by dividing by the so-called TEU factor. The

Table 1 Predicted number of 20 feet equivalent containers (TEU) per year at the fully developed MV2 stage. The TEU factor of Rotterdam is 1.7

	MV1	MV2	Total
TEU/year $\times 10^6$	8.7	8.6	17.3

TEU factor of a container batch is defined as $x+1$, where x represents the fraction of 40-ft containers in the batch.

Example A batch consisting of 100 40-ft and 60 20-ft containers represents 260 TEU. Here, $x = 100/160 = 0.625$, resulting in a TEU factor of 1.625.

In this work, the overall TEU factor of 1.7 of the Port of Rotterdam has been used.

The total number of real containers expected to be processed per year is 10.2×10^6, corresponding with 11.2×10^3 sea ships.

The existing terminals on the first Maasvlakte are modeled as they are. For the future MV2 terminals, three configurations are investigated:

- The compact configuration.
- The dedicated configuration.
- A combined configuration of dedicated and compact terminals. This configuration was based on the results of the simulation experiments with the compact and the dedicated case.

The compact and the dedicated configurations are two extreme situations. They are investigated to determine the range of interterminal transportation capacity in particular. The combined configuration is considered a realistic conceptual design that has to be further developed.

The contours of MV2 are determined by geographical and nautical considerations and are considered fixed in this study. The terminal configurations to be evaluated have been drawn up by experts of the Port of Rotterdam and other specialists. The combined configuration is shown in Fig. 5. The terminal 5 (MV2_III) is a typical compact terminal with a deep-sea quay (5.0), a gate area for road traffic (5.1), a barge terminal (5.2), and a rail terminal (5.3). All MV2 terminals have a truck handling facility to reduce ITT. Terminal 1 (Euromax) is still under development and will become a compact terminal. Terminal 2 (Delta)

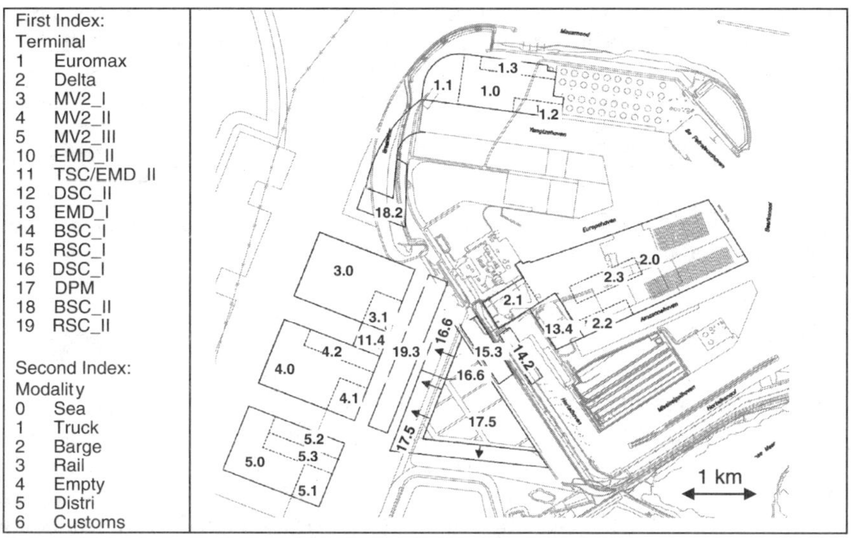

Fig. 5 Picture of the combined configuration. Courtesy of the Port of Rotterdam

represents the complete set of existing terminals at the ECT peninsula. TSC, BSC, and RSC in Fig. 5 are external dedicated truck-, barge-, and rail-service centers, respectively. Further, there are empty depots and the so-called Distripark for stuffing and stripping activities, and there is a customs area with security scan facilities.

5.2 Configuration data

The configuration data define the layout and equipment capacities available for the different elementary functions, as well as the interterminal road network.

Layout data include:

Number of terminals and for each terminal:

- Terminal name
- Location coordinates
- Number of quay cranes and the quay length available
- Inter- and intraterminal connections
- Composition of elementary functions and a decision whether the number of equipment units of the function is restricted or nonrestricted

The road network and for each road:

- Identification
- Coordinates, used as route points by the AGVs

Each deep-sea quayside has been assigned a finite unload and load handling capacity. These capacities are projected on a fixed number of quay cranes. The unload and load handling capacities, and thus the number of quay cranes, are tuned to cope with the workload under the condition that the sea ships are served in time by the transport system. For each quay, the length is set to the length available in the conceptual design of the terminal. The actually used quay length will follow from the simulation experiments.

5.3 Container flows

The year-based flows of Table 1 are resolved into ships and containers. Each container has modal split information and a dwell time.

5.3.1 Ships

A number of representative ship types have been defined. They are listed in Table 2. The actual ship types differ in call size, the number of cranes allowed simultaneously, and crane handling rates. For each ship, the actual call size is sampled from a distribution, taking into account the parameters of the corresponding ship type. The ship arrival times are sampled within a certain time window around the expected arrival times. An exception was made for the

F-type ships, also called "jumbo" ships. Here, it is anticipated that the arrival times are as expected. The jumbo ships are subdivided into five size classes from 8,000 to 12,500 TEU, with expected arrival times and destination terminals on a weekly schedule.

The destinations and origins of import and export containers, respectively, are obtained using modal split information. To that end, an "origin–destination" matrix for both import and export flows has been constructed. The input generator model creates a list of these deep-sea ships with container load information. The ship definition is shown in Table 3. For each ship, the numbers of import and export containers are sampled from distributions using the data from Table 2.

5.3.2 Dwell times and dwell time distributions

The dwell time of a container depends on its modal split connection. Import containers that have the same modal split connection and arrive with the same deep-sea ship are called a batch. On each batch, the corresponding dwell time pattern is applied. The same holds for export containers destined for the same deep-sea ship. In Table 4, an example of the definition of all possible modal split connections and the associated average dwell times are shown. In the model, any dwell time distribution may be defined [22]. Several dwell time distributions are tested. It appears that the shape of the distribution does not significantly influence the results. In this work, the actual dwell time of each container has been drawn from a uniform distribution between 0 and (2* dwell time average) associated with its modal split connection.

Containers that are transported with ITT usually stay in a marine stack and a landside stack. For each modality connection, a "stack-preference-factor" that prescribes the average fraction of the total dwell time in the marine stack has to be set. If the stack-preference-factor for a container batch, for example, equals 0.7, then the average stay time of the containers of that batch in the marine stack will be

Table 2 Definition of ship types

Ratio	L/U factor			Capacity (TEU)			Length	#QC	Type
	Min	Average	Max	Min	Average	Max	Max	Max	
0.3618	0.01	0.16	0.43	50	411	1,200	150	1	A
0.2276	0.02	0.29	0.93	600	1,693	3,000	200	2	B
0.0763	0.06	0.37	0.92	1,000	2,555	3,600	250	3	C
0.0719	0.06	0.32	0.80	2,700	4,599	6,600	300	4	D
0.2260	0.09	0.22	0.48	4,000	6,406	8,000	350	4	E
0.0364	0.30	0.375	0.45	8,000	10,000	12,500	450	6	F

Ratio fraction of total calls this ship type accounts for, *L/Ufactor* load/unload factor fraction of the capacity of this ship type that has the terminal as origin/destination, taken equal in the average, *Capacity* load capacity (in TEU) of this ship type, *Length* maximum length (in meters) of this ship type, *#QC* maximum number of quay cranes possible simultaneously on this ship type, *Type* ship-type indication (A–F)

For each ship, the L/U factor is sampled from a distribution

F-type is the biggest ship type. F-type ships are also called "Jumbo"

Table 3 Deep-sea ship definition

Ship name
Arrival day number
Arrival time
Arrival terminal
Ship category
Maximum number of quay cranes allowed
Number of container batches
Per batch:
Number of containers
Import or export ("I" of "E")
Originating/destination terminal and its modality

0.7× (total dwell time). Unless stated otherwise, the stack-preference-factor was set to 0.5.

5.3.3 Landside modalities and flows

The arrival of a deep-sea ship is anticipated by generating its export containers before the ship actually arrives, according to the modal split data of the export load and the appropriate dwell time. Export containers are generated in batches with a size representative for a transport load of the modality concerned. Import containers are collected at the landside terminal until a batch that represents a transport load of the modality concerned is formed. In Fig. 6, an example is shown of the generation of import container batches for a certain modality.

Table 4 Definition of the average dwell times of each pair of modality connections

From	To	Average (days)
Sea	Sea	3.4
Sea	Rail	5.2
Sea	Barge	3.5
Sea	Truck	6.4
Sea	Empty	3.0
Sea	DistriPark	3.0
Rail	Sea	3.7
Barge	Sea	3.2
Truck	Sea	3.7
Empty	Sea	3.0
DistriPark	Sea	3.0

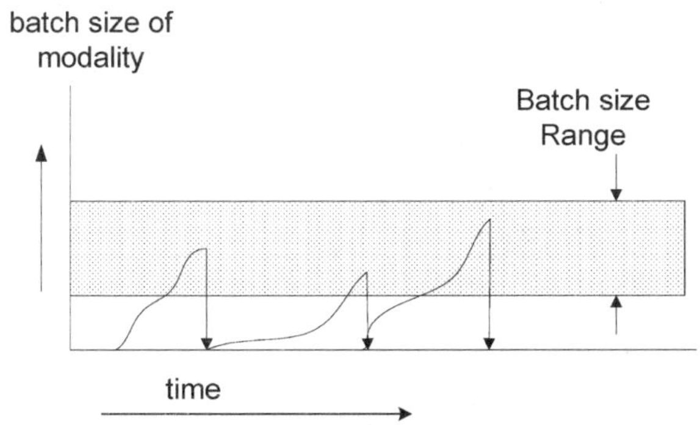

Fig. 6 Formation of import batches. As soon as the collected number of containers exceeds the batch size (sampled from a batch size distribution of the specific modality), the batch is supposed to leave the terminal with a transporter of the appropriate modality

6 Results

In this section, typical results of simulation experiments are shown. The run length of the simulation experiments was determined on 17 weeks. The first 4 weeks are found to be sufficient to stabilize the model. During this period, no statistics were collected. During the measuring period, about 2.7×10^6 containers are processed, transported by 3.0×10^3 sea ships. Every 15 min, the status of the model was monitored, resulting in time series of the equipment in use, the quay occupation, the stack content, and the traffic flows. With these data, the averages and the 95% percentile values are calculated.

To illustrate the variation of the measured values as a function of time, some typical parts of the time series will be shown as time plots. The statistics on the performance during the complete run length are presented in tables.

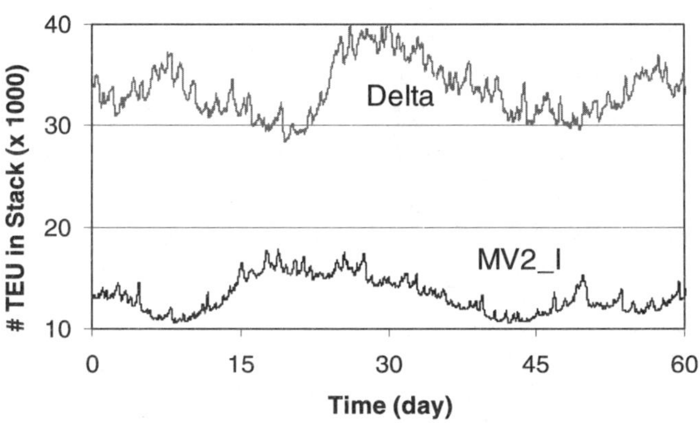

Fig. 7 Typical variation of the stack contents as a function of time for the terminals Delta and MV-III

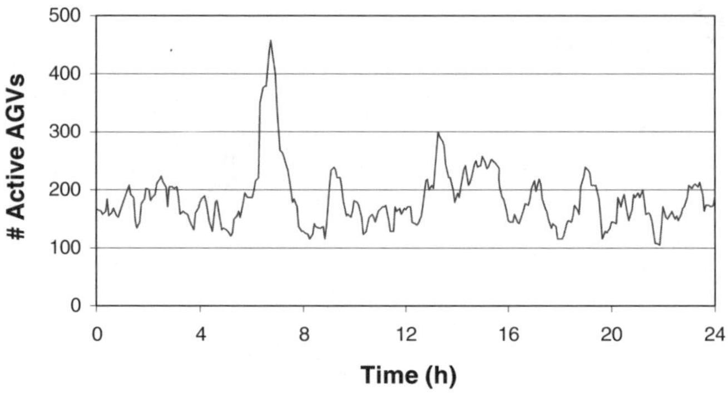

Fig. 8 Single-loaded interterminal AGVs, in use as a function of time, on a typical day

6.1 Time plots

Figures 7, 8 and 9 show typical time plots of the monitored data. Figure 7 gives the variation of the content of two marine stacks. Figure 8 shows the total number of interterminal AGVs in use, as a function of time representing the ITT transport demand. Only loaded vehicles are counted. Figure 9 shows the quay occupation for the MV2-III terminal for a period of 14 days.

6.2 Deep-sea quay occupation

Table 5 shows for each deep-sea quay the number of quay cranes and the quay length available (model-input) and the resulting quay occupation expressed in the average percentage, 95% percentile and maximum. The results for both compact and dedicated configurations are similar because the ship arrival patterns and ship load distributions are kept the same for all runs. The quay occupation perhaps

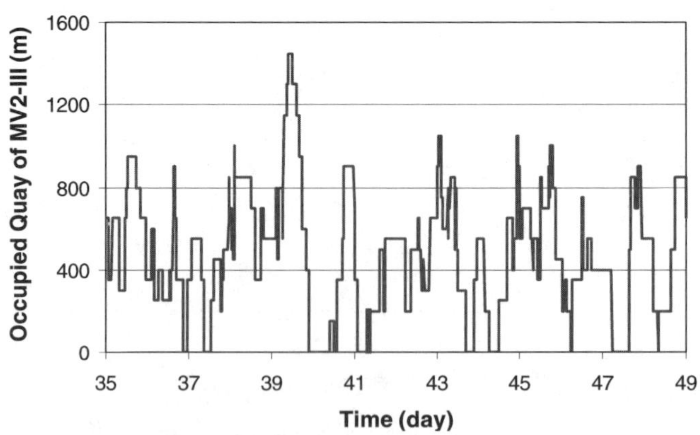

Fig. 9 Quay occupation of the MV2-III terminal during a period of 14 days

Table 5 The number of quay cranes and quay length available, average quay occupation, 95% percentile quay occupation, and maximum quay occupation for all deep-sea terminals

Deep-sea terminal	Number of quay cranes	Quay length (m)	Average quay occupation (%)	95% percentile quay occupation	Maximum quay occupation (%)
Euromax	13	1,800	29	58	86
Delta	34	5,250	26	41	52
MV2_I	15	2,000	29	59	92
MV2_II	15	2,000	32	64	90
MV2_III	15	2,000	28	58	85

seems low, but is rather common for service-oriented deep-sea terminals. The conclusion drawn from these data is that quay length will not be a bottleneck in the multiterminal operation.

6.3 Equipment use and stack volume

In Table 6, the ITT demand for all three configurations is shown. The numbers of AGVs in use in the compact and dedicated designs are in the proportion of 1:3. The number of AGVs in the combined configuration, as expected, is between the two extreme cases. The security scan requires almost 20% more ITT AGVs.

For the combined configuration, two values of the stack-preference-factor were tested. It appears that the resulting AGV needs for both values are nearly equal, indicating that the total AGV need has leveled out. Setting the stack-preference-factor to 1.0 requires "just in time" transportation of containers between the marine terminal and the proper inland modality.

Table 7 shows the stack content for two different stack-preference-factors. The total stack capacities used in both cases differ only by 0.4%, which is not significant. This was expected because the total dwell time of the containers did not change. Peak factors calculated using the 95% percentile values of Table 7 vary from 1.14 to 1.27. This is in line with the rule of thumb values used in practice.

Table 6 Number of single-loaded ITT AGVs in use for three configurations in the fully developed MV2 in 2025. In the combined configuration, the influence of the X-ray scan of 2% of all containers at a central point is also given

	Compact	Combined Normal/X-ray scan stack-preference-factor=0.5	Combined normal stack-preference-factor=1	Dedicated
Average # AGVs	104	173/205	173	318
95% percentile # AGVs	140	249/281	250	414

Table 7 Stack content in 10^3 TEU and peak factors for the combined configuration

Stack	Stack-preference-factor=0.5 stacks			Stack-preference-factor=1.0		
	Average (10^3TEU)	95% peak	Maximum peak	Average (10^3 TEU)	95% peak	Maximum peak
Euromax	13.6	1.27	1.41	14.2	1.26	1.40
Delta	33.0	1.15	1.24	36.3	1.15	1.24
MV2_I	12.9	1.15	1.25	16.2	1.11	1.21
MV2_II	15.9	1.15	1.25	17.9	1.14	1.23
MV2_III	13.7	1.20	1.31	14.8	1.20	1.30
Land side total	9.9			0		
Sum averages	99.0			99.4		

6.4 Interterminal traffic flows on the infrastructure

Figure 10 shows both the average and 95% percentiles of the traffic flows on the tracks of the terminal infrastructure. It shows that the 95% percentiles may be up to three times higher than the average flows. These flows are used to determine how many lanes and flyover constructions are necessary.

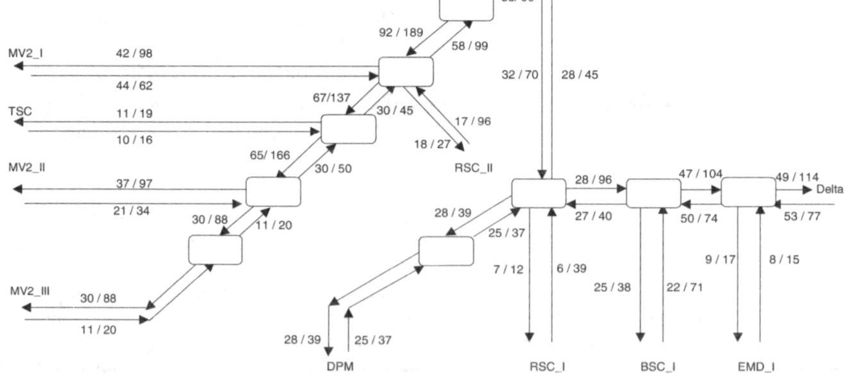

Fig. 10 Traffic flows (number of AGVs/hour) in the combined configuration expressed in loaded vehicles/hour. For each direction, the average and the 95% percentile of the traffic flow in AGVs/ hour is indicated

7 Conclusions

A concept was proposed to model a multiterminal system for container handling using a transport function, a transfer function, and a stacking function as basic building blocks. Each function has been modeled using the process interaction approach for simulation. The model structure allows zooming in on the elementary functions, giving possibilities for further tactical and operational extension of the model while preserving the container flows through the system. The model is applied to the set of existing container terminals in the Rotterdam port area and the conceptually designed second Maasvlakte, resulting in the determination of the requirements for the quay length, stacking capacity, handling and transport equipment, and the interterminal traffic flows on the transport infrastructure. The results of the study are used as a foundation for the further design of the MV2 infrastructure and the assessment of the cost.

7.1 Further developments

The design process of MV2 is still going on. Due to financial and nautical considerations in the meantime, the sea approach to MV2 has been diverted. As a consequence, the layout and the terminal configurations had to be adapted. Still, one of the main issues is to reduce ITT traffic. Possible improvements may be found in aiming at more compact terminals, optimizing empty depot locations, and applying container scan stations at various locations. A number of new projects are anticipated for detailed modeling and studying of parts of the MV2 complex. The main issue in these models will be the development of intelligent control systems and the optimization of equipment use and container throughput with a guaranteed service time for the various users.

Acknowledgement This work is a result of the research program "FAMAS.MV2" directed by the Port of Rotterdom. The simulation study has been carried out within the Delft Research School for Transport, Infrastructure and Logistics (TRAIL).

References

1. Birtwistle GM, Dahl OJ, Myhrhaug B, Nygaard K (1973) Simula Begin. Van Nostrand Reinhold, New York
2. Duinkerken MB, Veeke HPM, Ottjes JA (2001) Generator model MICL simulation. Report nr. 2001.LT.5555. TU Delft, fac. OCP, TRAIL Research School, Delft
3. Duinkerken MB, Ottjes JA, Lodewijks G (2002) The application of distributed simulation in TOMAS: redesigning a complex transportation model. Proceedings of the 2002 winter simulation conference, San Diego, December 2002, pp 1207–1213
4. Fishman GS (2001) Discrete event simulation. Modeling, programming, and analysis. Springer, Berlin Heidelberg New York
5. Fujimoto RM (2000) Parallel and distributed simulation systems. Wiley series on parallel and distributed computing. Wiley, NY
6. Grunow M, Günther HO, Lehmann M (2004) Dispatching multi-load AGVs in highly automated seaport container terminals. OR Spectrum 26:211–235
7. Healy J, Kilgore RA (1997) Silk:a java-based process simulation language. Proceedings of the 1997 winter simulation conference, December 1997, Atlanta, pp 475–482

8. in't Veld J (1998) Analysis of organization problems. Educatieve Partners Nederland BV, Houten
9. Kim KH (1997) Evaluation of the number of rehandles in container yards. Comput Ind Eng 32(4):701–711
10. Kim KH, Bae JW (1999) A dispatching method for automated guided vehicles to minimize delays of containership operations. Int J Manag Sci 5(1):1–25
11. Kim KH, Moon KC (2003) Berth scheduling by simulated annealing. Transp Res Part B Methodol 37:541–560
12. Little JDC (1961) A proof of the formula $L=\lambda W$. Oper Res 9:383–387
13. Liu CI, Jula H, Ioannou PA (2001) A simulation approach for performance evaluation of proposed automated container terminals. In: Proceedings of the IEEE intelligent transport systems conference, Oakland, CA, 25–29 August 2001, pp 565–570
14. Ottjes JA, Duinkerken MB, Evers JJM, Dekker R (1996) Robotised inter-terminal transport of containers: a simulation study at the Rotterdam port area. In: Proceedings of the 8th European simulation symposium, Genua, October 1996, pp 621–625
15. Park YM, Kim KH (2005) A scheduling method for berth and quay cranes. In: Günther HO, Kim KH (eds) Container terminals and automated transport systems. Springer, Berlin Heidelberg New York, pp 159–182
16. Rijsenbrij JC (2001) The impact of tomorrow's ships on landside infrastructure. In: Proceedings of the 26th annual terminal operations conference, Lisboa, June 2001, pp 1–17
17. Saanen YA (2004) An approach for designing robotized marine container terminals. PhD thesis, Delft University of Technology
18. Steenken D, Voss S, Stahlbock R (2004) Container terminal operation and operations research—a classification and literature review. OR Spectrum 26:3–49
19. Veeke HPM (2003) Simulation integrated design for logistics. PhD thesis, Delft University of Technology
20. Veeke HPM, Ottjes JA (2000) TOMAS: tool for object-oriented modeling and simulation. In: Proceedings of the advanced simulation technology conference, Washington, DC, April 2000, pp 76–81
21. Veeke HPM, Ottjes JA (2001) Applied distributed discrete process simulation. In: Proceedings of the 15th European simulation multiconference, Prague, Czech Republic, June 2001, pp 571–577
22. Veeke HPM, Ottjes JA (2002) A generic simulation model for systems of container terminals. In: Proceedings of the 16th European multi conference, Darmstadt, June 2002, pp 581–587
23. Vollmann TE, Berry WL, Whybark DC, Jacobs FR (2005) Manufacturing planning and control systems for supply chain management, 5th edn. McGraw-Hill, New York
24. Yun WY, Choi YS (1999) A simulation model for container-terminal operation analysis using an object-oriented approach. Int J Prod Econ 59:221–230
25. Zeigler BP, Praehofer H, Kim TG (2000) Theory of modeling and simulation, 2nd edn. Academic, San Diego, CA

Mark B. Duinkerken · Rommert Dekker · Stef T. G. L. Kurstjens · Jaap A. Ottjes · Nico P. Dellaert

Comparing transportation systems for inter-terminal transport at the Maasvlakte container terminals

Abstract In this paper, a comparison between three transportation systems for the overland transport of containers between container terminals is presented. A simulation model has been developed to assist in this respect. Transport in this study can be done by either multi-trailers, automated guided vehicles or automated lifting vehicles. The model is equipped with a rule-based control system as well as an advanced planning algorithm. The model is applied to a realistic scenario for the Maasvlakte situation in the near future. The experiments give insight into the importance of the different characteristics of the transport systems and their interaction with the handling equipment. Finally, a cost analysis has been executed to support management investment decisions.

Keywords Inter-terminal transport · Automated container terminals · Transport systems · Discrete event simulation

MSC Code 90B06 Transportation · Logistics

1 Introduction

Rotterdam's Maasvlakte complex has grown into a large complex of container terminals, both automated and conventional terminals (see Fig. 1). In 2004, more

---

M. B. Duinkerken · J. A. Ottjes
Delft University of Technology, Faculty of Mechanical, Maritime and Materials Engineering, Mekelweg 2, 2628 CD Delft, The Netherlands

R. Dekker (✉) · S. T. G. L. Kurstjens · N. P. Dellaert
Econometric Institute, Faculty of Economics and Business Economics, Erasmus University Rotterdam, Burgemeester Oudlaan 50, 3062 PA Rotterdam, The Netherlands
E-mail: rdekker@few.eur.nl

than 8 million twenty-foot equivalent units (TEU) were transferred in Rotterdam, while the coming years see a continuing growth.

Consequently, there will be an increased need for container transport between the various terminals and the various modalities (rail, road, barge, sea). Also, the transport between these modalities and other service centres (empty depots, DistriPark) will increase. This transport is called inter-terminal transport (ITT). At present, ITT is executed by means of the multi-trailer system (MTS). Such a system uses manned trucks, pulling trains of five trailers. The problem under consideration is whether this system is efficient enough to handle the large container streams predicted for the near future or whether other systems using automated guided vehicles would be more cost-effective. To this end, the project Inter-Terminal Transport was initiated. The project was commissioned by Incomaas (1994).

The objective of the study is to give a recommendation on the effectiveness and efficiency of three possible transport systems, viz, the present MTS system, a system based on automated guided vehicles (AGVs) and a system based on automated lift vehicles (ALVs). For this purpose, a simulation model of the inter-terminal transport (ITT) system at the Maasvlakte has been developed, including container handling at terminals, container transport between terminals and a control system.

Each transport system has different transport and handling characteristics. The MTS system uses manual drivers, and to increase efficiency five trailers are pulled. It has, however, problems in using that capacity efficiently. The AGV system transports only one load per trip and requires a crane to start or finish its job. Hence, it is very dependent on the capacity of the transshipment cranes. Finally, the ALV system works like an automated straddle carrier, although the one considered in this paper cannot stack containers on top of each other. It does not need cranes, as it can lift and drop containers itself. At the time of the study, it was a hypothetical system, but in the meantime, automatic straddle carriers have been developed by Kalmar,

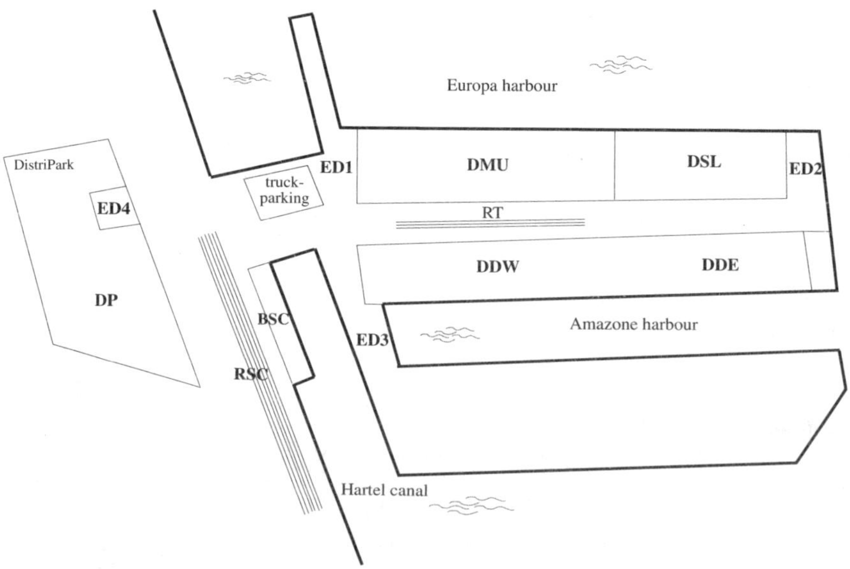

Fig. 1 General layout of the ECT peninsula at the Maasvlakte

the equipment manufacturer; they have been commissioned at the Patrick's container terminal in Brisbane, Australia as of Dec 1, 2005.

A lot of research on the processes and operations in container terminals has been done; Steenken et al. (2004) present an excellent overview. Research on an operational comparison of transportation systems, however, is scarce. Yang et al. (2004) compare through simulation an AGV system and an ALV system for loading and unloading containers from a ship to the stack at a hypothetical automated container terminal. They conclude that the ALV system is superior in performance because it does not have to wait for cranes to load/unload. The study by Vis and Harika (2004) is on a similar topic. They also conclude that the ALV system needs fewer vehicles than the AGV system.

In this paper, we consider inter-terminal transport rather than quay transport. We also include an MTS system and we consider the transportation over a network including a barge and rail terminal in detail with time-dependent demand. In the quay transport, the quay cranes are the bottleneck and the vehicle system should follow. In the inter-terminal transport, both the vehicle system and handling cranes are bottlenecks, and also a much more complex vehicle planning and control system needs to be used. Liu et al. (2004) have studied the effect of different layouts and traffic restrictions of AGVs in automated container terminals. Liu et al. (2002) compare four different terminal transport options: one using AGVs, one using a linear motor conveyance system, one using an overhead grid rail system and one using a high-rise automatic storage and retrieval system. They conclude that the AGV-based system performs best. Evers and Koppers (1996) emphasize the importance of AGV control for the system performance and present a modelling technique for traffic control at automated container terminals. Other related work is done on internal transport in warehouses; see e.g. Le Anh (2005) and Le Anh and De Koster (2006); the relations are very much in the planning and control of the vehicles. However, these systems do not consider the lifting of the loads explicitly. Finally, we like to mention Corry and Kozan (2006) who consider load planning of intermodal trains, which is important to understand our outcomes in the rail terminal.

In chapter 2, we present a simulation model with a very detailed modelling of the need for transport, e.g. created by departure of trains and barges, implemented by the use of double transport windows (viz, for departure and arrival). Furthermore, the influence of stochastic disturbances in handling and travel times is modelled. In chapter 3, results are given on the performance of the various systems. We provide the number of transportation units needed for a certain on-time performance.

1.1 Inter-terminal Transport (ITT)

The tasks of the ITT can be summarised as follows:

1. the punctual (neither early nor late) collection of containers from their point of origin
2. the punctual delivery of containers at the desired point of destination
3. the possible bridging of discrepancies in both these tasks by 'buffering on wheels' or in a transport-stack ('on ground')

An important performance criterion is the time it takes ITT-transported containers to reach their destination, including the handling at the destination terminal. If this completion time is later than the permitted latest arrival time for the container, this is considered as 'non-performance'. Non-performance as defined above is used as the most important criterion for the assessment of the ITT options. Another important performance criterion for ITT is the punctuality of departure of means of container transport such as trains and barges. In this case, 'non-performance' is defined as the late departure of a vessel or train as a result of the late delivery of one or more containers. Further performance indicators, which often take the form of averages and distributions, include vehicle occupation rates, number of empty trips, vehicle loading rate (percentage loaded, only MTS), number of vehicles waiting to load or unload and the equipment utilization at the terminals.

1.2 The area investigated

The area investigated consists of the Europe Container Terminals (ECT) Peninsula, with its marine terminals, and the peripheral service centres (Fig. 1). The system comprises the following terminals: the marine terminals (DDW, DDE, DSL and DMU) and the empty depots (ED) on the peninsula, the Barge Service Centre (BSC), the Rail Terminal (RT), the Rail Service Centre (RSC) and the DistriPark (DP). Each terminal has one or more handling centres (HC), indicated by a number behind its prefix. A handling centre is the origin or destination of an ITT move. In our approach, we created a transport network of handling centres in which the longest transport distance is almost 6,000 m.

1.3 Transport demand

Estimates for the container flows handled by the ITT are given in an origin–destination matrix in which the flows are shown on an annual basis, about 1.4×10^6 containers per year, or 27,277 per week; see the Appendix for details. To determine the dynamic effects of ITT, it is important to know these flows over a shorter time basis. To this end, the flow dynamics are generated with the aid of statistical distribution functions representing daily fluctuations. For the RT and the RSC, the incoming and outgoing flows are generated with the aid of service timetables. All the terminals except RT, RSC and BSC are equipped with a container stack, which is called the 'uncoupling stack'. ITT flows to and from marine terminals are uncoupled from the sea side at this stack. Also the DistriPark (DP) and the Empty Depots (ED) are equipped with a stack that uncouples the ITT from the activities inside these locations.

1.4 ITT vehicles

Three types of ITT vehicles are distinguished:

1. MTS, the multi-trailer system, consisting of a train of trailers on which containers (maximum ten TEU, Twenty-foot Equivalent Unit) can be carried, drawn by a manned traction unit (FTF); see Fig. 2

2. AGV, an automated guided vehicle, similar to the AGVs currently used by ECT; see Fig. 3
3. ALV, the automated lift vehicle, an automated guided vehicle equipped to pick up and set down containers; see Fig. 4

The MTS is the main system currently used for inter-terminal transport at the Maasvlakte. The AGV system is used in several automated terminals of the Maasvlakte, but not yet for inter-terminal transport. At the time of our study, the ALV system was still in the design stage; see Meeusen and Evers (1994) for a proposal. It is an interesting option because the way it operates uncouples the ITT from the equipment at the terminals.

2 Modelling

The four models that have been developed and the associated data files are shown schematically in Fig. 5. In the following paragraphs, the generator and the ITT simulation model are explained. The traffic density model has been used for a more detailed study of the traffic flows and will not be discussed in this paper. It uses the outcomes of the present study to investigate whether the envisaged flows and queues are also possible from a traffic physical point of view. The advanced planning model is used to generate a better planning for the MTS variant and is discussed in Kurstjens et al. (1996).

2.1 Generator model

The container flow data are generated by a separate computer model, the *Generator model*. In addition to the configuration data, the input data of this model includes the data relating to the container flows. For example, in the case of an RSC, this will relate the timetables of the trains and shuttles to the loading data. For marine terminals, empty depots and the DistriPark statistical distributions are used to derive the arrival times, the size, the origin and the destination of containers. The generator model takes into account the available transshipment, quay and batching

Fig. 2 The multi-trailer system (MTS) in use with ECT

Fig. 3 An AGV with a quay crane in use at ECT-DSL

capacity of the various terminals. The output of the generator model provides the input for the simulation model. In this way, experiments can be better controlled and the three transportation systems can be compared in a better way.

To generate well-defined ITT flows, a distinction is made between two types of flow called PUSH flow and PULL flow. If terminal A takes the initiative to send containers to terminal B, these are termed push containers from terminal A to B. If terminal B requests containers from terminal A, these are termed pull containers from terminal A to B. In both cases, containers are moving from A to B. Push containers have a time window varying between 2 and 24 hours, while pull containers one between 2 and 72 hours. The number of rush containers with a 2 hour transport time window varies per handling centre. It varies between 0 and 25% of the total number of containers for that handling centre. Because containers at RT, RSC and BSC are related to a train or barge, containers originating from these handling centres can only be pushed and containers destined for these handling centres only be pulled. In this way, a container from or to the RSC, BSC or RT can always be associated with a train or barge, and the (derived) performance of these carriers can be measured. In the case of rail and barge flows the generator

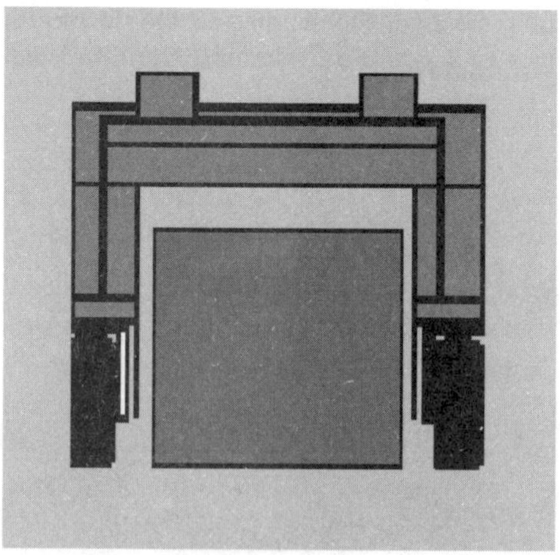

Fig. 4 A design for an ALV

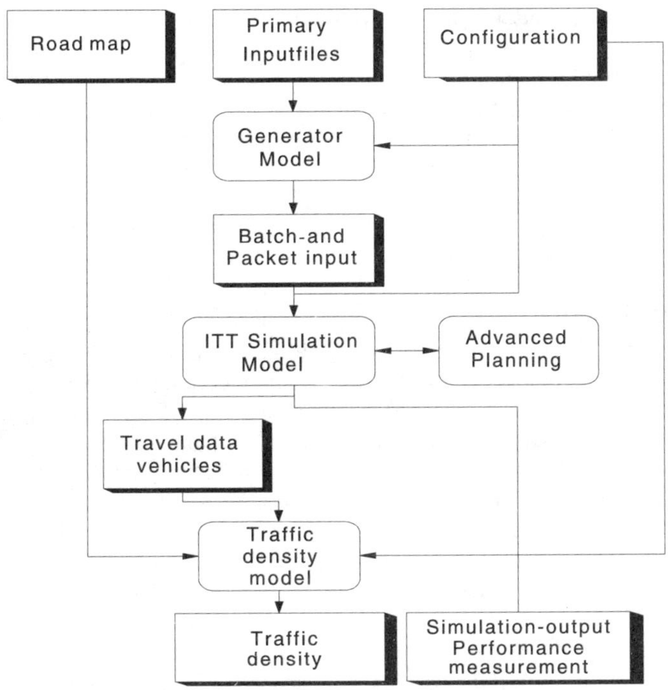

Fig. 5 Overview and relations between models and files (indicated by *rectangles*)

model takes care of allocation of trains to tracks and barges to quays. The timetables, as described in a position paper of the Dutch Railways, Hoenders (1994), are the basis of the allocation of the trains of suitable size on the available tracks of the Rail Service Centre (RSC) and the existing Rail Terminal (RT). From these timetables the time windows for the unloading and loading of a train can be used. The groups of containers to be unloaded and loaded are determined for each individual train and the position of each container is recorded as well as the time window within which handling must take place. In total we modelled per week 107 trains at the RT and 164 at the RSC. The short-stay trains are mostly assigned to the Rail Terminal, while the long-stay trains are assigned to the Rail Service Centre.

Barge arrivals and loads are generated on the basis of statistics. We modelled some 102 barges per week. In the model, a barge contains three layers of containers. As it is known in which layer each container is positioned, this can and will be taken into account during loading and unloading.

2.2 The ITT simulation model

This model simulates the entire ITT process: the handling of ITT vehicles at the terminals, the trip and waiting times of ITT vehicles, the control of the entire process and a planning mechanism. The control and planning processes in the model are 'rule-based'.

The simulation model is object-oriented and works according to the process description method (for a review, see Zeigler et al. 2000). The tool MUST is used for the implementation (Must 1992). The various objects contain the data structure of the components and, in the case of an active component, also the process description of the component. The main component classes applied in the model are described in the following paragraphs.

2.2.1 Containers

The Inter-Terminal Transport is simulated at container level. The object container serves as a model for the containers that are transported by the ITT. This means that during simulation a component is created for each container present in the system. Containers are grouped in packets. A packet is a group of containers with common ITT origin and destination. For containers going from or to trains and barges, packets are grouped in batches and assigned to a batch carrier. At the start of its departure time window the container is available for transport. Each container object keeps and updates the data that control the transport. When a container arrives at its destination, the container data indicate whether it is on time. After recording of the performance, the 'container' is removed from the model.

2.2.2 Terminals and handling centres

All the terminals except rail and barge service centres are equipped with a container stack, which is called the 'uncoupling stack'. ITT flows to and from marine terminals are uncoupled from the seaside at this stack.

Terminals are split into exchange points for ITT, the so-called handling centres (HCs). Each HC has transshipment equipment to transfer containers to and from the ITT vehicles. Several terminal types and related equipment are modelled. The marine terminal, the distribution centre and the empty depot make use of the standard HC that has 'standard equipment'. The rail and barge service centres each have their own type of HC, because at these handling centres the containers are related to batch carriers, which require special equipment and handling procedures. Every HC possesses all the data on vehicles that are present or on route to it.

2.2.3 Equipment for container handling

The object equipment is used to model loading/unloading equipment at HCs, such as straddle carriers, fork lift trucks and automatic stacking cranes. For a standard HC, the move time is drawn from a stochastic distribution. The standard capacity of most HC is some 35 moves/h during the week, while during the weekend, it is 25 moves/h. Empty depot HCs have different capacities (EDs varying between 10 and 45 moves/h, the DP 30 moves/h). We did not model individual equipment at each handling centre but considered the total as one super 'crane'.

As the handling capacity is likely an important factor in our experiments, we parameterised it by multiplying the standard capacity with a factor called CapFactor. The factor is not applied to the rail and barge handling centres, nor to the waterside terminals in the corresponding AGV case. In this way, we can also study the effects of varying the handling capacity. Each cycle of the equipment process begins with 'select action'. This selection is made on the basis of a decision tree, which can be summarised as follows: if there is only something to load or to unload, then do that; if both loading and unloading are possible, decide to unload a vehicle and start with the one most urgent, unless the time needed for this exceeds the planned execution time (plan time) of the succeeding load container.

For the equipment at rail and barge HCs, more detailed processes are modelled. Each rail HC comprises a number of rail tracks and vehicle tracks (termed 'bundle'), served by one rail crane. The control of the RSC and RT cranes takes into account the time that trains stay and the priority of the movement of individual containers. The movements of the crane and its trolley are modelled accurately, taking into account the positions of the containers on the train, the crane acceleration and speed and the trolley speed. Some details are max crane speed (2.0 m/s), crane acceleration (0.35 m/s^2), creep speed (0.2 m/s), trolley speed (1.3 m/s), container loading time (20±5 s), unloading time (on AGV, MTS, ground) (15±5 s). A barge handling centre consists of one quay with one BSC crane. It is assumed that a ship can only berth after the preceding ship has finished unloading and loading and has left. Therefore, a delay in the handling of one ship can cause a large non-performance, especially during the busy hours of a week.

In the model, all containers on a ship must be unloaded before the crane can start loading. The unloading of the ship starts at the top layer. Each layer must be finished completely before the crane starts handling the next layer. Loading starts with the bottom layer of the barge.

2.2.4 Vehicles for transportation system

In the model, a vehicle is a component that can transport containers from its loading point to its destination. For all three vehicle types, the travel times are calculated based on distance, average speed and a stochastic disturbance. The disturbance is drawn from a uniform distribution between zero and 30% of the nominal travel time.

An AGV is a vehicle that is loaded and unloaded by equipment at a handling centre and can travel from the loading point to its destination under its own power. In the model, the control assigns an idle AGV to the loading equipment of the handling centre (crane), which loads the AGV. After the crane activates the AGV, it travels to the destination with a speed of 5.0 m/s. At the destination, the AGV needs to position itself in 7.5 s, activates the crane at the destination and waits until unloading is completed. The AGV then is idle again.

An ALV is a vehicle that can both load and unload containers and travel from the loading point to its destination under its own power. In the model, the control activates an idle ALV. The ALV loads a container at its origin. Then it travels to the destination of the container at a 4.0 m/s speed, unloads the container in 30 s and is idle again.

A MTS is a train of coupled trailers that can be loaded or unloaded with one or more containers by handling centre equipment. An MTS is pulled by a manned traction unit (FTF) from its point of origin to its destination. An MTS object cannot carry out any process itself. A FTF couples and uncouples itself to and from an MTS in 65 s and rides from point of origin to destination. In the model, the control assigns an idle MTS to the equipment (crane), which loads the trailers with one or more containers. Then the MTS is put in a waiting queue at the handling centre. When a FTF arrives, the control can assign the FTF to the MTS. The FTF couples the MTS and travels to the destination of the containers on the trailers. At the destination, the FTF uncouples the MTS, puts it in a waiting queue and activates the crane at the destination. The MTS waits until it is unloaded, and then is idle again. The FTF speed is 7.7 m/s without MTS and 6.6 m/s with MTS.

Both planning and control generate empty trips for AGVs, ALVs, MTSs and FTFs when necessary. For an empty trip of an MTS, it is put in the waiting queue without a container load, and waits for an FTF. An empty trip for an FTF means that it does not pull an MTS.

2.3 The standard control and planning algorithm

The control of the simulation model determines for each vehicle (including the FTFs in the multi-trailer system) what the next action is after finishing a job. For the ALV and AGV systems, a job is finished when the destination is reached and a crane has removed a container if present. For the FTF system, a job is finished if it reaches a handling centre and is decoupled from a possible MTS it has pulled. For the MTS, a job is finished if it arrives at a destination and is unloaded of all containers. The next action for a vehicle is one of the following three options:

1. wait for loading at the current HC
2. go to another HC (AGV, ALV, FTF: start empty trip; MTS: wait for FTF)
3. remain idle

The standard planning algorithm generates extra empty trips of the vehicles based on the expected vehicle balance over a longer horizon. In the MTS option, an advanced planning system has been developed for the generation of all empty trips and the allocation of FTFs to trips. This module is reported in Kurstjens et al. (1996).

The type of control described in this paper is a way of centralised control with multi-attribute rules applied in a hierarchical way, with a look-ahead period. The review of Le Anh and De Koster (2006) gives some guidelines about when which control rules to use, in which our approach seems to fit. These recommendations are general, and the best type depends on the specific problem circumstances. Moreover, in every case, one has to optimise the control parameters as we did.

2.3.1 Vehicle control

The control heuristic is executed each time:

1. a vehicle or FTF at a handling centre becomes 'idle'
2. a handling crane has finished a loading or unloading action

This approach will result in frequent calls of the heuristic, several times per minute. It is expected that this behaviour can be optimised further.

The control heuristic contains the following steps:

1. Determine the vehicle requirement at each HC
2. Meet the vehicle requirement with idle vehicles
3. Activate all handling equipment if necessary

Step 1 In this step, it is determined whether a HC has a shortage or a surplus of vehicles. This balance is calculated based on the planned transport jobs, the number of vehicles in the waiting queues at the HC and the number of vehicles driving towards the HC.

For the planned transport jobs, a distinction is made between urgent and planned (non-urgent) transport jobs.

The degree of urgency is determined by:

1. BSC, the number of containers to be unloaded from a berthed barge
2. RSC, the number of containers that must be quickly unloaded from a train to prevent their late arrival at their destination (taking expected travel time into account)
3. Other HC, the number of containers where remaining time before non-performance occurs equals the expected travel time plus a certain safety margin

Only in the AGV variant the number of urgently required vehicles is artificially increased with a fraction (10%) of the total number of transport orders present. The result is that even when there are no urgent orders, empty trips are made to handling centres with many transport orders, which improves the performance substantially. The urgency requirement is limited to a specified maximum, depending on the type of handling centre and its capacity.

Besides the urgent vehicle requirement, there is also a 'normal' requirement for vehicles for the transport jobs that are not (yet) urgent. This number is also limited to a specific maximum, depending on the type of handling centre, its capacity, and for the AGV option on the length of the queue of loaded vehicles. This is an important feature, as it prevents long queues at the HCs.

Step 2 After the vehicle requirements for each HC are determined, the idle vehicles at each HC are assigned to an action. For each HC, the vehicle requirement is satisfied with vehicles from the local idle queue, or vehicles are sent from the queue of idle vehicles to elsewhere.

1. For AGVs and MTSs, the local requirement is first satisfied. First, the urgent requirement, then the normal requirement is satisfied with local vehicles, if available. After that, if there are still idle vehicles, the urgent requirement elsewhere is satisfied. The HCs have been put in an ordered list based on adjacency. For each HC with a surplus, one looks at the next HC with a shortage.
2. For ALVs, the urgent requirement at all HCs is satisfied first. Preference is for local ALVs, then ALVs from elsewhere are assigned. After satisfying the

urgent requirement, the normal requirement is satisfied, first with local ALVs, if available, and finally with ALVs from elsewhere.

Step 3 The control triggers the equipment to start loading the vehicles. This is necessary because a handling crane will become idle when there are no vehicles available for loading. Because the planning heuristic might have assigned vehicles to a crane, the crane must be activated again.

2.3.2 FTF control

In the MTS option, the control also allocates the FTFs. No distinction is made between urgent and non-urgent cases, but the number of FTFs needed at a HC is determined by the net requirement for FTFs. First, an attempt is made to meet the need for each HC from the FTFs present at the HC, which are in the 'idle' state. Next, FTFs are sent on an empty trip from the HC with the largest surplus to the HC with the greatest shortage of FTFs.

2.3.3 Standard planning

The standard planning module is always active for the AGV and ALV options and only for the MTS option if indicated in the configuration. The planning module is called up periodically (plan interval time). The planning determines the amount of ITT vehicles at the various handling centres and generates the empty trips to restore a balance. An empty trip is modelled as a transport job without container but with a certain plan time.

For each planning call, the following occurs:

1. Read all transport orders for which the plan time is within the coming two periods (plan interval time)
2. Calculate the vehicle surplus/shortage per handling centre: number of incoming orders minus number of outgoing orders
3. Generate empty trips from the handling centres with a vehicle surplus to the handling centres with a vehicle shortage, so that for each handling centre the resulting surplus or shortage is zero. The plan times for the empty trips are divided over the entire plan interval period.

2.4 Verification and validation

Verification and validation was done continuously during model development. Especially in the implementation phase of the advanced MTS planning all processes were verified thoroughly. The object-oriented modelling technique offers the opportunity to give close attention to the internal structure of the model. Important decisions concerning the number of handling centres at a terminal, the layout and the container handling at BSC and RSC were discussed in depth. Staff at ECT confirmed the validity of the modelling of the container streams, the terminal handling and the transport processes. All input data, like origin–destination matrices, distance tables, vehicle speeds and equipment capacity, originate from the specialists in the Incomaas project and are given in Celen et al. (1999).

3 Experiments and results

There are two reference layouts: the standard layout, termed the 'Landside', in which the ITT runs only via the central area of the Maasvlakte peninsula (see Fig. 1). This layout is used by the MTS, AGV and ALV variants. An alternative layout, termed the 'Waterside' is only used in an AGV variant where the AGVs drive along tracks at the waterside of the stack. In the first layout, vehicles at the DSL, DDE and DDW terminals are served by straddle carriers, as these also serve trucks for import and export of containers. The capacity of these straddle carriers is modelled through the CapFactor variable. In the second layout, the AGVs at the DSL, DDE and DDW terminals are handled by automatic stacking cranes, which also serve the AGVs used for transport to the quay cranes (but which are not modelled in our study). The results are split up into performance measurements, vehicle characteristics and equipment characteristics. This paper contains a sample of the results and shows typical phenomena observed over all experiments.

The models that have been developed are used for an extensive experimental programme, the objective of which was to determine the characteristics of the various ITT options.

The five investigated systems are:

1. MTS/FTF with control and standard planning
2. MTS/FTF with control and advanced planning
3. AGV landside
4. AGV waterside
5. ALV

Note that the MTS systems employ batching to improve the productivity of the driver in the FTF. As the FTF can be decoupled from the MTS, much less FTFs are needed than multi-trailers. As in this case the planning is important, we put a lot of work in developing both the standard planning and an advanced planning (described in "Kurstjens et al. 1996"). Although an advanced planning may also be interesting for AGVs, it was not developed primarily because vehicle dispatching rules are known to work quite well in stochastic circumstances; see Le Anh (2005).

The two AGV options differ in the fact that the AGVs are being loaded/unloaded in the stack. In the waterside case, the loading/unloading could be somewhat more easily done during ship operations, as during those times the stacking cranes will be close to the waterside. That capacity is, therefore, cheaper. We did model larger fluctuations in the waterside handling capacity, as these cranes are also used for loading and unloading ships.

A distinction can be drawn between runs, which serve for the tuning of a number of parameters, runs to determine the run length and production runs that determine the ITT performance as a function of a number of factors. The tuning of the model is necessary to determine the setting of a number of parameters. These parameters relate to the control and planning; for example: under what conditions is a transport order urgent, how long before the physical arrival of a container is a transport order planned, how long is the permitted pre-work time for a transport job. Separate tuning is required for the railway handling of MTSs. We found that a 1-h cut-off time to define urgency in case AGV and ALV systems worked best while the MTS system needed a 2-h cut-off time.

Before starting the experiments for each system, the optimal control-parameters were determined. We did ten replications of a reference run of 10 weeks, which revealed a relative standard deviation of some 15% (0.1% in absolute terms) in the non-performance outcome of 0.8%, which was considered to be acceptable. Hence, the length of the runs was set at 10 weeks of operations, preceded by a running-in period of 1 week. Stochastic variation was reduced by the use of the generator file and using always the same seed. In the production runs, the available ITT and the handling capacity of the handling centres are varied. This variation is effected by multiplying the nominal capacity by a capacity factor, shown in the graphs as CapFactor. All in all, some 1,000 runs of different settings have been carried out, each taking between 1 and 3 h (with the advanced planning) of computer time.

3.1 Performance measurement

Figures 6, 7, 8 and 9 show the non-performance as function of the number of vehicles. In Fig. 10, the non-performance as a function of the CapFactor is shown for the AGV landside variant.

Even with extreme high numbers of vehicles, the performance of the MTS system remains clearly poorer than that of AGV and the ALV systems. The number of late trains is almost eight times greater for the MTS option than for the AGV option and 18 times greater than for the ALV option. This can be easily explained, particularly in the case of train handling. AGVs and ALVs travel as closely as possible to the destination or point of origin of their container on the train, resulting in shorter crane move times compared to the MTS variant. This is explained later in Figs. 18 and 19: at the RSC, the average crane move time for the AGV case is 0.024 h and for MTS it is 0.034 h. By definition, the time lost by containers mounts up as a result of the batch-type work method of the MTS option. Furthermore, in the control and planning, concessions are made with regard to the departure times to attain a better MTS loading rate. Containers can be kept back to be loaded on an MTS with containers that become available later. As a result of peak loading at the

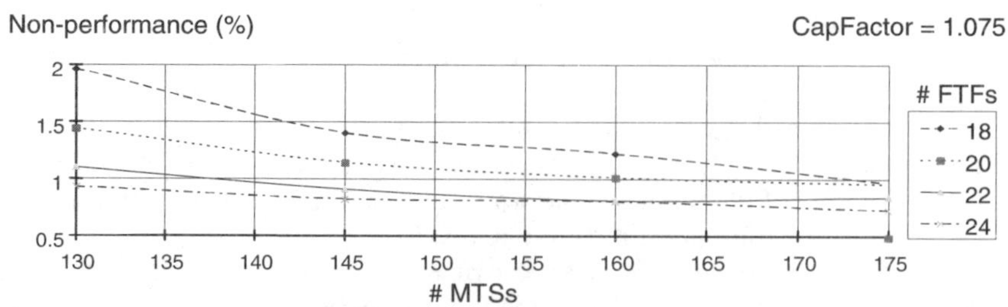

Fig. 6 Non-performance as a function of the number of MTSs

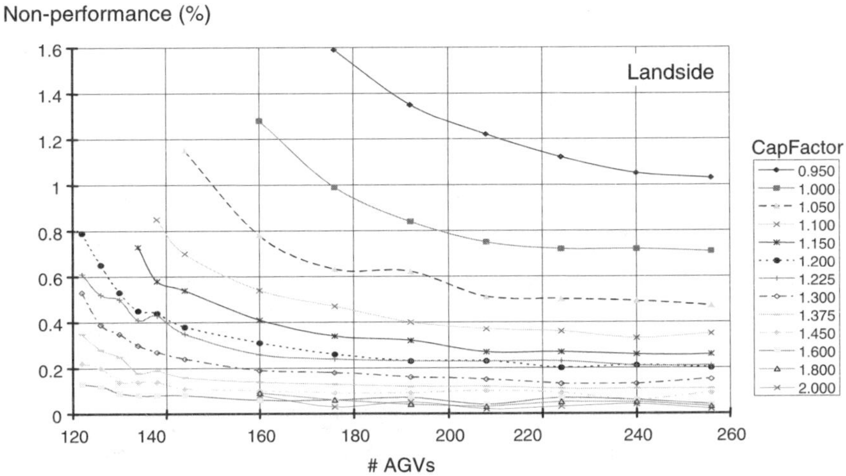

Fig. 7 Non-performance as a function of the number of AGVs—landside

destination terminal, this loss of time may lead to non-performance. This is a typical disadvantage of batch-processing.

Figure 11 shows the irregular occurrence of non-performance. It is not evenly spread over the weeks, but in peaks, as it is strongly related to workload peaks, such as departures of trains and barges.

3.2 Vehicle characteristics

In this section, we show figures about the number of active vehicles. Although the figures are for one particular case and particular week, they are exemplary for all

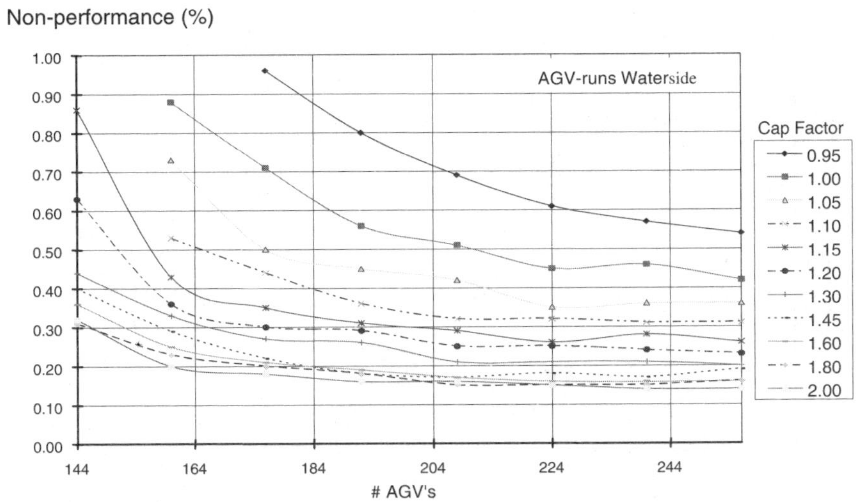

Fig. 8 Non-performance as a function of the number of AGVs—waterside

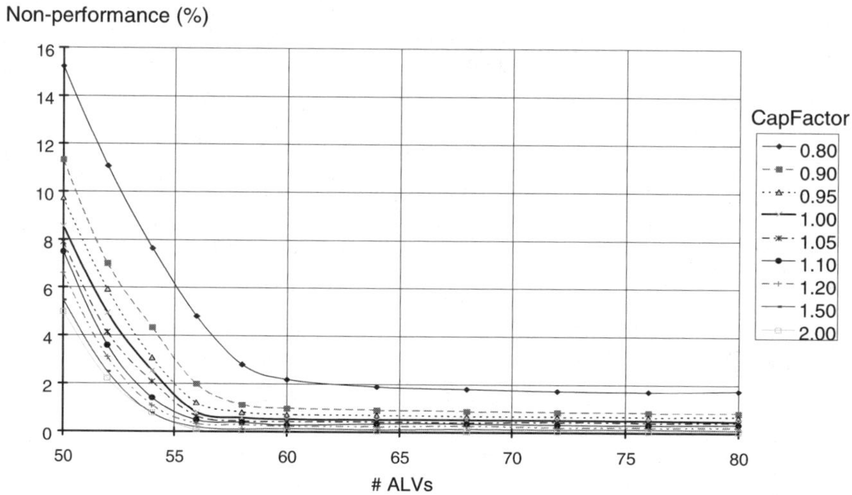

Fig. 9 Non-performance as a function of the number of ALVs

weeks and cases. First, we show the variations in the number of active AGVs in Fig. 12, next we show the number of busy ALVs in Fig. 13.

Notice that the use of the ALVs is much more constant in time than the use of the AGVs. This is presumably due to the fact that they can work independently of the cranes, while the AGVs are often engaged in waiting, which gives much more fluctuations. In Fig. 14, we show the number of active MTS and FTF in case of standard and advanced planning.

Notice the difference between the MTS case with and without advanced planning. In the first case, more combinations between forward and return trips are made, with a lower utilisation of the FTFs as a result. In the case of control and standard planning, the FTFs are continuously sent somewhere: they are running

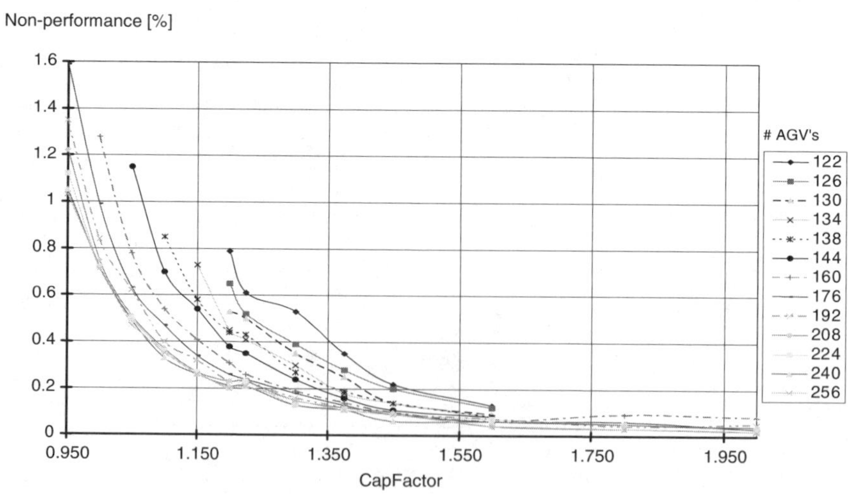

Fig. 10 Non-performance as a function of the CapFactor—landside

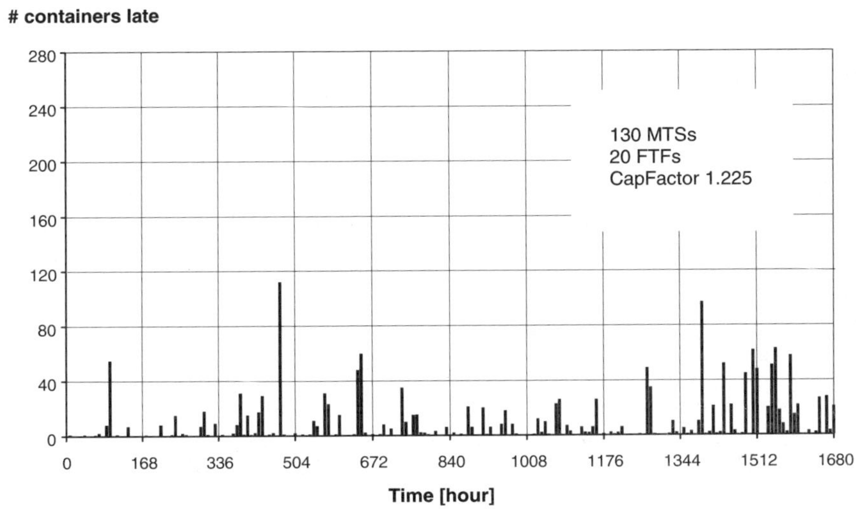

Fig. 11 Number of late containers per shift

behind their jobs all the time. Using an advanced planning does reduce the number of MTSs needed somewhat, but the number of FTFs needed cannot be reduced, presumably because the transport peaks determine this number.

The average number of MTSs in use is about 50% of the number available. However, during some hours, many more vehicles are used (see Fig. 14). Reducing the number of MTSs below 130 or the number of FTFs below 20 leads to an unacceptable high level of non-performance. The ratio between loaded trips and empty trips is 3:2. The utilisation of FTFs is rather high, with an average of 70%,

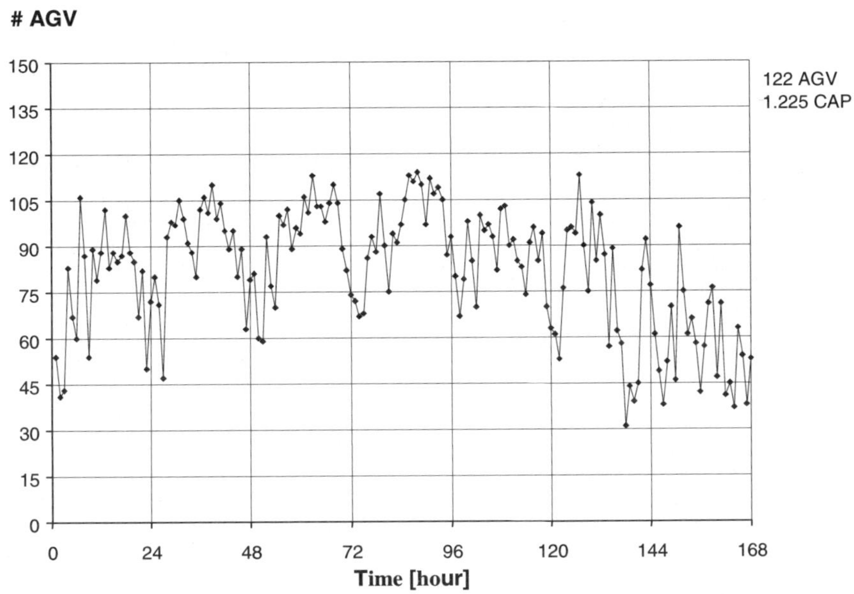

Fig. 12 Number of active AGVs per hour during a week (AGV landside case)

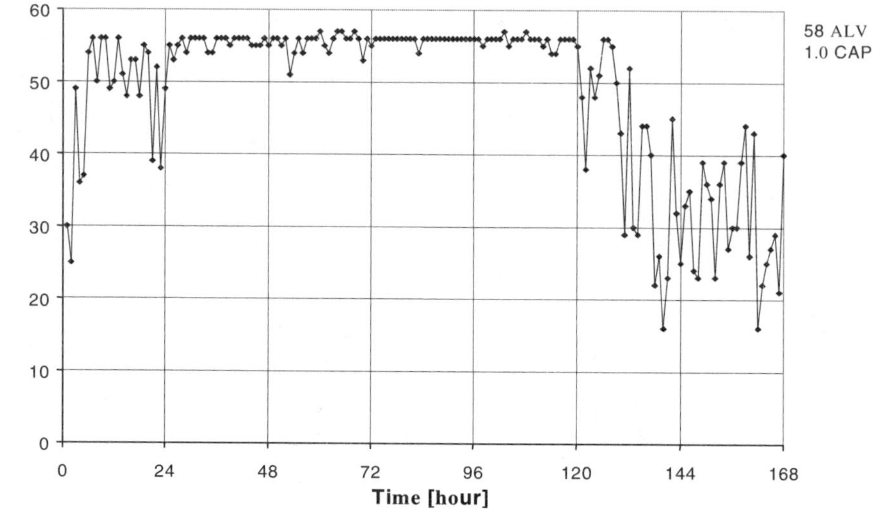

Fig. 13 Number of busy ALVs per hour in a week

with a ratio between pulling an MTS and not pulling of 13:1. The average number of AGVs in use is rather constant, about 90. In peak situations, some 130 AGVs are needed to avoid an unacceptable high level of non-performance. The ratio between loaded trips and empty trips is 50:50. Notice the very large fluctuations in the number of waiting AGVs in Fig. 15.

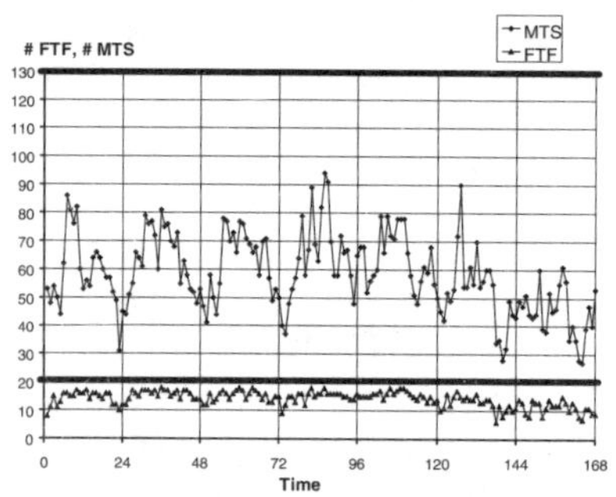

Utilisation of vehicles (FTFs and MTSs), with adv. planning

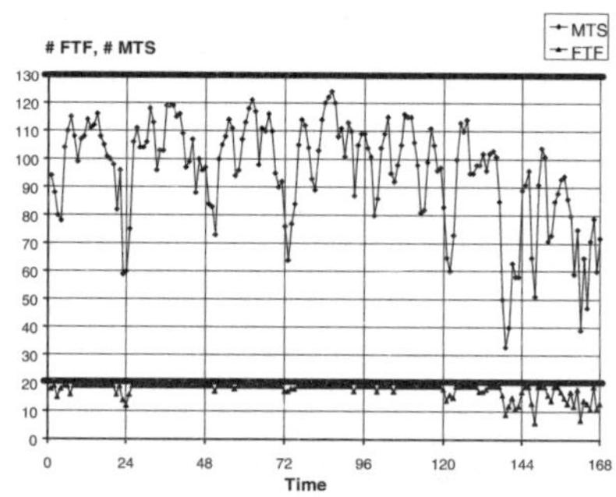

Utilisation of vehicles, without adv. planning

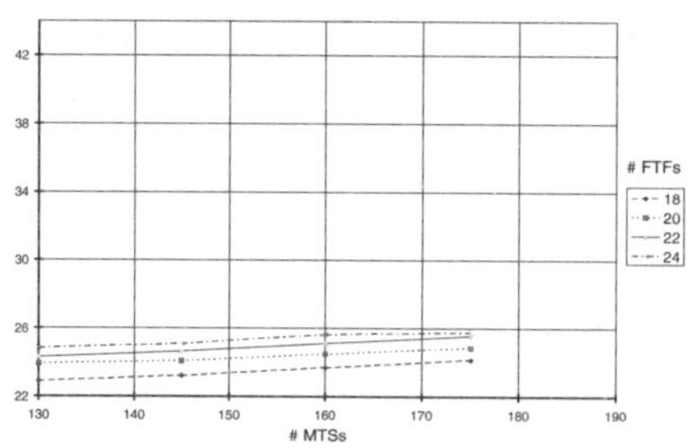

Waiting queues of MTSs to be unloaded, with adv. planning

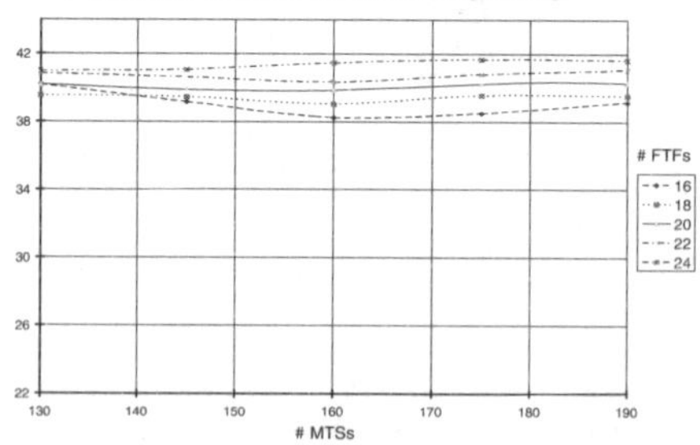

Waiting queues of MTSs to be unloaded, without adv. planning

Fig. 14 Number of active MTSs and FTFs (average per hour) in a week with and without advanced planning

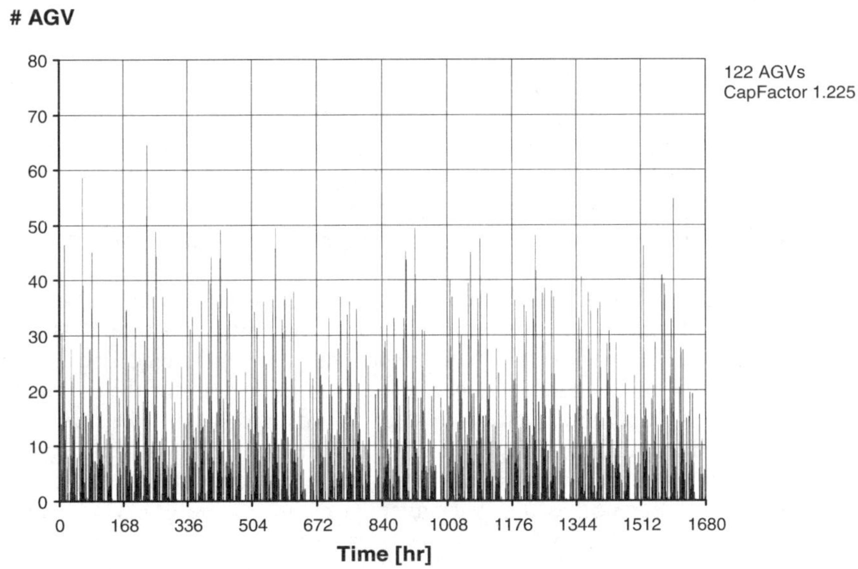

Fig. 15 Number of waiting AGVs (average per hour) at the RSC

The average number of busy ALVs is about 50, which is much less than the number of AGVs needed for the same performance. The main difference with AGVs is that ALVs are not kept in a waiting queue at the handling centres, like AGVs (see Fig. 15). The loaded-trip-to-empty-trip ratio for ALVs is 5:3.

Figure 16 shows the distribution of the load utilisation (number of loaded containers units/load capacity) of MTSs. A peak is shown at 10, which means a 100% loading utilisation. The peak at 2 is due to urgent containers (two TEU correspond to a single 40-ft container), which are loaded on an MTS and then sent away, while the other load positions at the MTS remain empty. The average load utilisation is only 62%, which is caused by the impossibility to combine containers with different destinations on the same MTS.

```
Total                        Excluding zero             Minimum            1.000
 Entries         70194        Entries        70194      90% Quantile       9.711
 Mean                6.175    Mean               6.175  95% Quantile       9.856
 Std.Deviation       3.422    Std.deviation      3.422  Maximum           10.000

        Range Numb Perc Cum %  |  10|  20|  30|  40|  50|  60|  70|  80|  90| 100|
                               |
 <=     0.00     0  0.0   0.0  |
 <=     1.00 5289  7.5   7.5  |
 <=     2.0011523 16.4  24.0  |
 <=     3.00 4704  6.7  30.7  |
 <=     4.00 7023 10.0  40.7  |
 <=     5.00 3253  4.6  45.3  |
 <=     6.00 4991  7.1  52.4  |
 <=     7.00 2437  3.5  55.9  |
 <=     8.00 3468  4.9  60.8  |
 <=     9.00 3192  4.5  65.4  |
 <=    10.0024314 34.6 100.0  |
  >    10.00    0  0.0 100.0  |
```

Fig. 16 Load factor MTS

3.3 Equipment characteristics

To investigate the possibility of sharing terminal equipment, the moves of several handling centres at a terminal have also been added up. Figure 17 shows the aggregate moves at the DMU terminal. Note the lower level of activities during the weekend. It is clear that levelling out of peaks occurs, although the remaining peaks are substantial. This is due to deadlines for all kind of departing transport means. It also means that we cannot save much by exchanging equipment between handling centres.

Figures 18 and 19 show the difference in cycle times of a rail crane when using MTSs or AGVs. In the latter case, AGVs travel to the correct position next to the train, while the MTS has to position itself with a number of containers along the train. We did assume that the loading sequence at the train is random for the dispatching handling centre, as the position on the train depends on many factors like size, weight and destination of a container, implying that not all containers of a handling centre are likely to be next to each other on the train; see e.g. Corry and Kozan (2006). Therefore, the RSC crane has a shorter average cycle in case of AGV transport (0.024 h) than for the MTS case (0.034 h), resulting in higher productivity.

3.4 Cost characteristics

In Table 1, the average and maximum numbers of vehicles in use are given for a realised non-performance of 1.0%; the results in this table are the cost-optimal solutions for each variant; hence, the differences in the value of the CapFactor. In the multi-trailer system, the MTS is a trailer combination with a capacity of ten TEU, while in the AGV and ALV systems, the capacity of a vehicle is one container (1 or 2 TEU). A much larger transport capacity (in TEU) in the MTS case is needed than in the AGV and ALV systems to reach the same non-performance target.

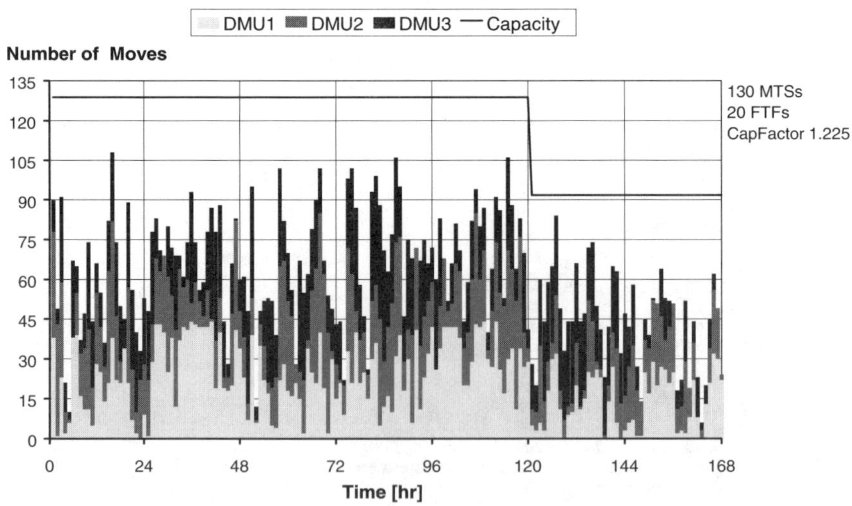

Fig. 17 Aggregate number of moves per hour at DMU (1, 2 and 3) in the first week

Total		Excluding zero			
Entries	19601	Entries	19601	Minimum	0.008
Mean	0.034	Mean	0.034	90% Quantile	0.052
Std.Deviation	0.015	Std.deviation	0.015	95% Quantile	0.061
				Maximum	0.111

	Range	Numb	Perc	Cum %
<=	0.005	0	0.0	0.0
<=	0.010	0	0.0	0.0
<=	0.015	1	0.0	0.0
<=	0.020	506	2.6	2.6
<=	0.025	2426	12.4	15.0
<=	0.030	3095	15.8	30.8
<=	0.035	3077	15.7	46.5
<=	0.040	2909	14.8	61.3
<=	0.045	2332	11.9	73.2
<=	0.050	1789	9.1	82.3
<=	0.055	1270	6.5	88.8
<=	0.060	704	3.6	92.4
<=	0.065	458	2.3	94.7
<=	0.070	286	1.5	96.2
<=	0.075	174	0.9	97.1
<=	0.080	110	0.6	97.6
<=	0.085	119	0.6	98.2
<=	0.090	100	0.5	98.8
<=	0.095	76	0.4	99.1
<=	0.100	67	0.3	99.5
<=	0.105	43	0.2	99.7
<=	0.110	38	0.2	99.9
<=	0.115	19	0.1	100.0
<=	0.120	2	0.0	100.0

Fig. 18 Move times in hours at HC 19 (RSC1) when using MTSs

Total		Excluding zero			
Entries	19601	Entries	19601	Minimum	0.009
Mean	0.024	Mean	0.024	90% Quantile	0.038
Std.Deviation	0.015	Std.deviation	0.015	95% Quantile	0.057
				Maximum	0.112

	Range	Numb	Perc	Cum %
<=	0.005	0	0.0	0.0
<=	0.010	0	0.0	0.0
<=	0.015	8	0.0	0.0
<=	0.020	2242	11.4	11.5
<=	0.025	9587	48.9	60.4
<=	0.030	3456	17.6	78.0
<=	0.035	1349	6.9	84.9
<=	0.040	707	3.6	88.5
<=	0.045	443	2.3	90.8
<=	0.050	316	1.6	92.4
<=	0.055	222	1.1	93.5
<=	0.060	227	1.2	94.7
<=	0.065	207	1.1	95.7
<=	0.070	148	0.8	96.5
<=	0.075	117	0.6	97.1
<=	0.080	118	0.6	97.7
<=	0.085	88	0.4	98.1
<=	0.090	82	0.4	98.6
<=	0.095	74	0.4	98.9
<=	0.100	56	0.3	99.2
<=	0.105	63	0.3	99.5
<=	0.110	57	0.3	99.8
<=	0.115	30	0.2	100.0
<=	0.120	4	0.0	100.0

Fig. 19 Move times in hours at HC 19 (RSC1) when using AGVs

Table 1 Overview used transport vehicles and capacity factor for a service level of 99%

	MTS standard planning	MTS advanced planning	AGV land side	AGV waterside	ALV
Average vehicle use	FTF 16.5 MTS 106.4	FTF 13.5 MTS 64.2	84.8	95.0	50.0
Maximum	FTF 18 MTS 145	FTF 18 MTS 145	122	144	60
CapFactor	1.23	1.15	1.23	1.15	0.90

In Table 2, we show the utilisation of the vehicles. The differences between MTS control and MTS planning are caused by differences in utilisation rate and waiting times; MTS-planning is more efficient. However, the peak levels for both variants are identical. When using AGVs, much less vehicles are needed and the utilisation rate of the vehicles is higher. In the waterside variant, more AGVs are needed than in the landside variant because the distances are longer. In the ALV variant, the number of needed vehicles is the lowest, their utilisation is very high and their waiting times are minimal. From a logistic point of view, this seems to be the best choice.

A cost analysis by Incomaas (1994) resulted in the cost curves shown in Fig. 20. Actual costs were confidential. The service level (100% minus the non-performance) is shown at different costs for the MTS, AGV and ALV options. The costs are directly related to the CapFactor (capacity at the handling centres), the number of vehicles and the usage of the vehicles. At each cost level, the combination of CapFactor and number of vehicles with lowest non-performance is determined for all options. These points are used to draw the lines in Fig. 20.

It can be concluded that the robotised ITT (AGVs and ALVs) achieves the best service level at the lowest costs. Notice that for both the AGV and the ALV approaches, the service level rapidly increases to a plateau level if their number increases. The level of this plateau depends on other bottlenecks in the system, e.g. limited transfer cranes, or peaks in the demand for transport. The MTS solutions seem

Table 2 Utilisation (as % of time) of the transport vehicles and equipment given capacity from Table 1

% time	MTS std. planning	MTS adv. planning	AGV land s ide	AGV waterside	ALV
Idle	26.6	55.7	30.5	34.0	16.7
Trip fully loaded	5.4	5.3	19.3	18.2	49.1
Trip empty	4.0	3.5	17.2	16.4	29.9
Loading	3.1	3.3	3.4	2.9	2.2
Unloading	3.1	3.3	3.4	3.0	2.2
Waiting for loading	8.9	8.2	4.0	3.6	0.0
Waiting for unloading	25.1	15.4	22.2	22.1	0.0
Waiting for FTF	23.5	5.2			
Handling equipment idle	50.7	46.8	50.6	37.4	32.8

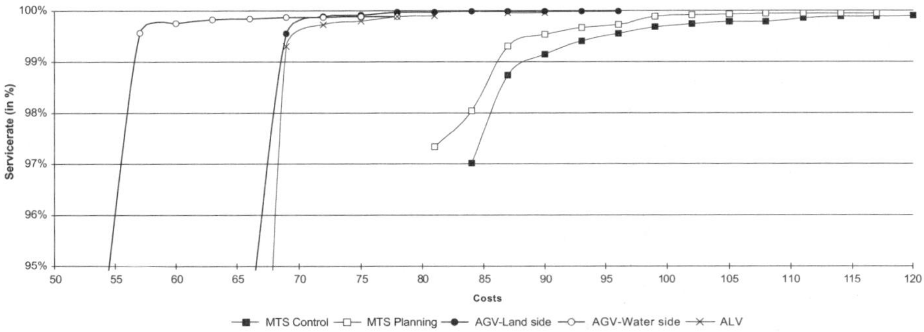

Fig. 20 Service rate vs costs

to converge slower to a somewhat lower plateau. This may be due to the combination of the MTS with the trucks as well as in the lower flexibility of this system.

4 Conclusions

Stochastic fluctuations in the workload determine the number of ITT vehicles and equipment required, and in that sense it is too expensive to ban out all non-performance, in contrast with studies that focus on vehicles needed for quay transport (Vis and Harika 2004 and Yang et al. 2004). A non-performance of less than 1% can be achieved for all the options investigated. The lowest non-performance rates attained with the AGV and ALV options are smaller than those attained by the MTS option. The number of times that non-performance occurs at rail or barge terminals when MTS is used is considerably higher than it is when AGV and ALV options are used.

For each option, an estimate of the minimum number of ITT vehicles needed to provide an acceptable performance can be given. These numbers, however, still depend on the terminal capacities. The most economic combinations of numbers of vehicles and terminal handling capacities are derived from calculations of costs.

For all options, the first aim is to achieve the lowest costs at acceptable levels of non-performance. With the MTS option, it appears that to attain an acceptably low non-performance, it is necessary to put a great deal of effort into the control and planning of the ITT vehicles and terminal equipment. The handling of barge and rail traffic is a complicating factor, largely as a result of the batch transport nature of MTS processing.

The lower limit on the number of MTSs required is high because the MTSs are also used as buffers on wheels. There appear to be occasional peaks in the number of MTSs waiting to be unloaded. Moreover, the occupation rate achieved by MTS is only of the order of 50%. A reduction in the number of MTSs seems to lead only to an unacceptable non-performance. This is caused by the fact that there are a number of periods during which a much higher number of MTSs is required and that to be available for use, an MTS must not only be free, but must also be at the right place. This last requirement becomes increasingly important as ITT flows become less balanced. Further research into this phenomenon is recommended.

The number of AGVs needed is strongly influenced by the buffer function that the AGVs fulfil at the terminals and, thus, is closely related to the terminal capacities. The utilisation rate is approximately 70% on the land side and 65% at the waterside. About half of the trips are made by loaded vehicles, but when fewer vehicles are used, the non-performance is greatly increased. Here, too, in peak situations, almost all the vehicles are required, and the use of a free AGV is largely determined by the place where it is located and the time at which it will be required.

The number of ALVs needed is less than half the number of AGVs. Here, the role of the terminal capacities is smaller than that of the other options because an ALV can load and unload itself and can, thus, work independently of the terminal equipment. A problem that arises is that there is limited space for containers to be placed on the ground or picked up by ALVs. The utilisation rate of ALVs is 85%, the percentage of loaded trips being 60%. The ALV option can be regarded as setting a benchmark for individual automated container transport.

The utilisation of the terminal equipment is very variable. In most cases, the average utilisation rate is lower than 50%, but the peak occupation rates are frequently 100%. Combination of the work loads at different handling centres and terminals shows that levelling out of peaks occurs. This may create opportunities to share equipment. By extending the model, it will be possible to conduct further research into this aspect.

The handling strategies used in the model for the RT and the RSC are reasonably satisfactory. However, they are complex problems on their own, requiring more research. It is difficult to serve the RSC and BSC with the MTS. The loading rate of an MTS remains low and the numbers of late barges and trains are much higher than those of the AGVs or ALVs.

Acknowledgements The research described in this paper was coordinated and partially funded by the research school TRAIL. The authors are grateful to Ruud van der Ham, Anko Nagel from ECT and Frank Nooijen, from Nooijen Consultancy for their contributions and to the referees for useful comments. They further acknowledge the financial support from Connekt.

Appendix

OD-matrix on weekly basis

	DMU	DSL	DDE	DDW	ED1	ED2	ED3	ED4	BSC	RT	RSC	DP	
DMU	0	288	348	348	154	0	288	730	1,557	366	807	536	5,422
DSL	288	s	248	248	76	0	134	346	270	442	192	288	2,532
DDE	348	248	0	0	116	0	212	558	384	366	462	424	3,118
DDW	348	248	0	0	116	0	212	558	384	366	462	424	3,118
ED1	77	38	58	58	0	0	0	0	77	19	77	0	404
ED2	0	0	0	0	0	0	0	0	58	19	58	0	135
ED3	154	76	116	116	0	0	0	0	173	38	135	0	808
ED4	385	192	288	288	0	0	0	0	423	115	365	0	2,056
BSC	1,786	308	442	442	19	19	38	115	0	0	0	58	3,227
RT	519	326	326	326	6	4	19	19	0	0	0	96	1,641
RSC	807	384	616	616	17	13	38	96	0	0	0	308	2,895
DP	326	192	288	288	0	0	0	0	173	154	500	0	1,921
	5,038	2,300	2,730	2,730	504	36	941	2,422	3,499	1,885	3,058	2,134	27,277

References

Celen HP, Slegtenhorst RJW, Ham RTh van der, Nagel A, Berg J van den, Vos Burchart R de, Evers JJM, Lindeijer DG, Dekker R, Meersmans PJM, Koster MBM de, Meer R van der, Carlebur AFC, Nooijen FJAM (1999) FAMAS—NewCon: Phase 1: Starting points; Phase 2: Architecture integrating information system, CTT publicatiereeks 32 (in Dutch)

Corry P, Kozan E (2006) An assignment model for dynamic load planning of intermodal trains. Comput Oper Res 33:1–17

Evers JJM, Koppers SAJ (1996) Automated guided vehicle traffic control at a container terminal. Transp Res A 30(1):21–34

Hoenders CGA (1994) Position paper, RSC-Rotterdam, Rotterdam

Incomaas (1994) Report definition phase. Rotterdam (in Dutch)

Kurstjens STGL, Dekker R, Dellaert NP, Duinkerken MB, Ottjes JA, Evers JJM (1996) Planning of inter terminal transport at the Maasvlakte. Working paper. TRAIL Research School, Delft

Le Anh T (2005) Intelligent control of vehicle-based internal transport systems. Ph.D. Thesis, Erasmus University Rotterdam

Le Anh T, de Koster MBM (2006) A review of design and control of automated guided vehicle systems. Eur J Oper Res 171(1):1–23

Liu C-U, Jula H, Ioannou PA (2002) Design, simulation, and evaluation of automated container terminals. IEEE Trans Intell Transp Syst 3(1):12–26

Liu C-U, Jula H, Vukadinovic K, Ioannou PA (2004) Automated guided vehicle system for two container yard layouts. Trans Res C 12:349–368

Meeusen KP, Evers JJM (1994) Verkenning operationele inzetbaarheid van een IT-hefvoertuig. TRAIL Onderzoekschool, Delft (in Dutch)

MUST (1992) Simulation software. Upward systems, The Netherlands

Steenken D, Voss S, Stahlbock R (2004) Container terminal operation and operations research—a classification and literature review. OR Spectrum 26:3–49

Vis IFA, Harika I (2004) Comparison of vehicle types at an automated container terminal. OR Spektrum 26:117–143

Yang CH, Choi YS, Ha TY (2004) Simulation-based performance evaluation of transport vehicles at automated container terminals. OR Spectrum 26:149–170

Zeigler BP, Praehofer H, Kim TG (2000) Theory of modelling and simulation, 2nd edn. Academic, San Diego

Rajeeva Moorthy · Chung-Piaw Teo

Berth management in container terminal: The template design problem

Abstract One of the foremost planning problems in container transshipment operation concerns the allocation of *home berth* (preferred berthing location) to a set of vessels scheduled to call at the terminal on a weekly basis. The home berth location is subsequently used as a key input to yard storage, personnel, and equipment deployment planning. For instance, the yard planners use the home berth template to plan for the storage locations of transshipment containers within the terminal. These decisions (yard storage plan) are in turn used as inputs in actual berthing operations, when the vessels call at the terminal. In this paper, we study the economical impact of the home berth template design problem on container terminal operations. In particular, we show that it involves a delicate trade-off between the service (waiting time for vessels) and cost (movement of containers between berth and yard) dimension of operations in the terminal. The problem is further exacerbated by the fact that the actual arrival time of the vessels often deviates from the scheduled arrival time, resulting in last-minute scrambling and change of plans in the terminal operations. Practitioners on the ground deal with this issue by building (capacity) buffers in the operational plan and to scramble for additional resources if needs be. We propose a framework to address the home berth design problem. We model this as a rectangle packing problem on a cylinder and use a sequence pair based simulated annealing algorithm to solve the problem. The sequence pair approach allows us to optimize over a large class of packing efficiently and decomposes the home berth problem with data uncertainty into two smaller subproblems that can be readily handled using techniques from stochastic project scheduling. To evaluate the quality of a template, we use a dynamic berth allocation package developed recently by Dai et al. (unpublished manuscript,

---

Part of this work was done when the second author was at the SKK Graduate School of Business, Sungkyunkwan University, South Korea.

R. Moorthy · C.-P. Teo (✉)
Department of Decision Sciences, NUS Business School, National University of Singapore,
1 Business Link, 117592, Singapore
E-mail: bizteocp@nus.edu.sg, rajeeva.moorthy@gmail.com

2004) to obtain various berthing statistics associated with the template. Extensive computational results show that the proposed model is able to construct efficient and robust template for transshipment hub operations.

Keywords Container logistics · Transshipment hub · Sequence pair · Project management

1 Introduction

Mega container terminals around the world routinely handle more than 10 million TEU of cargo and serve thousands of vessels in a year. Efficiency of container operations (along berth and within yard), to certain extent, determines the competitiveness of the terminals within the global shipping network. This depends on a delicate coordination of various expensive resources, including the deployment of quay cranes and crews, allocation of prime movers and drivers, planning and deployment of yard resources etc.

Port operations planning can be broadly classified into the following categories:

- *Strategic* planning deals with long-term issues, such as strategic alliances with shipping lines, infrastructure development to support volume growth, etc. A major exercise in this phase is to identify proper allocation of major/feeder services to different terminals or various sections within a terminal to ensure quick vessel turnaround and transship containers in short time windows.
- *Tactical* planning deals primarily with midterm berth and yard planning issues. A berth template and an associated yard template are usually drawn so as to minimize berthing delays and operational bottlenecks. The tactical plans follow the general guidelines laid out in the strategic plans and is a primary driver of the operational planning phase.
- *Operational* planning involves more detailed equipment and manpower deployment plans, taking into consideration real time operational constraints.

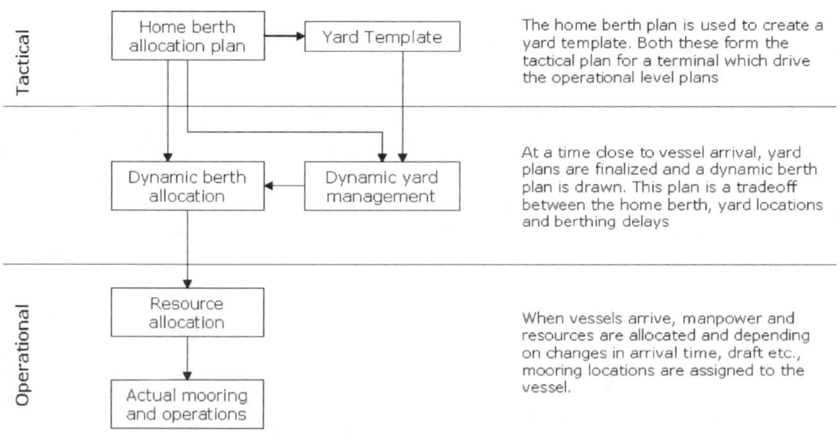

Fig. 1 Tactical and operational planning before mooring a vessel

These plans broadly follow the tactical plans and changes are dynamically made so as to satisfy customer service demands.

In this paper, we study a tactical problem motivated by the operation in a large container terminal, where close to 80% of the containers handled are designated for transshipment to other destinations. Figure 1 shows the major activities and the influence of the tactical plans on the operational level planning and execution.

Planning the yard is critical because the containers, after being unloaded from a vessel, will be moved to an area in the yard to wait for the arrival of a connecting vessel. The designated storage location in the yard and its distance from the mooring positions of the connecting vessels along the quay determine to a large extent the workload needed to carry out the transshipment operations. In general, finding a proper storage plan for the transshipment containers is the main challenge confronting the yard planners in the terminal. Designing these storage plans, however, requires prior knowledge of the mooring locations and time-of-arrival of the vessels. To address this issue, the terminal operator currently assigns a *home berth* location for each vessel calling at the terminal on a weekly basis. Note that in the current operational environment, almost all vessels calling the terminal follow a fixed cycle of 7, 10, or 14 days, of which majority arrive on a weekly schedule. The storage plan for the transshipment containers is designed by assuming that these vessels will be moored at the designated home berths upon arrival.

Figure 2 shows two possible solutions for the home berth allocation problem. The horizontal axis shows the scheduled time of call, which depends on the ship's schedule (thus not entirely within the control of the terminal operator) and scheduled departure time of a vessel, whereas the vertical axis shows the berthing location assigned. There are two groups of vessels (i.e., $\{3,4\}$ and $\{1,2,5\}$) belonging to two shipping lines that call at the terminal. In this example, let us assume that containers are exchanged only between vessels within a group. The key question we would like to address in this paper is this: *Which one of the two templates should we use for the home berth allocation*?

To minimize the amount of work for container operations, the template on the left is desirable because vessels belonging to the same group are moored in close proximity to each other. The transshipment containers in each group can thus be

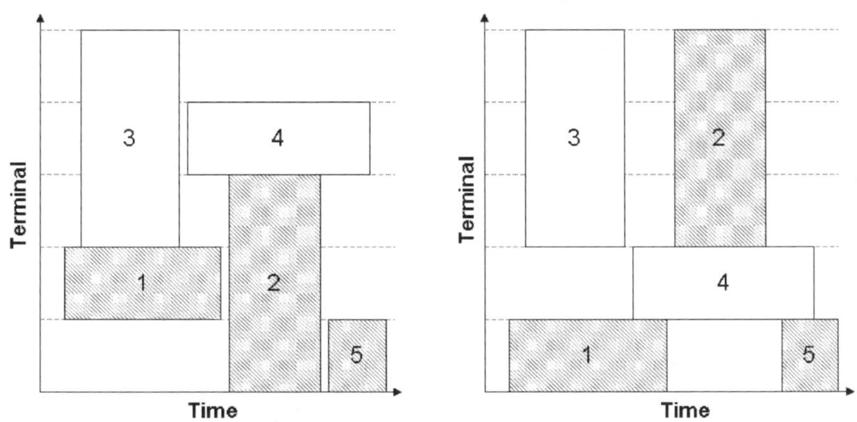

Fig. 2 Two solutions to the home berth allocation problem

stored in the same area of the yard, reducing the amount of work and distance covered by the prime movers. However, in reality, the actual time of call of each vessel will normally deviate from its scheduled time, and the processing time needed to service a vessel may vary. This results in the necessity to adjust the mooring locations of the vessels along the quay. For instance, if the departure of vessel 1 is delayed for an extended time, the operation for vessel 2 will be adversely affected. The delays may propagate and spread to other vessels (such as 5). In practice, the planners may have to moor a vessel far away from its home berth, just to cut down the waiting time of the vessel. In this case, the template on the right of Fig. 2 may be more desirable, as there are sufficient gaps (buffers) in front of the scheduled time-of-call of each vessel. The template is thus more *robust* to unexpected delays experienced by the vessels, and this leads to a more stable and reliable mooring plan, even though it may result in higher container handling cost.

Finding a good home berth template is clearly a difficult combinatorial problem, because we have to search through exponentially many different ways to assign a location to each vessel. Note that the time-of-call of the vessels are preannounced by the shipping lines and cannot be changed by the template designer. We also need to address the associated problem of finding a good way to evaluate the *robustness* of a home berth template. To this end, we identify two primary objectives used in container operations:

- *Service level-waiting time*. This is defined as the time elapsed between the actual time-of-call at the port and the beginning of the mooring operation along the berth. A vessel is said to be *berthed-on-arrival* (BOA) if the mooring operation commences within 2 h of arrival. The BOA statistics is often used as a proxy to gauge the quality of service provided by the port operator.
- *Operational cost-connectivity*. The actual movement cost (containers move from quay to yard storage location and, subsequently, from storage to quayside to be loaded to the connecting vessel) is difficult to estimate and depends also on the storage plan of the containers. As a proxy, we approximate the movement cost with the following: Let x_i and x_j denote the berthing locations of vessels i and j (measured with respect to the midpoint of the vessels) and c_{ij} the number of containers to be exchanged between vessels i and j. If vessel i arrives before j, then c_{ij} denotes the number of containers that need to be transferred from i to j. On the other hand, if vessel j arrives before i, then c_{ij} denotes the number of containers that need to be transferred from j to i. The connectivity cost is defined to be $c_{ij} \times d(x_i, x_j)$, where $d(\cdot,\cdot)$ is a properly selected distance function. In reality, the effort required to transport containers depends on their storage locations in the yard. However, as a policy, most of the containers are stored close to the berth where the vessel on which they should be loaded will be moored. Hence, using the berthing locations of the vessels to compute connectivity cost is acceptable in reality.

We use the two opposing objectives to develop an approach to evaluate the relative performance of different home berth templates. Figure 3 shows the efficient frontiers for two different templates: B is clearly a better design compared to A, as it attains a higher service level at a lower operating cost.

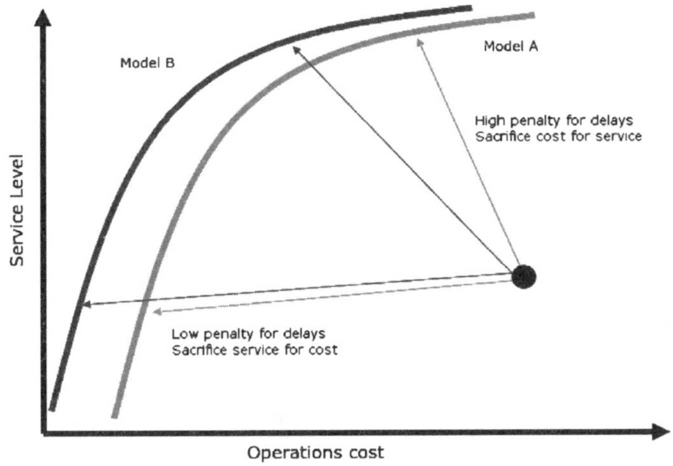

Fig. 3 The efficient frontier for two different templates

In the rest of this paper, we develop a methodology to design a good home berth template, and we use the dynamic berth allocation planning package developed in Dai et al. (unpublished manuscript, 2004) to estimate the efficient frontier attained.

1.1 Literature review

For an introduction to the terminal operations, we refer to Murty et al. (2005). The operational issues involved in managing a container terminal are vividly described, and they identify a number of operational planning problems. The foremost of these is the berth assignment problem. It is also highlighted that data uncertainty plays a critical role in decision making, and the authors specifically recommend techniques to solve some of the operational issues. The paper, however, deals only with operational issues in the terminal and furthermore does not solve the berth planning problem.

Most of the papers in the berth planning literature focus on the combinatorial complexity of the static berth planning problem, where the objective is to obtain a nonoverlapping berth plan within a given scheduling window (cf. Brown et al. (1994); Lim (1998); Chen and Hsieh (1999) and the references therein). In all these papers, it is assumed that some preferred berth location is already known. The focus is to penalize deviation from the preferred locations and to penalize for excessive waiting times. The connectivity cost component is not considered. Dai et al. (unpublished manuscript, 2004) solve the berth planning problem for the dynamic case when vessels arrive over time. They provide a set of stability conditions when the arrival information is random using the tools from stochastic processing networks. They also provide a local search method for obtaining good berth plans using the sequence pair approach.

The berth template problem is also related to a variant of the two-dimensional rectangle packing problem, where the objective is to place a set of rectangles in the plane without overlap so that a given cost function will be minimized. In typical rectangle packing papers, the objectives are normally to minimize the height or area used in the packing solution. Imahori et al. (2003), building on a long series of work by Murata et al. (1996, 1998), Tang et al. (2000) etc., propose a local search method for this problem, using an encoding scheme called sequence pair. Their approach is able to address the rectangle packing problem with spatial cost function of the type $g(max_i p_i(x_i), max_i q_i(t_i))$ where p_i, q_i are general cost functions and can be discontinuous or nonlinear, and g is nondecreasing in its parameters. Given a fixed sequence pair, Imahori et al. (2003) showed that the associated optimization problem can be solved efficiently using a Dynamic Programming framework.

The technique in this paper builds on the sequence pair concept in this series of works. We extend the concept to handle the situation of rectangle packing on a cylinder (instead of the plane). We use this technique to handle the complexity inherent in the combinatorial explosion of the number of different possible template designs. We also integrate this method with results in stochastic project scheduling, which allows us to analyze the impact of arrival time uncertainties on waiting times of the vessels for a class of templates. Note that the problem of evaluating expected delay in a project network can be cast as one of determining the expected longest path in a network with random arc lengths. The problem is well-studied in the project management community.

2 The sequence pair approach

In this section, we describe how we structure the search over all possible template designs, using an encoding scheme called sequence pair (represented by a pair of permutations of the vessels). Each pair of permutation corresponds to *a class* of templates, where the connectivity cost and waiting time objectives can be evaluated efficiently. We use the standard simulated annealing scheme to search the space of all sequence pairs.

Note that a container terminal is divided into a number of linear stretches of berthing space called wharfs, which are further subdivided into sections, called berths, depending on the draft.

In the template, the berthing time is fixed at the scheduled arrival time of the vessels. The decision variables in the home berth problem is essentially x_i, the berthing location of vessel i within the terminal. However, as we are also interested in the delays experienced by each vessel, we let t_i (another decision variable) denote the *planned* berthing time of vessel i, where t_i may be larger than $\mu(r_i)$, the scheduled (expected) arrival time of the vessel. The difference between t_i and $\mu(r_i)$ is the planned delays for vessel i. The actual delays experienced by each vessel are more complicated, as they depend on the (random) arrival time of several other vessels calling at the terminal. We will develop a technique in the next section to capture the expected delays due to the home berth allocation decisions. (See Table 1 for the list of notations used in this paper.)

A priori, it is not clear whether we can design a home berth template where all the vessels are nonoverlapping, because there may be instances where the demand for terminal space exceed the total available space within the terminal. Another complication lies in the periodic schedule operated by the vessels, as we need to ensure that the packing near the two boundaries are properly aligned to avoid excessive overlapping of terminal space, as one move from the end of 1 week to the beginning of the next.

Table 1 Notations used in the paper

l_i	Length of vessel i
p_i:	Expected length of port stay upon berthing by vessel i
w_i:	Cost for delaying vessel i. This can be interpreted as the vessel's priority class.
P:	Number of hours in the planning horizon (we use 24×7 h for a weekly template)
M:	Number of berths in the terminal
L_i:	Length of berth i, $i = 1, \ldots, M$

If we ignore the inherent periodicity of the weekly arrival schedule and the issue of overlapping vessels in the template, the layout of the home berth template (cf. Fig. 2) can be conveniently viewed as a packing of rectangles in a two-dimensional (time–space) plane, where the berthing time of the vessel i is taken to be $\mu(r_i)$, the scheduled (expected) arrival time of the vessel. However, the periodicity of the schedule introduces added complexity to the problem—the template is more suitably viewed as packing rectangles on a cylinder with circumference P.

For ease of exposition, we will first review the sequence pair concept for packing on a plane. Consider the template of two vessels as shown in Fig. 4. We define a pair of permutations H and V associated with the template and construct them with the following properties:

- If vessel a is on the right of vessel b in H, then vessel b *does not see* vessel a on its LEFT-UP view.
- Similarly, if vessel i is on the right of vessel j in V, then vessel j *does not* see vessel i on its LEFT-DOWN view.

It is clear that, given any (nonoverlapping) template, we can construct a pair (H, V) (need not be unique) satisfying the above properties. For the rest of the paper, we write $a <_H b$ (and $a <_V b$) if a is placed on the left of b in H (respectively in V). For any two vessels a and b, the ordering of a, b in H, V essentially determines the relative placement of vessels in the packing.

- If $a <_H b$ and $a <_V b$, then a does not see b in LEFT-DOWN or LEFT-UP, i.e., vessel b is to the right of vessel a. In other words, vessel b can only be berthed after vessel a leaves the terminal.

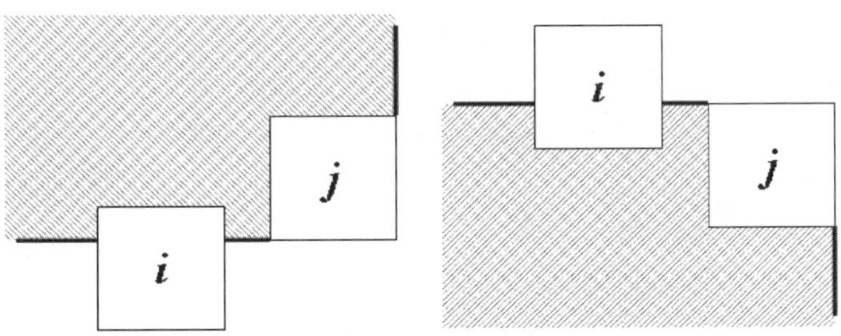

Fig. 4 The figure on the *left* shows the LEFT-UP view of vessel j, whereas the figure on the *right* shows the LEFT-DOWN view of vessel j in the time–space plane. In the former, j cannot be on the right of i in H, whereas in the latter, j cannot be on the right of i in V

– If $a <_H b$ and $b <_V a$, then a does not see b in LEFT-UP and b does not see a in the LEFT-DOWN view, i.e., vessel b is berthed below vessel a in the terminal.

For any H and V, either one of the above holds, i.e., either vessel a and vessel b do not overlap in time (one is to the right of the other) or do not overlap in space (one is on top of the other).

Note that every sequence pair (H, V) corresponds to a *class* of templates satisfying the above properties. The constraints imposed by the sequence pairs splits into two classes: constraints of the type $x_i + l_i/2 \leq x_j - l_j/2$ (in the space variables) or of the type $t_i + p_i \leq t_j$ (in the time variables). In this way, finding the optimal packing in this class, given a fixed sequence pair, decomposes into two subproblems: *space* and *time* (cf. Fig. 5). In the space and time graphs, each vessel is represented by a node, and the constraints imposed by the sequence pair are represented by directed arcs.

This decomposition provides the flexibility to address the stochastic issues in the time problem and the connectivity cost minimization in the space problem separately. This feature turns out to be extremely useful for our problem, as the delay experienced by each vessel, for a fixed sequence pair, no longer depends on the home berth location decision, but on the precedence constraints imposed by the time-constraint graph.

With this insight, we obtain the optimal packing by searching among all permutations of H and V. For each sequence pair (H, V), the procedure for solving the time problem is described in the "Estimating the expected delays" section. "Estimating the connectivity cost" section describes a model to solve the space problem. The periodicity of vessel arrivals introduces additional complexity in solving the template problem and we introduce the virtual wharf mark technique in the "Rectangle packing on cylinder" section to address this issue. In "Searching over the sequence pairs," a simulated annealing-based search procedure is described to obtain the optimal sequence pair.

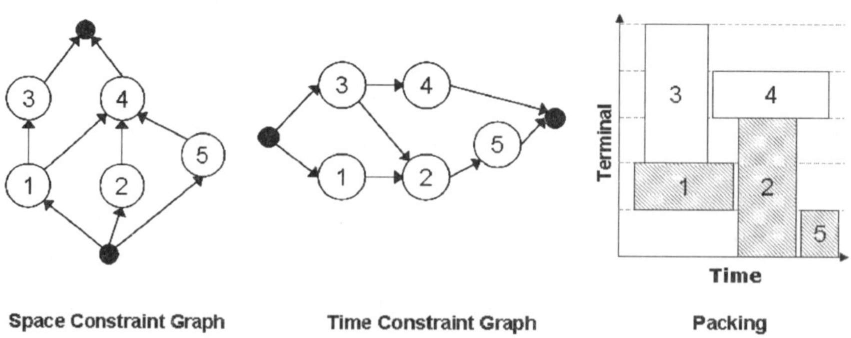

Fig. 5 Directed graphs on the space and time variables along with a packing arising from the sequence pair, H:{3,4,1,2,5} and V:{1,3,2,5,4}

3 Evaluation of the template

3.1 Estimating the expected delays

In this section, we present a derivation for estimating the expected waiting time in a general project management network with release time constraints. The associated precedence relationship in time is shown as the *time-constraint graph* (cf. Fig. 5).

Let r_i and p_i represent the release time and processing time, respectively, for each job in the network. We assume that both r_i and p_i are normal r.v., with mean $\mu(r_i)$, $\mu(p_i)$, and variance $\sigma(r_i)^2$, $\sigma(p_i)^2$. Let S_i represent a minimal set of vessels that should complete before vessel i can be processed. Note that S_i depends on the template obtained. In our example in Fig. 5, $S_1 = \emptyset$, $S_3 = \emptyset$, $S_2 = \{1, 3\}$, $S_4 = \{3\}$, and $S_5 = \{2\}$.

The planned berthing time t_i is easy to determine in this case: it is simply the (random) earliest possible starting time of vessel i. The value t_i depends on the completion time of the jobs in S_i, as

$$t_i = \max\left(r_i, \max_{j \in S_i} (t_j + p_j)\right).$$

The expected waiting time for job i is thus $E(t_i - r_i)$, where

$$t_i - r_i = \max\left(0, \max_{j \in S_i} (t_j + p_j) - r_i\right).$$

Finding the exact distribution of the maximum of a multivariate distribution with an arbitrary covariance structure is a difficult computational problem. In the project management area, a common technique, called the Project Evaluation and Review Technique, identifies a critical (longest) path in the network and uses certain, carefully chosen distribution (such as Beta or Normal distribution) to approximate the longest path duration in the stochastic network. The parameters in the distribution for $t_i - r_i$ are chosen with mean $\mu(t_i - r_i)$ and variance $\sigma^2(t_i - r_i)$,

where

$$\mu(t_i - r_i) = \max\left(0, \max_{j \in S_i} (E(t_j) + E(p_j)) - \mu(r_i)\right), \tag{1}$$

and $\sigma^2(t_i - r_i) = 0$ if $\mu(r_i) \geq \max_{j \in S_i} (E(t_j) + E(p_j))$, otherwise

$$\sigma^2(t_i - r_i) = (\sigma^2(t_{j^*}) + \sigma^2(p_{j^*})) + \sigma^2(r_i) \text{ with } j^* = \operatorname{argmax}\left(E(t_j) + E(p_j) : j \in S_i\right). \tag{2}$$

This approach works well when there is a dominant longest path, i.e., a unique solution to Eq. 1 exists, and this path attains the largest value in most realization of the stochastic longest path problem. However, the variance estimation of the longest path is too conservative, especially if there are many paths with mean path length close to $\mu(t_i - r_i)$ in Eq. 1. Unfortunately, this is normally the case in the template design problem, especially in a heavily congested environment. It is crucial to distinguish between templates where there are only one vs several potential paths, which may delay the berthing of a particular vessel. For this purpose, we need a more refined estimate of the variance of the longest path distribution.

When the underlying distributions are Gaussian, the following result is well-known: Let $X = (X_1, \dots, X_n)$ be a multivariate normal distribution, with identical mean $E(X_j) = m$, and variance $\sigma^2(X_j)$ for all $j = 1, \dots, n$. Let $\max(X)$ denote $\max(X_1, \dots, X_n)$.

Proposition (Borel's Inequality) For all $\lambda > 0$,

$$P(\max(X) - E(\max(X)) \geq \lambda) \leq P(Z \geq \lambda),$$

where $Z \approx N(0, \max(\sigma^2(X_1), \dots, \sigma^2(X_n)))$.

Note that the result does not hold for multivariate normal distribution if the mean values are not identical. It suggests that the maximum variance can be used to approximate the spread of $\max(X)$ above the mean. We use the above insight to improve our estimation of the variance for the longest path.

Fixed a buffer length L, define $S_i(L) \subset S_i$ with

$$k \in S_i(L) \text{ if and only if } (E(t_k) + E(p_k)) \geq \max\left(\mu(r_i), \max_{j \in S_i} (E(t_j) + E(p_j))\right) - L.$$

$S_i(L)$ consists of those paths whose expected length is within the buffer L from

$$\max\left(\mu(r_i), \max_{j \in S_i} (E(t_j) + E(p_j))\right),$$

the earliest time vessel i can be berthed at the terminal. It corresponds to those jobs that may potential block the berthing operations of vessel i. We refine our estimate of the variance of the longest path by:

$$\sigma^2(t_i - r_i) = \begin{cases} 0 & \text{if } S_i(L) = \emptyset \\ \max_{j \in S_i(L)} \left(\sigma^2(t_j) + \sigma^2(p_j)\sigma^2(r_i)\right) & \text{otherwise.} \end{cases} \quad (3)$$

Note that the refined estimate for $\sigma^2(t_i - r_i)$ is larger in Eq. 3 than Eq. 2 and grows with the number of elements in $S_i(L)$. This is desirable as it provides a convenient way to penalize against template where there are many other vessels blocking (within buffer of L hours) the berthing operation of another vessel.

For the rest of this paper, we will approximate the distribution of $t_i - r_i$ by a normal distribution $N(\mu(t_i - r_i), \sigma^2(t_i - r_i))$, with the mean and variance given by Eqs. 1 and 3. Hence,

$$E(t_i - r_i) \approx \int_0^{\infty} (x) \frac{1}{\sigma(t_i - r_i)\sqrt{2\pi}} \exp\left\{ -\frac{1}{2}\left(\frac{x - \mu(t_i) - \mu(r_i)}{\sigma(t_i - r_i)}\right)^2 \right\} dx$$

$$= \int_{\frac{\mu(r_i)-\mu(t_i)}{\sigma(t_i-r_i)}}^{\infty} (y(\sigma(t_i - r_i)) + \mu(t_i) - \mu(r_i)) \frac{1}{\sqrt{2\pi}} \exp\left\{ -\frac{1}{2} y^2 \right\} dy$$

$$= \sigma(t_i - r_i) L\left(\frac{\mu(r_i) - \mu(t_i)}{\sigma(t_i - r_i)}\right) + (\mu(t_i) - \mu(r_i))\left(1 - F\left(\frac{\mu(r_i) - \mu(t_i)}{\sigma(t_i - r_i)}\right)\right)$$

where $L(\cdot)$ and $F(\cdot)$ are the standard unit normal loss and cumulative density function and $\sigma(t_i - r_i)^2 = \sigma(t_i)^2 + \sigma(r_i)^2$. This value can thus be easily computed to a high degree of accuracy by evaluating a simple integral, provided the value $\mu(t_k)$ and $\sigma(t_k)$ can be determined a priori for $k \in S_i$.

3.2 Estimating the connectivity cost

Given a sequence pair (H, V), let G_S be the directed graph associated with constraints involving the berthing location variables x_i . Let $L_i, U_i, \quad i = 1, \ldots, W$, denote the position of the lower and upper end of wharf i in the terminal. The connectivity-cost problem can be formulated as:

$$\begin{aligned}
(P_C) \quad & \min \sum_{i=1}^{N} \sum_{j=i+1}^{N} c_{ij} d(x_i, x_j) \\
s.t. \quad & x_k + l_k/2 \le \sum_{i=1}^{W} U_i y_{ik} \quad \forall \quad k = 1, \ldots, N \\
& x_k - l_k/2 \ge \sum_{i=1}^{W} L_i y_{ik} \quad \forall \quad k = 1, \ldots, N \\
& \sum_{i=1}^{W} y_{ik} = 1 \quad \forall \quad k = 1, \ldots, N \\
& x_i + l_i/2 \le x_j - l_j/2 \text{ if } (i,j) \in G_S \\
& y_{ik} \in \{0, 1\}, \quad i = 1, \ldots, W, k = 1, \ldots, N \\
& x_i \le 0, \quad i = 1, \ldots, N.
\end{aligned}$$

The first three constraints ensure that the vessels are not berthed across different wharfs. By definition, each wharf corresponds to a stretch of linear space along the quay in the terminal. Note that $y_{ik} = 1$ implies that vessel i is berthed in wharf k. The complexity of this problem depends on the structure of $d(\cdot)$. Furthermore, we need an extremely efficient routine to estimate the minimum connectivity cost, because we have to solve this problem repeatedly over a large number of sequence pairs. To this end, it will be convenient if x_i is chosen to be the *smallest* value satisfying the constraints in (P_C), so that a solution to (P_C) can be obtained easily from the space precedence graph G_S, in a recursive manner: For each vertex j, suppose $\max_{i:(i,j)\in G_S}(x_i + l_i/2) - l_j/2 \in [L_k, U_k)$, then

$$x_j = \begin{cases} \max_{i:(i,j)\in G_S}(x_i + l_i/2) + l_j/2 & \text{if } \max_{i:(i,j)\in G_s}(x_i + l_i/2) + l_j < U_k \\ L_{k'} + l_j/2 & k' \text{ is the nearest wharft above } [L_k, U_k), \text{ with } U_{k'} - L_{k'} \geqq l_j. \end{cases} \tag{4}$$

Unfortunately, choosing x_i in the above manner may produce a very bad solution with respect to the objective function $\sum_{i=1}^{N} \sum_{j=i+1}^{N} c_{ij} d(x_i, x_j)$. Furthermore, the periodicity in the vessel schedule introduces additional complexity into this problem. We introduce the notion of virtual wharf mark to address both these issues in the next subsection.

3.3 Rectangle packing on cylinder

We first discuss how the estimation of delays can be modified. Consider Fig. 6, where the template on the left ignores the wrap around effect (periodicity) of the schedule; hence, we can treat $S_4 = S_1 = S_2 = \emptyset$. After computing the values $\mu(t_i)$ and $\sigma(t_i)$ for $i \in \{1, 2, 4\}$, we can proceed to compute $\mu(t_3), \sigma(t_3)$, and $\mu(t_5), \sigma(t_5)$, using $S_3 = \{1\}$, $S_5 = \{1, 2, 4\}$. However, if we factor into the periodicity of the schedule, we find that vessel 5 blocks the berthing operation of vessel 1 and 4 (in the following week).

We address this problem by iteratively computing the first and second moment values using the layout of the template on a plane over a few periods (i.e., few weeks in our case). The overlap and blocking situation in Fig. 6 is thus captured if we look at the layout in the second period (week). We stop the computation after a few iterations so that the wrap -around effect is propagated into the computation. We use this modification as the time cost objective while evaluating the waiting time objective of a particular template.

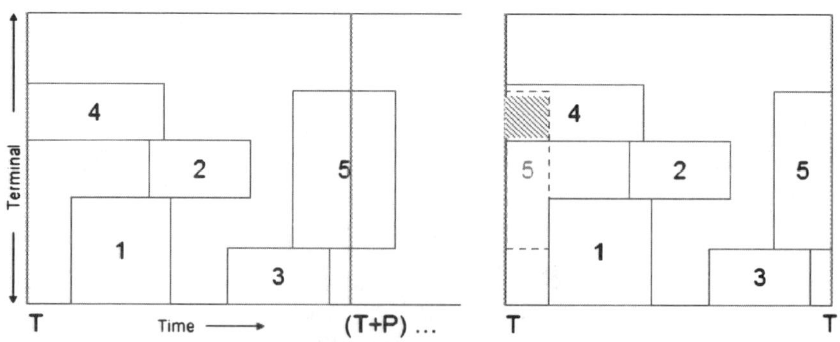

Fig. 6 Template on a plane vs template on a cylinder

We note that, for any nonoverlapping berth template (i.e., packing on a cylinder), it is possible to find a partition such that it can be *unwrapped* to an associated static berth plan (i.e., packing-on-a-plane). Figure 7 shows one such partition obtained by unwrapping. The converse, however ,is not true because we cannot guarantee that the vessels would not overlap when wrapped. Figure 6 shows one such example.

To ensure that vessels can be propped appropriately in the associated static berth plan, we introduce the notion of "virtual wharf mark" (VW). VW's are essentially additional vessels with appropriately chosen r_w, $l_w = 0$ and $p_w = 0$ and with additional lower-bound constraint on the berthing location.

The problem of overlaps due to *wrapping* of the layout in the plane can be avoided by using the virtual wharf mark technique. For example, Fig. 8 shows how the overlap between vessel 2 and 4 (after wrapping around) can be eliminated through the introduction of a virtual wharf mark w, with appropriately chosen lower bound on the berthing location, and arrival time (in this case, T) for w. In Fig. 8, vessel 4 can be kept propped by maintaining $(4, w)$ and $(w, 4)$ in H and V sequences, respectively.

In general, obtaining the appropriate wharf mark set and their associations with the vessels is difficult, as it changes with the sequence pair. Furthermore, a large number of wharf marks may be required to prop all the affected vessels to appropriate height in the layout. Obtaining the set of virtual wharf marks and appropriate lower bounds in the optimal solution is nontrivial, and our strategy is to introduce VW as and when required. However, the dynamically introduced VW aids in enlarging our search space, thereby allowing us to search over a larger neighborhood. More details on implementation involving the virtual wharf mark is presented in "Searching over the sequence pairs."

Because the berth template is tactical in nature, there may be instances where overlaps (in the template) cannot be avoided. These are instances where many vessels need to be serviced at the same time according to the vessel performa. To take care of this issue, we can also allow for vessel overlaps in the packing and try to minimize them by adding a large penalty term to the objective function.

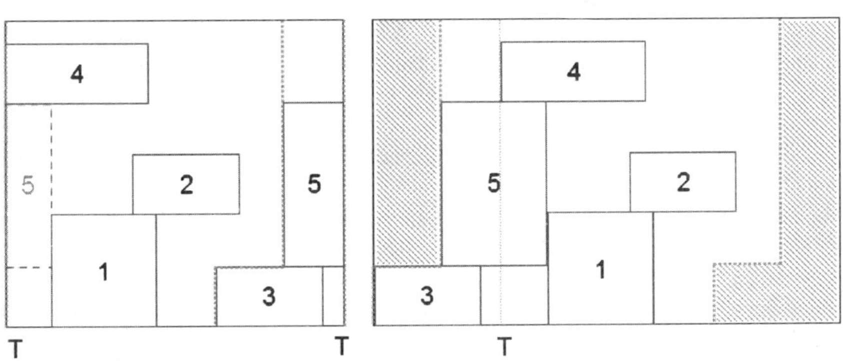

Fig. 7 Unwrapping a feasible template

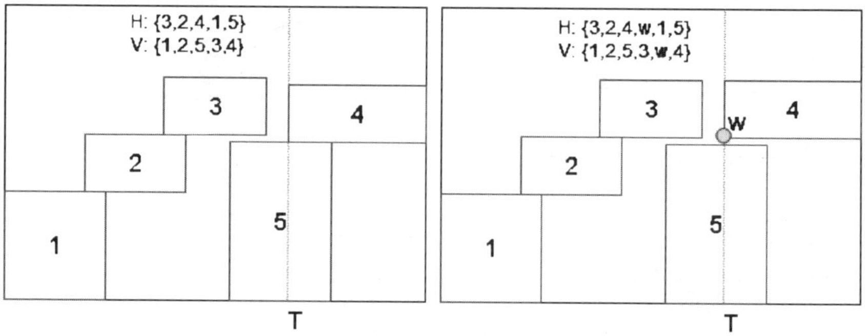

Fig. 8 A virtual wharf mark can be used to add constraints for propping vessels as shown

4 Searching over the sequence pairs

Searching for the optimal sequence pair is nontrivial, and we use a simulated annealing-based local search procedure similar to the one proposed in Dai et al. (unpublished manuscript, 2004). We employ the standard swap and shift neighborhoods and a modification of the greedy neighborhood presented in Dai et al. The greedy neighbor simply represents all the templates that can be obtained by visually manipulating the vessel locations while keeping the berthing time fixed. While manipulating the location, we ensure that the vessels do not overlap even after wrapping the template.

The critical aspect for getting good solutions is to define an appropriate cooling schedule and a sufficiently large neighborhood that can be explored efficiently. With regard to the latter, we use the following standard structures (cf. Dai et al., unpublished manuscript, 2004):

(a) *Single swap* This is obtained by selecting two vessels and swapping them in the sequence by interchanging their positions. Single swap is defined when the swap operation is performed in either H or V sequence.
(b) *Double swap* Double swap neighborhood is obtained by selecting two vessels and swapping them in both H and V sequences.
(c) *Single shift* This neighborhood is obtained by selecting two vessels and sliding one vessel along the sequence until the relative positions are changed; i.e., if $i, j, \ldots, k, l$ is a subsequence, a shift operation involving i and l could transform the subsequence to $j, \ldots, k, l, i$. There are many variants of this operation depending on whether vessel i (or l) is shifted to the left or right of vessel l (or i). We define single shift as a shift operation along one of the sequences.
(d) *Double shift* This defines the neighborhood obtained by shifting along both H and V sequences.

Figure 9 shows examples of the above operations and their impact on the packing obtained. The above neighborhoods described are simple perturbations of the sequence pair, but they result in remarkably different packing when compared visually. Our next neighborhood structure, however, is obtained using visual manipulation of the rectangles in the packing.

(e) *Greedy neighborhood* Given a sequence pair H and V and the associated packing $\mathcal{P}$, we evaluate all possible locations that vessel i can take, with the

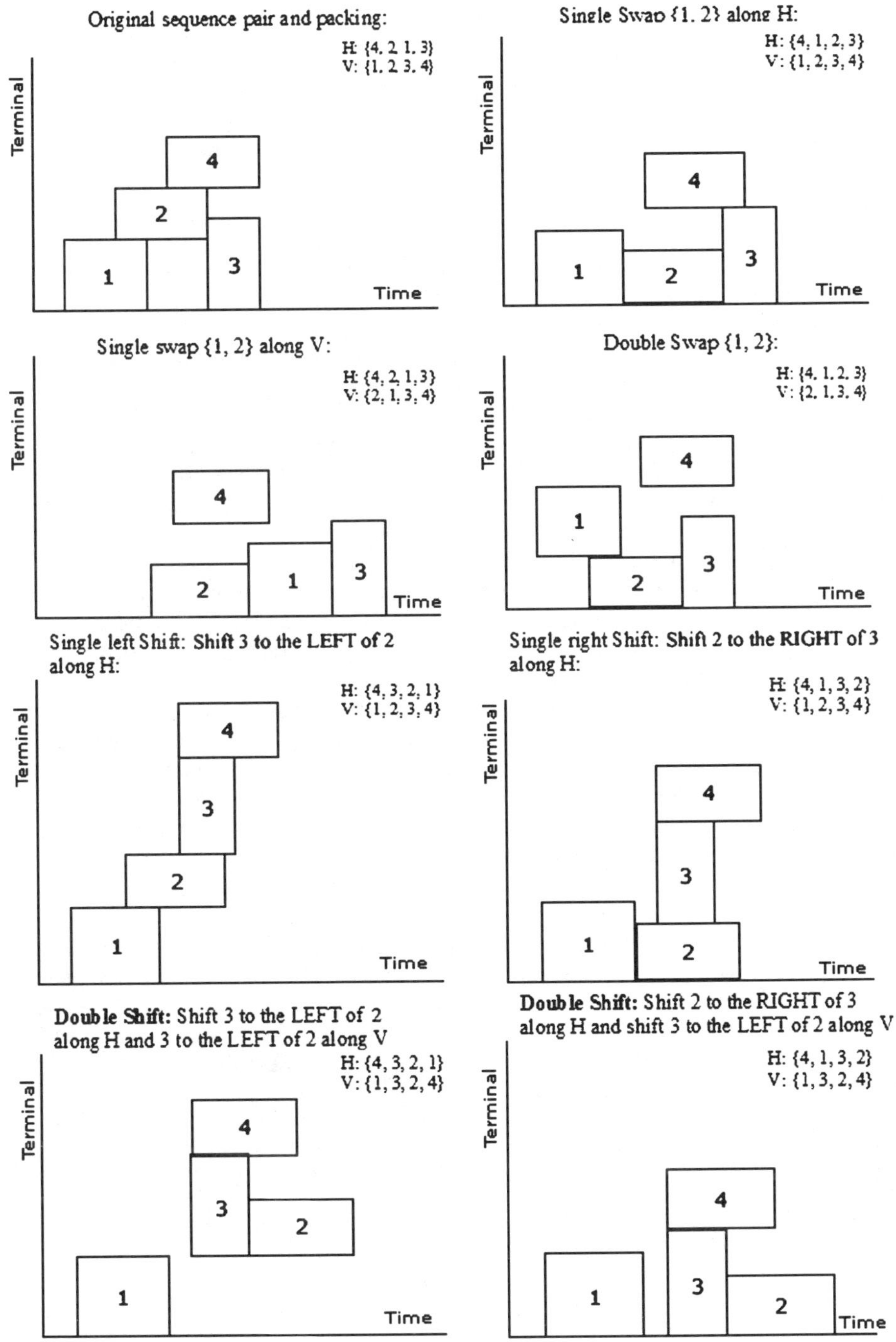

Fig. 9 Examples of swap and shift neighborhoods

rest of the vessels fixed in their respective positions. If there is a better location for the vessel, then we set the berth location of i to its new location. Note that because the time is kept constant, it is easy to check whether there is an overlap along the space dimension. Once the vessel is placed in the new location, we repeat the procedure for the rest of the vessels, until no improvement is possible.

Figure 10 shows the packing and the corresponding sequence pair obtained from a simple greedy neighborhood. The greedy neighborhood artificially modifies the position of the vessel along the space dimension, while keeping the berthing time decision fixed. Once we find the best berthing location for a vessel and obtain the corresponding sequence pair, we can then proceed to the next iteration of the simulated annealing algorithm. However, because we confine ourselves to packing where the berthing location decisions are obtained via a greedy manner (cf. Eq. 4), we need to ensure that the packing obtained from the greedy neighborhood exploration is suitably propped in the packing with additional VWs. New VWs are thus dynamically added and dropped from the search procedure as we explore the various neighborhoods in the simulated annealing algorithm.

Implementation incorporating virtual wharf mark The problem in adding virtual wharf marks as additional vessels in the search space is that it increases the problem size and, hence, the computation time. Here, we propose a cost-effective way of implementing the approach by employing dynamic lower bounds.

Note that the vessels need to be propped after we employ the greedy neighborhood. Instead of adding virtual wharf marks, the idea is to (dynamically) set lower bounds for the berthing location for those vessels that need to be propped by a virtual wharf mark in the packing.

We retain these lower bounds while exploring the neighborhood using operators (a)–(d) and change the lower bounds only when operator (e) changes the packing and introduces new virtual wharf marks. Searching the greedy neighborhood is computationally more expensive than simple sequence pair manipulation; hence, for the experiments, the operators (a)–(d) are employed successively, while the operator (e) is used whenever the other operators get stuck in a local optima.

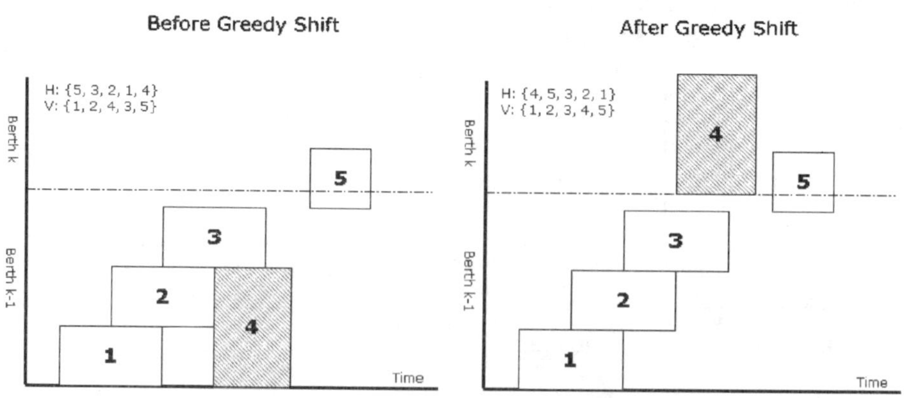

Fig. 10 An example for greedy neighbor

The dynamic lower bounding technique described above is equivalent to adding a virtual wharf mark w_k to i (with w_k coming immediately after i in the H sequence, and w_k immediately before i in the V sequence) and performing all neighborhood searches treating iw_k in H and $w_k i$ in V, as *virtual* vessel. Note that swapping or shifting iw_k with j in H or swapping or shifting $w_k i$ with j in V has the same effect of swapping or shifting i with j in the original H sequence, but maintaining a lower bound (determined by virtual wharf mark w_k) on vessel i in the neighborhood.

5 Computational and simulation results

In the previous section, we described a procedure where good home berth template can be constructed, using the sequence pair manipulation. In this section, we evaluate the usefulness of this approach, by simulating the performance of the template using a dynamic berth allocation package developed recently by Dai et al. (unpublished manuscript, 2004). The simulation allows us to track the berthing performance of the vessels over several months. We use the BOA statistics (percentage of vessels berthed on arrival) and the connectivity cost (based on actual berthing location of the vessels in the simulation) to construct the efficient frontier for the template.

Throughout our simulation, we use the following observation to create realistic input parameters for our model: In a typical port, the connectivity profile normally exhibits the following feature. On the average, each vessel exchanges containers with about 30% of the rest of the vessels calling at the port. Fifteen percent of these container exchanges are in the range $[0, 100]$, and another 15% in the range $[100, 1000]$. We use $d(x_i, x_j) = |x_i - x_j|$ in our experiment.

To simplify our simulation, we also make the simplifying assumption that port stay time is deterministic, and we focus on the importance of capturing the variability of the arrival time of the vessels at the port. The variability of the port stay time is within the control of the terminal management and can be controlled or influenced by deploying appropriate amount of resources. The vessels' arrival time, however, is not within the control of the terminal; hence, it is deemed to play a more important role in the berth management process.

The general simulation environment for the experiments is as follows:

- We use a data set that represents a vessel arrival pattern at a typical port. The expected arrival time of vessels is the same as used in the template. It is assumed that the berth planner does not get any updates on the vessel arrival time between the time the template is drawn and when the actual deployment is done.
- Based on the vessel location in the template, a stepwise constant space cost is generated for the problem. We call the berth that the vessel is expected to be moored in (according to template) as the *preferred berth*. The space cost is considered to be constant within each berth in the terminal. The cost of allocating a vessel in any berth other than the preferred berth is set to be proportional to the distance between the two berths.
- During dynamic deployment, the berth planner plans the berth allocation to a set of vessels arriving within a scheduling window. Of course, as time rolls by, the

scheduling window rolls forward too, hence, including newer vessel arrivals. For the experiments, we set the scheduling window to *48 h*.

- To prevent last-minute reshuffling of resources, the berth plan is frozen for two shifts (i.e., 16 h) from the current time. Any changes in the vessel's arrival time will be accommodated within this window, but the vessel's berth location will remain fixed. This rule is of practical importance for the port operator as it eases the resource bottlenecks.
- We assume that the vessels update their exact arrival time around *8 h* from the current time. For sensible comparison, we generate the actual arrival information (from a normal distribution with mean r_i , and a small variance) beforehand for all the experiments.
- We compute the container movement cost between vessels once the simulation is over. In this step, we assume that only the vessels within the same period are *connected*. Specifically, for a vessel berthed at time t_i' , we assume it to be *connected* only o the vessels berthed between $t_i' - \frac{P}{2}$ and $t_i' + \frac{P}{2}$.
- In dynamic deployment, we consider a vessel to be berthed-on-arrival if it is berthed earlier than $(\hat{r}_i + 2)$ hours. Here, $\hat{r}_i$ is the actual arrival time. BOA directly corresponds to the service levels that a port guarantees to the vessels, and maximizing BOA is a prime concern during deployment.
- The statistics on the percentage of vessels allocated to the preferred berth is collected. This is a measure of the amount of replanning the berth planner has to do when using a template.

5.1 The impact of variability on berthing performance

In practice, the home berth template is usually designed assuming that the vessels' time-of-call is precise and does not fluctuate from the scheduled time of arrival. Furthermore, the planners normally try to design a layout with as few overlaps as

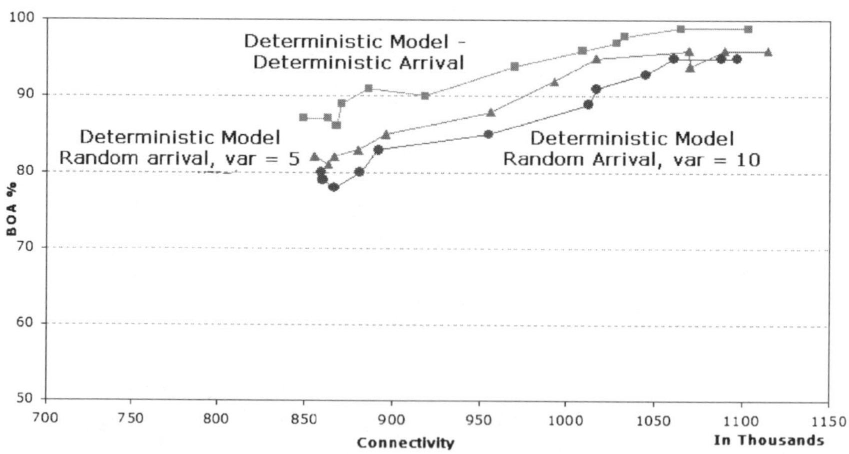

Fig. 11 Plot shows a steady drop in performance as variance in arrival data is increased

possible. We can incorporate these considerations in our home berth model, using $\sigma^2(r_i) = 0$ and a huge, lumpy cost to penalize for overlaps in our template.

Using the template obtained this way, Figure 11 shows the efficient frontiers obtained under three different scenarios: (1) $\sigma^2(r_i) = 0$, (2) $\sigma^2(r_i) = 5$, and (3) $\sigma^2(r_i) = 10$. As expected, the efficient frontier drops steadily from case 1 to 3, as the variability of the arrival data slowly erodes the performance of the home berth template.

5.2 Evaluating the template obtained from the robust model

We design another template, using the robust model outlined in the previous sections. Instead of using lumpy cost to minimize the number of overlapping vessels in the layout, we use the connectivity and delay cost estimation methods proposed in this paper to distinguish between templates. By varying the weights on the connectivity and waiting time cost components, we actually obtain different home berth templates. The planners can then choose the right template to balance the two objective functions in berthing operations.

Sensitivity to parameters Figure 12 shows the variation of connectivity and waiting time cost components (obtained from our model) when the weights of delays (time penalty), with respect to connectivity cost, are increased steadily. We benchmarked against the template obtained using a deterministic approach, i.e., ignoring the information on the variance of arrival time data.

The templates obtained from the stochastic model performs consistently well in terms of waiting time cost component for all time penalty weighing factors. By increasing the weights for time penalty, the model is able to find new templates with slightly better waiting time performance, but at the expense of a large increase in connectivity cost.

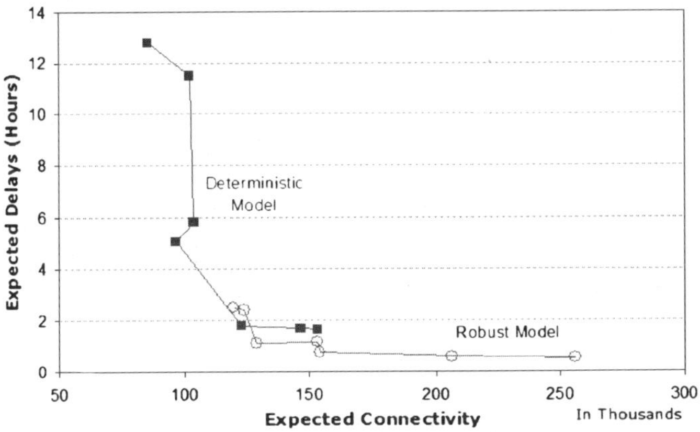

Fig. 12 Comparing robust templates with varying penalties on time. For similar connectivity cost, one can expect to have smaller expected delays while using a robust measure for the waiting time objective

The templates obtained from the deterministic model, however, is very sensitive to the weights used for the time penalty cost function. This suggests that it is very important that the parameters should be set correctly to ensure desirable results from the proposed model. This behavior arises, we believe, because the deterministic model is not able to capture the effect of delays propagation on the rest of the vessels.

Resource conflicts Ideally, the home berth template should have a minimal number of overlap between vessels. This is important because too many overlaps (though quantified in the form of delays) would eventually make the real time vessel deployment problem tough. Figure 13 shows the overlaps obtained from templates produced from the robust model, while we vary the weights on the time penalty cost function.

The robust model, as it is seen, is successful in keeping the number of overlaps down. This is an inherent quality of the model rather than an exception since, while evaluating the waiting time objective using a robust measure, the concerns posed by overlaps are captured implicitly in the waiting time estimation in the model.

Performance of the robust model in simulation Figure 14 shows an efficient frontier comparison for *good* templates. For the sake of fair comparison, we choose the home berth plan that balances the connectivity cost and expected delay, one from each model. The results indicate that the robust model is the better choice. One could choose a minimum service level expected and could pick a template that can achieve the service levels with smaller connectivity cost. For example, the minimum connectivity cost that can be achieved with a service level of 90% is 836,446 and is possible using the robust template.

If one compares the delays experienced by an individual vessel, another inherent property of the robust model is revealed. In the presence of traffic congestion, the robust model would ensure that vessels are distributed so as to minimize the delay propagation. The deterministic model fails to employ the information from the immediate past. This means that specific vessels may suffer recurring high delays when the deterministic home berth plan is used. This observation is corroborated by Fig. 15. The delay experienced by individual vessels fluctuates widely while using the deterministic measure for waiting cost objective, whereas the fluctuation is more controlled while employing the robust measure in evaluating a home berth template. The average delays in the

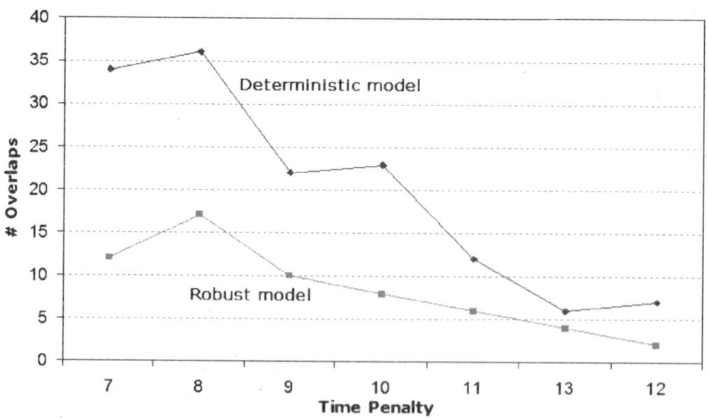

Fig. 13 Variation of overlaps vs weights in time penalty function

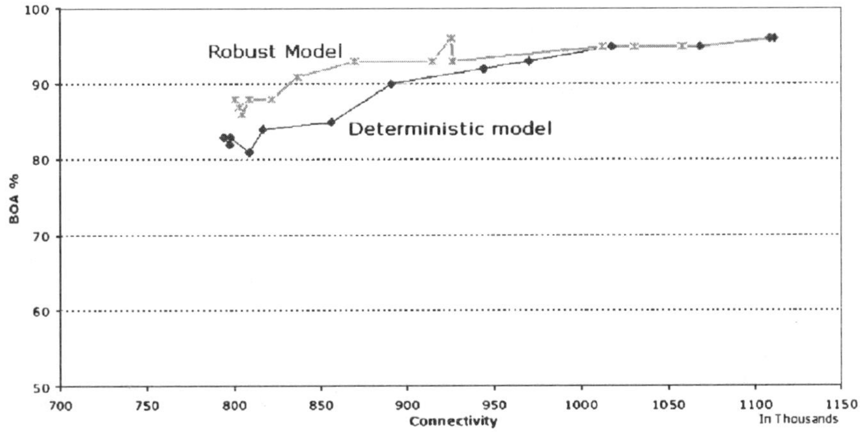

Fig. 14 Efficient frontiers for "good" templates

deterministic model is 1.65 h with a variance of 2.75, whereas the average delay in the robust model is 0.75 h with a variance of 1.13. Furthermore, 27 vessels in the robust model's template have an expected delay of 0 h, as opposed to 13 vessels in the deterministic model.

5.3 Comparison to "optimal" template

It is desirable to obtain an absolute measure comparing the solution produced by the robust model to the optimal template or with respect to a tight bound, as it is important to understand how good the proposed robust model performs. To address this concern, we create an artificial problem instance wherein a real-life berth template is chosen and the connectivity data are modified so that the given template is near optimal (i.e., vessels in close proximity will have higher number of container exchanges). Vessels in the same berth have high connectivity (1,000), vessels in adjacent berths have low connectivity (300), and the connectivity is 0 for the rest.

We did not modify the arrival time information or adjust the layout of the template. As the template has been deployed in practice, the planners must have already visually inspected the layout and are satisfied that all potential bottlenecks have been resolved from the template. Using these data, we create another home berth template using the robust model.

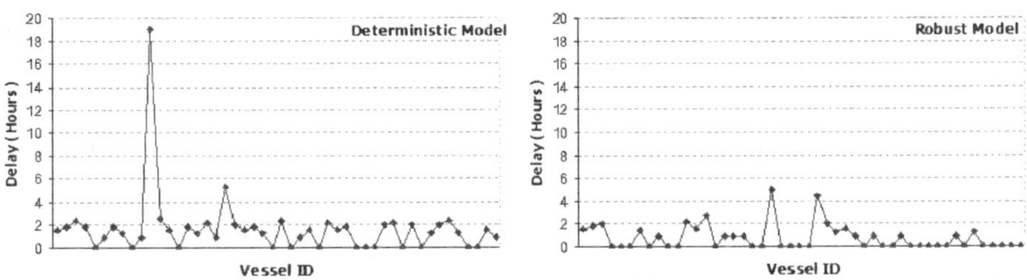

Fig. 15 Expected delays for each vessel based on the templates

The connectivity optimal template and the robust template are as shown in Fig. 16. The respective efficient frontiers are as shown in Fig. 17. We observe that even in this scenario, the efficient frontier obtained from the robust model slightly outperforms the optimal template. The results are promising and one can expect to obtain better dynamic performance in the actual deployment of the robust model.

6 Conclusion and extensions

In this paper, we have addressed an important tactical planning problem in container terminal management, the home berth allocation. The problem is modelled as a bicriteria optimization problem. We study and provide a framework to deal with the trade-off between the operational cost and service levels demanded by customers.

The combinatorial nature of the home berth problem poses a primary challenge and we address it by modelling the problem as one of packing rectangles on a cylinder. We motivate the use of sequence pair for defining the search space and provide a series of extensions to adapt the approach for the problem presented herein. The sequence pair approach naturally decomposes the home berth problem into a space and time subproblem, and we use this insight to address the temporal problem with random inputs. Because the home berth problem is tactical in scope, the temporal data, specifically the vessel arrival time, is stochastic. To address this inherent randomness, we borrow techniques from stochastic project scheduling and explicitly derive an expression for expected delays.

In brief, we address both the combinatorial and stochastic nature of the problem in our proposed framework and create robust home berth allocations, which translates to better service levels and better resource management during actual operations.

In addition, the paper provides extensive computational experiments and simulations to analyze the effect of solving the problem in the robust home berth allocation framework. The results show that the model cannot only effectively

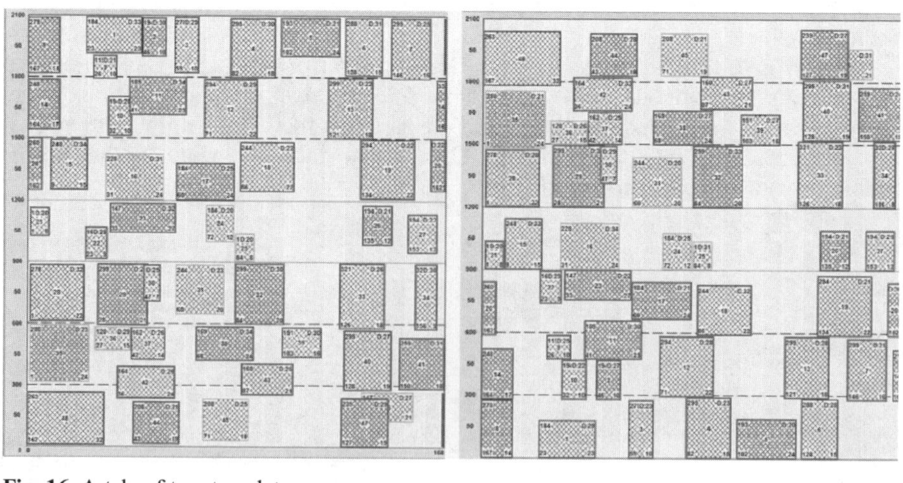

Fig. 16 A tale of two templates

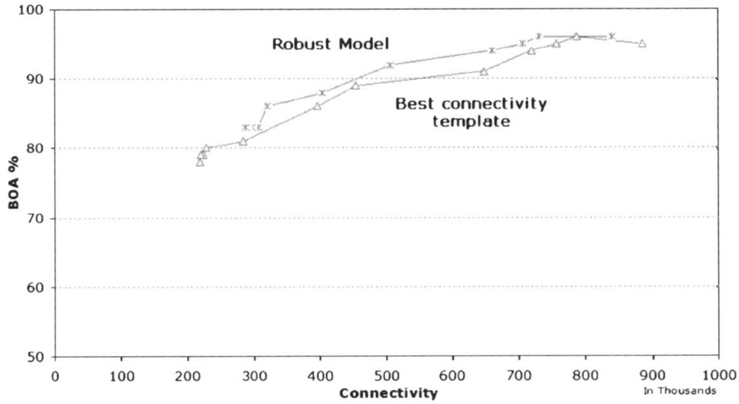

Fig. 17 Efficient frontier analysis shows that the robust model outperforms the "optimal" template

measure the dynamic impact of the randomness but also has other advantages like minimizing overlaps in the template.

To the best of our knowledge, this is the first paper that solves a berth template problem and analyzes the impact of the template on the real time berth allocation. However, our approach is limited by the fact that the template is relevant only when a substantial number of vessels arrive periodically and within the same period. An extension of the problem to create a template for vessels with different periods would be interesting and we leave it for future research. We do not address the problem of crane allocation in this paper. We leave this problem as an extension and note that the crane allocation decision affects the port stay of vessels, and it requires a more sophisticated model and careful study to analyze its impact on the template and during dynamic deployment.

Recent trends in port management has been toward making the operations flexible. This means that megaports would have to frequently redraw customer contracts and appropriately change internal operations viz, yard plans. In such a situation, flexibility in drawing reliable home berth allocation becomes ever more important. We hope that with the model described in the paper, a berth operator can create berth templates that can then be follow with minor modifications during actual deployment.

References

Brown GG, Lawphongpanich S, Thurman KP (1994) Optimizing ship berthing. Nav Res Logist 41:1–15

Chen CY, Hsieh TW (1999) A time–space network model for the berth allocation problem. In: 19th IFIP TC7 conference on system modeling and optimization, Cambridge, England, 12–16 July 1999

Imahori S, Yagiura M, Ibaraki T (2003) Local search algorithms for the rectangle packing problem with general spatial costs. Math Program Ser B 97:543–569

Lim A (1998) On the ship berthing problem. Oper Res Lett 22(2–3):105–110

Murata H, Fujiyoshi K, Nakatake S, Kajitani Y (1996) VLSI module placement based on rectangle packing by the sequence pair. IEEE Trans Comput-Aided Des Integr Circuits Syst 15(12):1518–1524

Murata H, Fujiyoshi K, Kaneko M (1998). VLSI/PCB placement with obstacles based on sequence pair. IEEE Trans Comput-Aided Des Integr Circuits Syst 17:60–68
Murty KG, Liu JY, Wan YW, Linn R (2005) A decision support system for operations in a container terminal. Decis Support Syst 39(3):309–332
Tang X, Tian R, Wong DF (2000) Fast evaluation of sequence pair in block placement by longest common subsequence computation. In: Proceedings of the conference on design,, automation and test in Europe, pp 106–111

Erhan Kozan · Peter Preston

Mathematical modelling of container transfers and storage locations at seaport terminals

Abstract This paper models the seaport system with the objective of determining the optimal storage strategy and container-handling schedule. It presents an iterative search algorithm that integrates a container-transfer model with a container-location model in a cyclic fashion to determine both optimal locations and corresponding handling schedule. A genetic algorithm (GA), a tabu search (TS) and a tabu search/genetic algorithm hybrid are used to solve the problem. The implementation of these models and algorithms are capable of handling the very large problems that arise in container terminal operations. Different resource levels are analysed and a comparison with current practise at an Australian port is done.

Keywords Scheduling · Heuristics · Containers · Seaports · Genetic algorithms · Tabu search

1 Introduction

The introduction of containerisation caused some dramatic changes to the layout at seaport terminals. These changes include alterations to the storage area and the introduction of specialised container handling equipment. However, that main change has been in the storage area. Storage methods have undergone significant modifications to take full advantage of the container-stacking ability. This means more cargo can be stored at the port requiring a smaller area of land.

Berthing time of a container-carrying ship accounts for a considerable proportion of its journey, which concerns shipping lines who wish to minimise the waiting time and berthing time of the ships at the port. Decreasing the

---

E. Kozan (✉) · P. Preston
School of Mathematical Sciences, Queensland University of Technology, G.P.O. Box 2434, Brisbane, Qld. 4001, Australia
E-mail: e.kozan@qut.edu.au, p.preston@qut.edu.au

turnaround time at port would reduce the total travelling time of ships, thus, reducing the cost of transporting containers.

There are no suitable tools available to assist the management in obtaining the optimal efficiency of container terminals. It is necessary to develop a technique that allows managers to better control the terminals by ensuring that container transfers allow maximum throughput, taking into account operating constraints and service reliability. Efficient transfers can defer or eliminate the need for significant infrastructure investment. After achieving optimal efficiency, infrastructure changes and investments can be more accurately considered.

Kozan and Preston (1999) use genetic algorithm (GA) techniques to reduce container handling/transfer times at the multimodal terminals. When containers are stacked to multilevels or high levels, more handling time is needed to retrieve a container at the lower level of the stack. Total throughput time of containers as a function of cranes, forklifts/highstackers and terminal transfer trucks are used to measure the performance of the system. Kozan (2000) discusses the major factors influencing the transfer efficiency of container terminals and a network model is designed to analyse container progress in the system to minimize the total handling and travelling time of containers. The author studies the flow of containers between various locations (ships, berths and different sections of storage yard) of a multimodal terminal and the expected number of handling due to the height of container stacks. The paper considers various types of handling and transfer equipment and the location of containers in the yard.

Kim and Kim (1999) study the routing of a single straddle-carrier in the storage yard. A model is proposed to minimise the distance travelled by the straddle-carriers between yard bays. Kim and Kim (2001) estimate the cost of terminal operation. This paper also suggests a way to estimate the travel time of a transfer crane between yard bays. Bish (2003) investigates the case when import containers are loading to a ship and export containers are unloading to another ship at the same time. The author proposes the "transshipment problem-based list scheduling heuristic" for large size problems. Vis and Koster (2003) give a comprehensive review on the literature relating into recent research on container terminal. The authors suggest that future research needs to extend models for simple cases to a more "realistic situation". Steenken et al. (2004) give classification of problems surrounding terminal operations and suggestions for future research. They also point out that stacking and storage logistics are becoming increasingly important as a result of growth in container traffic and are also getting more complex and sophisticated.

Preston and Kozan (2001a) determine an optimal storage strategy for various container-handling schedules. They minimise the ship turnaround time of container ships by genetic algorithms and design a scheduling model and apply it to container terminals taking into account factors such as container handling equipment, labour resources, storage capacities and terminal layout. Major factors influencing container transfer efficiency are analysed to optimise resource usage resulting in lower operating costs while achieving a desired level of customer service. Tabu search (TS) and genetic algorithm heuristics are used to compare the benchmark of the Fisherman Island Container terminal in Australia. Similarly, optimising the storage location to match a particular transfer schedule is developed by Preston and Kozan (2001b) in a later study, and some improvement could be gained.

Seaport systems are very complex and due to the dynamic nature of the environment, a large number of timely decisions have to be continuously reviewed in accordance with the changing conditions of the system. This complex system, comprising many interrelated subsystems, was individually modelled with relative ease in previous studies. It would be impractical to develop a comprehensive model that incorporates each of the subsystems and the interactions between them. Our approach models each subsystem separately with the addition of a structured feedback system where the output of one model becomes the input of another, thereby, capturing the interactions between them and the integration of two such subsystems. This technique uses small initial improvements and provides feedback to the other half of the problem, gradually increasing the accuracy of the solution as the algorithm progresses. This is where an integrated model is used to combine two separate models developed by Preston and Kozan (2001a,b). The aim of the model is to simultaneously optimise both the storage locations and the handling scheduling. This paper presents an innovative approach to solving an integrated model for seaport systems.

Improvements are needed in these areas to reduce transportation time and streamline the whole container transportation industry. This research examines the sea interfaces, specifically the transfer of export containers from the storage area to ships. However, the formulation and implementation of the model and algorithms are capable of handling the very large container export and import problems that arise in container terminal operations.

2 The problem

When a container vessel calls to port, the containers on board must be unloaded and stored at the port until they are transported further by rail or road. The containers must be stored in a manner so as to minimise the amount of handling needed to place a container in the storage area and to remove it when needed. Therefore, the problem being investigated is minimising the total throughput time which is the handling time for all the containers from ships at berth and the transferring time of the containers to the storage area. When dealing with export containers, the problem would be reversed. That is, the handling time of the containers from when it first arrives at the port until the ship carrying the containers departs from the port.

The method of assigning containers to yard machines for loading/unloading of ships most widely employed is the use of “gut instinct” or heuristics. These approaches may seem to be effective but may actually increase the berthing time of the ships. A better assignment technique may involve an analytical model. Many papers have considered analytical models to replace “gut instinct” methods of loading and unloading containers.

The containers that are remaining must be placed in storage areas until they are needed. The company does not know when or in what order the containers will be called for loading or unloading. Therefore, they must stack the containers in a manner so as to minimise the time taken to retrieve a container by considering the storage area constraints. In the case of exports, the stevedoring company usually knows when a container will depart as it arrives. The stevedoring company charges a fee for containers that are delivered too early in respect to the departure time, and after cut-off times, no containers are received.

After the containers have been unloaded from the container vessel and placed in the marshalling area, they are moved to the storage areas, A_{js}, the multimodal terminal truck area or road intermodal terminal as seen in Fig. 1. The containers are moved from the storage areas to the road intermodal terminals by a multimodal terminal truck for transferring to the hinterland. For use in the model, each of the storage areas are segregated into s parts each, such as $A_{11}, A_{12}, A_{13} \ldots, A_{1s}, A_{21}, A_{22}, A_{23} \ldots, A_{2s} \ldots\ldots\ldots\ldots\ldots\ldots$, and $A_{j1}, A_{j2}, A_{j3}, \ldots.. A_{js}$.

3 The model

When a container ship arrives at the port, management allocates a number of yard machines to service it. (i.e. transfer the import containers to storage or road/rail links and the export containers from the storage area to the berth). The problem is to determine the schedule in which to transfer the containers. The Container Transfer Model (CTM) is used to handle this problem. The objective of CTM is to minimise berthing time of the ship by considering the setup and travelling time for each container. Therefore, CTM minimises the completion of the transfers from a storage area to ship and/or ship to the storage area and determine the optimum schedule.

The containers must be stored in a manner so as to minimise the amount of handling needed to place a container in the storage area and to remove it when needed. Therefore, the Container Location Model (CLM) is designed to determine

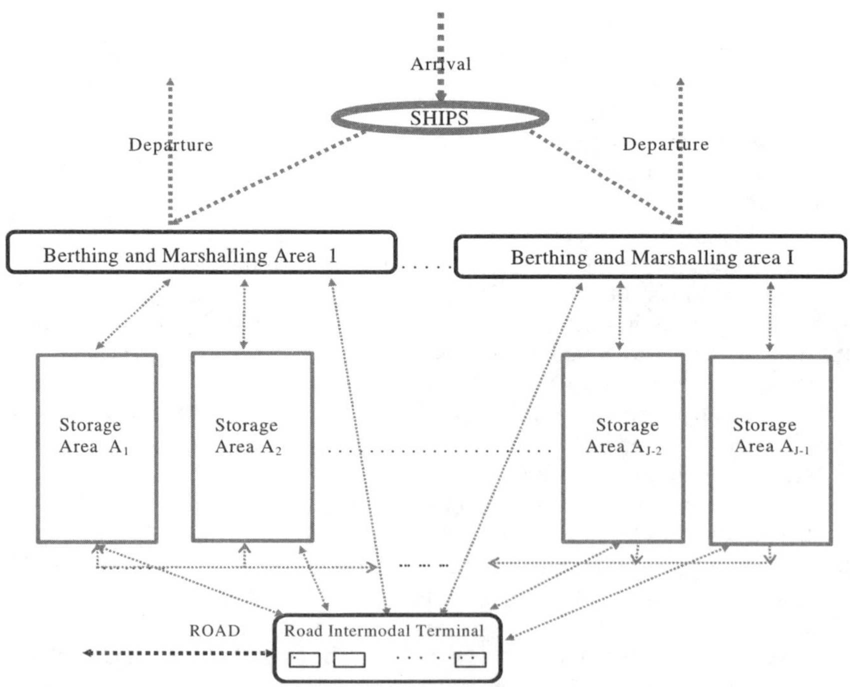

Fig. 1 A layout of a Multimodal Container Terminal

the optimal storage locations for various container-handling schedules by minimising the total throughput time of containers.

The rationale behind integrating CTM and CLM is that it is pointless optimising one when the other is far from optimal. While optimising the transfer schedule for a given storage location, assignment will reduce loading time; it will not provide the best solutions. Similarly, whilst optimising the storage location to match a particular transfer schedule will offer some improvement, more could be gained. This is where an integrated model is used. The aim of the model is to simultaneously optimise both the storage locations and the handling scheduling. The main advantage of using the integrated model over a single model is the solution space. By solving the models iteratively as opposed to that in a single model reduces the feasible search space. The high dependency of the variables and parameters in the two reduced models would also lead to greater complexity of a single model.

In essence, the container transfer and container location models are a decomposition of the real problem. The approach is used to solve the decomposed problem successively. The problem in solving the two subproblems independently is that the decision variables for one are problem parameters (input) for the other. Firstly, CTM is solved for container transfers using random initial storage locations. The output, handling schedule, is then used as input for CLM. The optimal locations of containers determined are then subsequently used as input to CTM. This continues iteratively until a stopping criterion is reached. This technique uses small initial improvements and gradually increases the accuracy of the solution as the algorithm progresses.

The notation of the parameters and variables of the model are detailed below:

cw, rw	The width of a column and row, respectively, in the storage area
t_i	Time at which container *i* is scheduled for handling (movement)
x_i	The row of the storage area partition where container *i* is stored
y_i	The column of the storage area partition where container *i* is stored
$z_{i,t}$	The vertical storage position of container *i* stored at time t_i; this is measured as the number of containers stored on container *i* which delays access by handling equipment
lock	This parameter is defined as the time required by the yard machines to "lock on" to a container before picking it up; it is assumed that the time to "unlock" a container after moving it, is the same
move	When a container is stored below one or more of the others and is required for loading, the upper container/s is/are moved to a temporary storage location to remove the desired container; *move* is the time required moving containers to the adjoining temporary position
mac_i	The yard machine, container *i* is scheduled to be transferred by
ship_i	The ship, container *i* is scheduled to depart on
depart^s	Departure time of ship *s*
arrive^s	Arrival time of ship *s*
v^m	Velocity of the yard machine *m*

travel_i	The time required to transport container *i* between the storage area, marshalling area, track area and/or intermodal terminal
$Travel_i = lock + \frac{x_i * rw + y_i * cw}{v^m} + lock.$ *where* $m=\text{mac}_i$.	This equation defines the travelling time for all containers; the travelling time includes the time to lock and drop off the container at the start and end of the journey
setup_I	This is the time required to move the container(s) stored above the next scheduled container; if the desired container is on top, there is no setup time. If the desired container is not on top of the other containers, the setup time incorporates the time required to move these containers to an adjoining position $setup_i = \begin{cases} 0 & If\ z_{i,t} = 0 \\ z_{i,t} * (4 * lock + 2 * move) + 2 * lock + move & Otherwise \end{cases}$ The setup time has two components one of these is "lock–unlock" time and the other is moving time; in this case, the yard machine must first move the top containers to a temporary storage position, then move the desired container from the storage area and finally, return the top containers from the temporary storage position back to the storage area

Both separate models, CTM and CLM, use the same mathematical model; however, the decision variables and input parameters change. In CLM x_i, y_i and $z_{i,t}$ are decision variables and mac_i, and t_i are input parameters and vice versa for CTM.

The objective of this model is to minimise the time ships spend at the berth. We will minimise time spent transferring containers from a storage area to ship or ship to the storage area. This transfer time is the sum of the setup and travelling time for each container. As the idea is to minimise the completion of the transfers, we want to find the minimum time for the yard machine that is in use the longest by adding the transfer times of the containers allocated to it.

$$Minimise \max_{Mac} \sum\nolimits_{\{i|mac_i=mac\}} (travel_i + setup_i) \tag{1}$$

This equation is designed to find the maximum time any yard machine is in service and minimise this value. This will minimise the time the ship spends at the port and also minimise the total working time of all yard machines.

The location constraints in Eq. 2 are used to satisfy the physical condition that only one container can be stored in a given storage position. The solution program will either move one of the containers to another position or store them on the top of the other. The initial storage locations are also checked for feasibility of the height parameter, ensuring if $z_{i,t}>0$; then, there is, in fact, containers occupying those positions.

$$If\ x_i = x_{i'}\ and\ y_i = y_{i'}\ then\ z_{i,t} \neq z_{i',T}.\ \forall i \neq i' \tag{2}$$

Equation 3 defines the machine constraints which are used to satisfy the physical condition that each yard machine can only be scheduled to handle only one container at a time. The inverse that each container is scheduled exactly once is covered in the definition of the neighbourhood for tabu search.

$$If\ mac_i = mac_{i'}\ ,\ then\ t_i \neq t_{i'}.\ \forall\, i \tag{3}$$

Equation 4 is a modelling constraint used to modify the parameter, $z_{i',t}$, when container i is stored above i' and is scheduled to be loaded before container $i \neq i'$ at time t_i.

$$If\ x_i = x_{i'}\ and\ y_i = y_{i'}\ and\ z_{i,t} < z_{i',t}\ and\ t_i < t_{i'},\ then\ z_{i',t'} = z_{i',t} - 1\ \forall\, t_{i'} \geq t_i, i \neq i' \tag{4}$$

The ship constraint, Eq. 5, ensures that each ship has a time window [time interval (arrive^s, depart^s)] within which loading and unloading service should begin and end.

$$arrive^s + \sum\nolimits_{\{i|ship_i=s\}} (travelling\ time_i + setup\ time_i) \leq depart^s\ \forall s \tag{5}$$

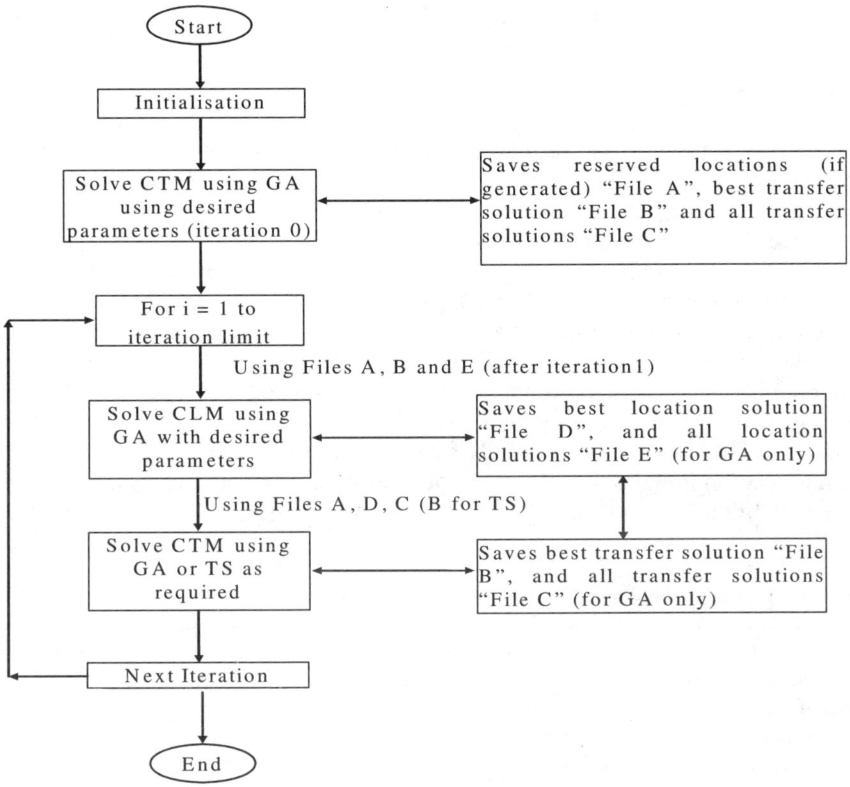

Fig. 2 Flow chart showing feedback for integrated iterative algorithm

Time windows can be fixed where ships will arrive at a certain time and must depart at a certain time or are flexible where ships will arrive at a certain time but the departure date can be negotiated depending on the number of imports and exports amongst other factors. Generally, the time window is fixed because this allows the shipping company to know exactly what port costs are. With flexible time windows, the shipping company may decide to pay extra to allow their ship to be serviced (imports unloaded and exports loaded) quicker or to allow them to stay at port longer.

While the model may seem similar to many job shop machine scheduling, it is greatly different in the way the setup time is determined. The setup time in general job shop machine scheduling problems is dependent only on the job immediately preceding the job in question. In this model, the setup time is a dependent order of scheduling of the containers (if any) initially stored on the top of the container in question. For this reason, the solution is dependent on the order of the whole sequence, not just the immediate predecessors of certain jobs.

The problem is known to be NP-hard; thus, its computation complexity increases exponentially with the number of containers in the schedule. This makes it difficult to solve in reasonable time with the current exact solution techniques, (i.e. branch and bound or tree searches). This implies that for large-size, real-life problems, heuristic techniques have to be used. Genetic algorithm has been applied previously to this problem by Kozan and Preston (1999) with promising results, but the solution times were found to be quite large in some cases.

4 Solution techniques

The two separate models are integrated into a single-solution algorithm and are solved iteratively in an attempt to find a storage arrangement and handling schedule to minimise the turnaround time. This iterative approach allows both models to be optimised, thus, giving a better overall solution.

Two iterative techniques are applied to solve the integrated models. The first one is the nonincreasing (normal) algorithm which has the same number of "generations" within each iteration, and the second one is the increasing algorithm which has the increasing number of "generations" within each iteration. The reason to use nonincreasing and increasing algorithms is to speed up the process. The basis behind this is that the first iteration uses a random initial handling schedule, and there seems to be a little point in finding the best storage locations for this obviously suboptimal schedule only for it to change dramatically after the first iteration. Rather, this technique searches for smaller improvements that gradually

Table 1 The benchmark setup for comparison

Containers for export	500 TEU
Containers in storage area	500 TEU
Storage capacity	2,306 TEU
Yard machines used	10
Storage levels	3
Storage policy	Fixed

Table 2 Mean and standard deviation with varying numbers of chromosomes

Chromosomes	Mean	Standard deviation	CPU Time(s)
10	694.97	7.6697	2,065
20	687.62	8.5158	4,704
26	696.63	7.1774	6,285
50	696.97	7.6934	13,778
100	694.93	8.0800	28,298

increase with each iteration to save needless fine-tuning of solutions in early iterations. This procedure is shown in the following algorithm.

Algorithm

Step 0 The distance matrix of distances from various storage locations (in Cartesian coordinates) to the berth space is generated. The storage allocation is randomly generated. A random schedule for container handling is input ensuring that only one container is scheduled for a particular yard machine at a time. In the case of using genetic algorithm for CTM, a number of these transfer schedules (chromosomes) are generated. The iteration counter is initialised.

Step 1 Run CTM for N iterations. It would be N generations if GA is used rather than tabu search.

Repeat

Step 2 CLM is then run using the best handling schedule found from Step 3 (Step 1 when i=1), Step 1 as the fixed handling schedule. This is run for $M*i$ generations, where i is the iteration number.

Step 3 Run CTM for $N*i$ iterations. (It would be N generations if GA is used rather than TS). This uses the best storage locations found as input and modifies the previously determined solutions. If using GA, then it uses the previous chromosomes as the starting point while TS uses the best solution found in the previous iteration.

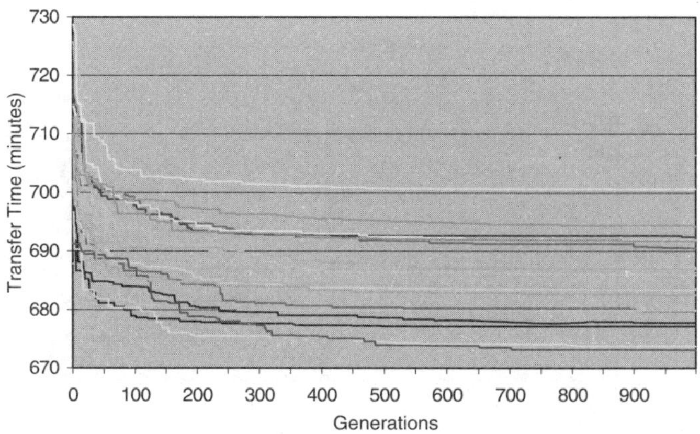

Fig. 3 Transfer time vs number of generations for 20 independent simulations

Table 3 Mean and standard deviation with varying numbers of chromosomes

Chromosomes	Mean	Standard deviation	CPU Time
10	679.39	5.5976	5,283
20	669.02	6.0675	12,751
26	667.49	6.2084	16,887
50	658.63	6.3207	32,751
100	648.55	6.6755	63,321

Step 4 The generation counter (i) is incremented.
Until either the iteration limit is reached or the successive solutions of CTM and CLM have converged.
END

The use of M^*i and N^*i generations in Step 2 and Step 3 is to increase the accuracy of the solution with further iterations while limiting the use of large amounts of CPU time early in the process. The use of feedback from previous iterations is demonstrated in Fig. 2.

The reserved locations (file A), best transfer solution (file B) and all transfer solutions, i.e. chromosomes (file C), are saved after iteration 0. During the iterative procedure, CLM saves the best location solution (file D), and all location chromosomes (file E), and CTM saves the best transfer solution (file B) and all transfer chromosomes, if using GA, (file C). For each iteration, CLM reads and uses files A, B (as the fixed transfer schedule) and E (to continue with the same chromosomes after iteration 1 and for subsequent iterations). Conversely, CTM reads and uses files A, D (as the fixed storage locations) and C, if using GA (to continue with the same chromosomes for subsequent iterations), or B, if using TS, to continue with the same solution string for subsequent iterations.

Using data supplied by the Port of Brisbane it is estimated that an average of 486 containers is exported with each ship. It is found that each of these 'average' ships have a transfer time of 673 min. With this in mind and using current resources and storage practices of the port, the CTM benchmark is given in Table 1.

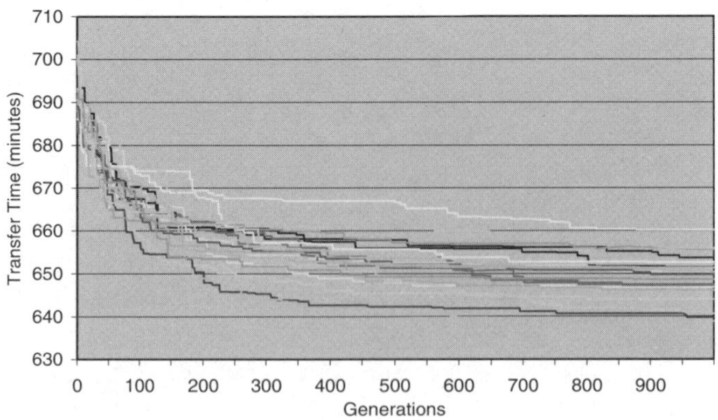

Fig. 4 Transfer time vs number of generations for 20 independent replications

The key to a successful application of GA or any heuristic, for that matter, is to ensure that the solution parameters are optimised. This means that the crossover rate, mutation rate, number of chromosomes and generation limit need to be examined to find the best values to ensure that good solutions are found using a minimum CPU time.

The number of chromosomes GA uses in each generation is critical. Too few chromosomes will not allow enough of the search space to be examined while too many will mean there are more initial solutions generated, crossover's, repairs and objective functions calculated, all of which increase the CPU time. Another consideration is that more chromosomes mean that each chromosome has a smaller probability of selection for reproduction. This is because roulette wheel selection is used where each chromosome has a probability inversely proportional (due to a minimisation problem) to its objective function value.

The model was simulated 50 times with varying number of chromosomes and the results were compared. The mean of the best solutions and standard deviations were calculated and are tabulated in Table 2, along with the CPU time for a replication.

As seen in Table 2, 20 chromosomes is what are best to use as it has a lower mean (but higher standard deviation) and requires less CPU time than all but ten chromosomes. It is thought, however, that ten chromosomes are too few, so, it is disregarded.

Analysis was also performed on the number of generations required and demonstrated in Fig. 3. The figure shows transfer time vs number of generations for 20 independent replications. It is seen that the curves flatten out (as expected) after about 500 generations, and it is felt that the little improvement found after that point does not justify spending more time to find the solution. With this in mind, it was decided that 500 generations would be used as the generation limit.

The crossover rate was the next parameter examined. Crossover rates were trialled with 0.005 increments in the range 0.05–0.99. It is clearly seen that a crossover rate of 0.50 provides, on average, the lowest solution. However, there isn't a significant difference amongst any of the solutions. It was decided, however, to use a crossover rate of 0.50.

The other parameter examined is the mutation probability. The mutation parameter was varied in the range, 0.01–0.99. Although little overall difference was observed for mutation rates above 0.25, the value of 0.99 provided the best overall average.

Table 4 Mean and standard deviation with varying crossover rate

Crossover rate	Mean	Standard deviation
0.05	673.61	5.7913
0.10	668.96	6.9172
0.25	662.97	7.0655
0.50	658.38	6.7707
0.75	653.59	7.5916
0.90	652.61	6.2084
0.95	654.83	6.0882
0.99	654.48	6.2459

The CLM was run 50 times using the benchmark parameters for CLM with varying number of chromosomes. The average of the best solution found in each replication and standard deviations were calculated and are tabulated in Table 3, along with the CPU time for a replication. Therefore, using 100 chromosomes is significantly better as it has a lower mean but requires more CPU time.

Analysis was performed on the number of generations required. To demonstrate this, plots of transfer time vs number of generations for 20 independent replications is shown in Fig. 4.

It is seen that the curves flatten out (as expected) after about 300–500 generations. With a view to the CPU time, however, it is decided that for further analysis, 350 would be the generation limit for the 20 replications shown in Fig. 4. The mean after 350 iterations is 654.6 while after 1,000 iterations it dropped to 649.2, but this required over 11 more hours of CPU time.

Crossover was the next parameter to be optimised, with an examined rate between 0.05 and 0.99. These are tabulated in Table 4, which shows a steady decrease in the range, 0.05–0.75 and a flattening from 0.75–0.99. The best results were those with a crossover rate of 0.9, and thus, this was selected to be used.

The final parameter examined is the mutation probability. The mutation rate parameter was varied in the range, 0.01–0.99 as shown in Table 5. It was found that there is a little difference between a mutation of 0.05 and 0.25; however, 0.05 gave marginally better solutions so the mutation rate was set at 0.05.

5 Results

This technique has been coded in C++ and is setup to use either tabu search or genetic algorithm to optimise CTM and genetic algorithm for the CLM. There is the option of inputting a file containing "reserved locations" that cannot be used for storage. These are locations that either currently have a container stored there or reserved for another incoming container that will not be exported on the ship under consideration. In this paper, the reserved locations are randomly selected.

Table 5 Mean and standard deviation with varying mutation probability

Mutation probability	Mean	Standard deviation
0.01	655.17	7.0394
0.05	651.45	6.1557
0.10	652.05	5.9576
0.25	652.61	6.2084
0.50	655.05	5.7171
0.75	655.73	5.8564
0.90	656.60	6.2172
0.95	657.27	5.7169
0.99	658.04	5.7517

5.1 Genetic algorithm for CLM and CTM

Preliminary results indicate that about ten iterations of each of the CLM and CTM are needed before the solutions settles to a minimum value. Also, the algorithm ends with a CTM because its fitness function is higher, due to not including the setup time for containers "shifted" during mutation for the current generation. This was done to reduce CPU time by not having to shift down (i.e. drop containers so they are not floating) and recalculate storage depth as often. In effect, this gives a lower bound fitness for the solution. Figure 5 shows ten replications of the normal

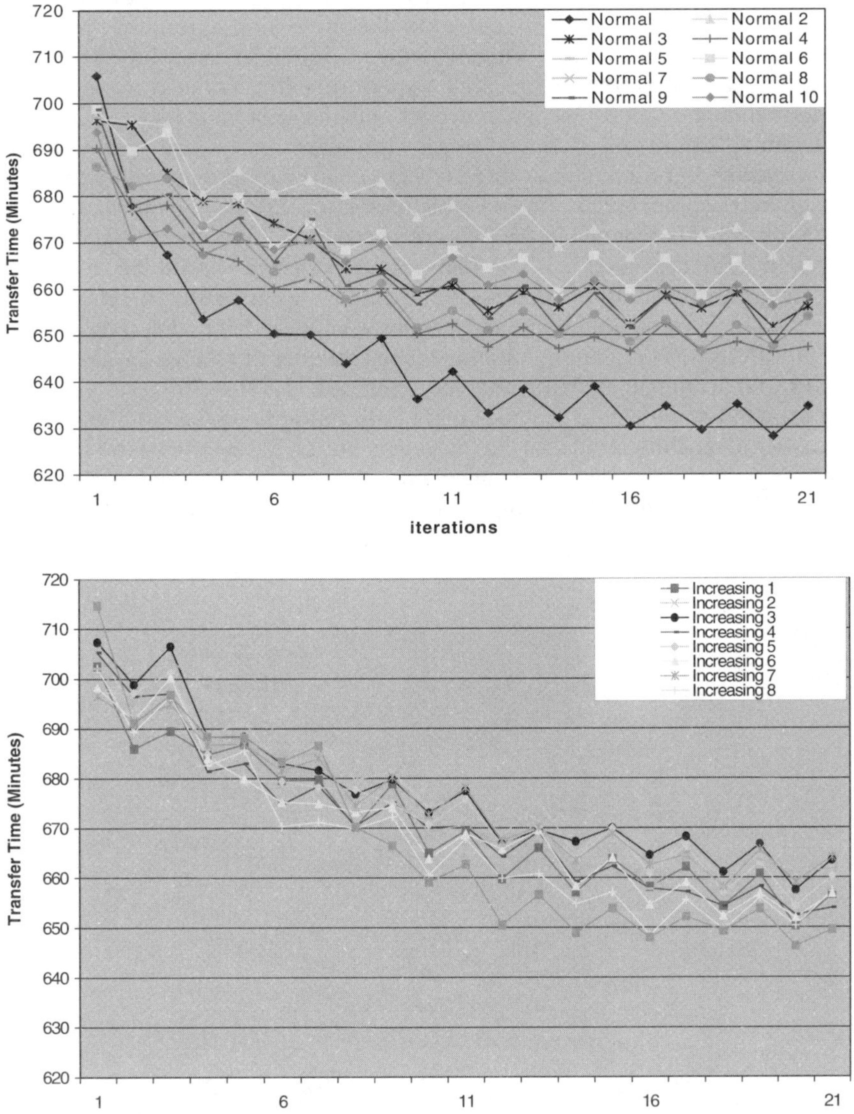

Fig. 5 Ten replications of the benchmark problems

and increasing algorithms for the integrated CLM–CTM model using the benchmark problem.

It is seen that the final solution is generally in the range, 650–670 with a few outliers. The outliers are most likely due to using different random "reserved locations" for each replication. This was done to see how the algorithm would perform in a variety of situations. Overall, the increasing algorithm had a mean of 660.44 and a standard deviation of 4.871, and the normal algorithm had a mean of 657.80 and a standard deviation of 8.607.

As with the decomposed models, analysis was performed to assess changes to the port infrastructure on the solution time window. Firstly, a comparison was made varying the maximum height. The results are shown graphically in Fig. 6. Figure 6 shows two distinct curves. The curve for the increasing algorithm produced an exponential decrease as the maximum storage height decreased. Intuitively, this is because it reduces the chance of having to perform extra container moves to access the desired container. In practice, however, this may not be feasible, as it requires much more additional storage area. The other curve for the normal or nonincreasing algorithm shows three levels to be the minimum with an increased average for the other levels. One possible reason for this may be the possible range of solutions that is reduced in the two-level scenario (1,536 total storage positions of which 500 are reserved), so the solution strings would be more alike. Consequently, the solution becomes more similar faster so that it gets stuck in a genetic hole in early iterations and cannot escape in later iterations. The only method of escape is mutation, but the reduced number of storage spaces limits the number of feasible mutation alternatives to just 536 (1,536—500 reserved—500 in use), many of which would be undesirable from an improvement point of view. The increasing algorithm avoids this, as it would not go as deep into the hole in the earlier iterations. This phenomenon would also explain the higher variation in solutions for the nonincreasing algorithm, as it has a greater tendency towards very good or very bad solutions with less mid-range solutions.

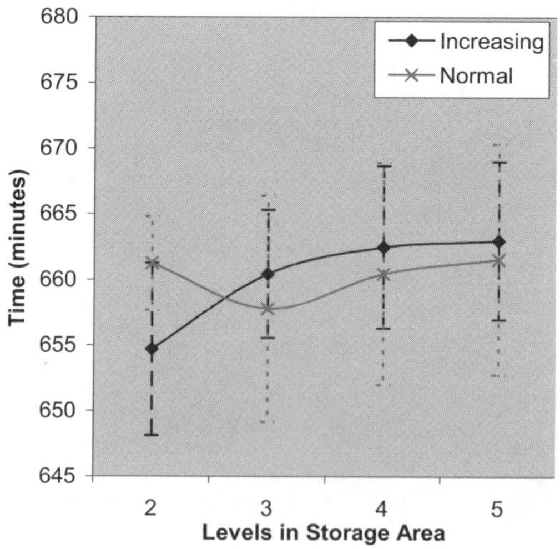

Fig. 6 Variation due to maximum storage height

Sensitivity analysis was also performed to analyse what effect changing the number of assigned yard machines has on the loading time. Figure 7 shows, graphically, these results. This shows an exponential growth in the average loading time as the number of available yard machines is reduced. This figure also shows a little difference in the averages between the increasing and normal algorithms, but the standard deviation of the increasing algorithm is generally half that for the nonincreasing algorithm.

5.2 Genetic algorithm for CLM and tabu search for CTM

This integrated model is similar to that outlined in the previous section but uses tabu search to solve the Container Transfer Model. Iteration 0 (see Fig. 2) still uses GA, however, to "seed" the initial iteration.

Figure 8 plots ten replications of this integrated model for increasing and normal algorithms. Figure 8 shows that most of the final solutions were again in the range, 650–670; however, there was less variation compared to the technique used above. Once again, solutions were found to converge after ten iterations (steps 11–20 in the graph). A typical execution would require around 4 h CPU time.

Once more, sensitivity analysis was performed to analyse the differences due to the changing of the maximum storage height. The results are graphed in Fig. 9. This shows that for the increasing algorithm, the average solution time increases slightly as the maximum storage height increases. The nonincreasing iterative algorithm produced a curve showing that the averages decreased as the maximum storage height decreased until the two-level storage that again provided a higher average. This is due to less storage locations, resulting in a smaller solution space for CLM and the algorithm zooming in too quickly in early iterations.

The effect of changing the number of yard machines is shown in Fig. 10. This shows a polynomial increase as the number of yard machines decrease. Obviously the more yard machines available decreases the amount of work to be done by each machine, but halving the number of machines requires less than twice the time.

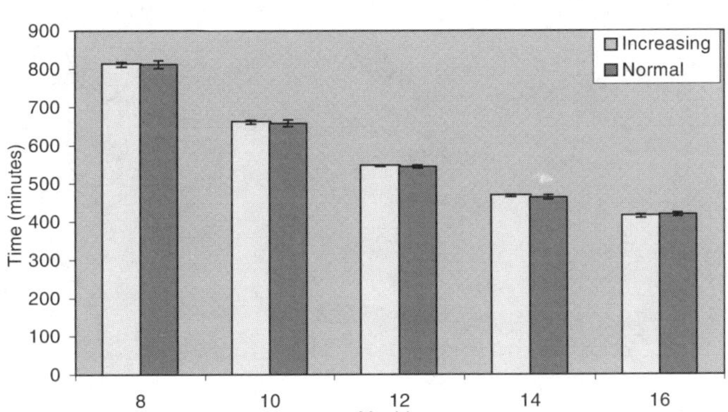

Fig. 7 Average time with varying yard machines

6 Comparison

To assess the merits of the seven algorithms covered and to determine the best approach a comparison of the three individual (i.e. disaggregated) models and the four integrated methods. For the following tables, GA for CLM is designated as approach 1, GA for CTP as approach 2, TS for CTP as approach 3, increasing iterative search using GA as approach 4, normal iterative search using GA as approach 5, increasing iterative search using GA and TS as approach 6 and normal iterative search using GA and TS as approach 7.

Table 6 provides results for the benchmark problem for the various solution approaches. Table 6 shows that the integrated algorithms had significantly less variation (observed by smaller standard deviation) and all averaged around

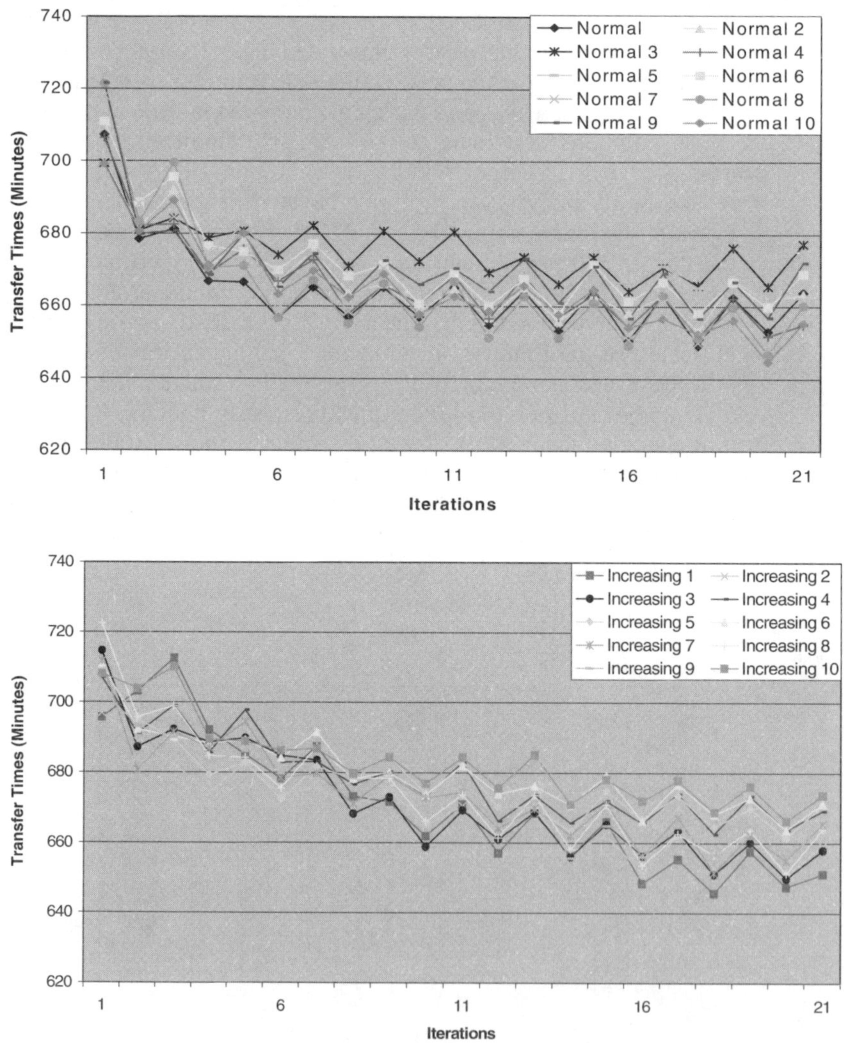

Fig. 8 Ten replications for the benchmark model

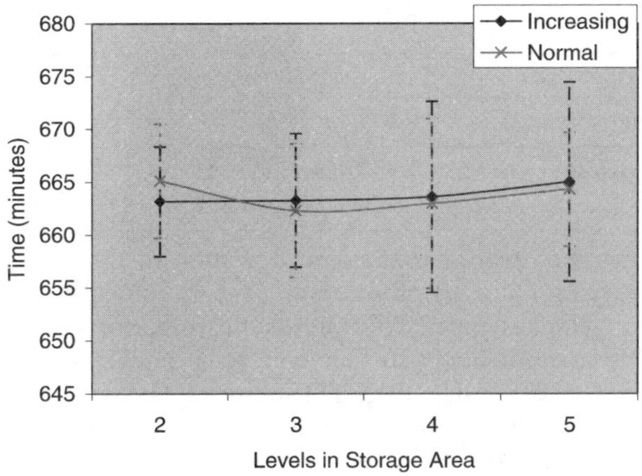

Fig. 9 Maximum storage height variation

660 min. The GA for CLM model provided better average, but like all the separated models, the standard deviation was much greater. The varying of maximum storage height had a different effect for each of the solution techniques. These are tabulated in Table 7.

Generally, the integrated algorithms found better solutions than the single model solutions, with the exception of GA for CLM with three levels of storage and TS for CTM with two levels of storage. The integrated algorithms also had much less variation in solutions with the standard deviation less then 10 in all cases, while it was often over 20 for the separated models.

This is mostly due to the integrated algorithms being able to work with a poor initial random location configuration and improve this to match in with the transfer schedule and vice versa, with a poor initial transfer schedule. In a sense, when the integrated methods started with bad initial solutions, they were able to find more improvement than the separate models. Of the iterative searches for the integrated models, GA solving both CLM and CTM provided better solutions.

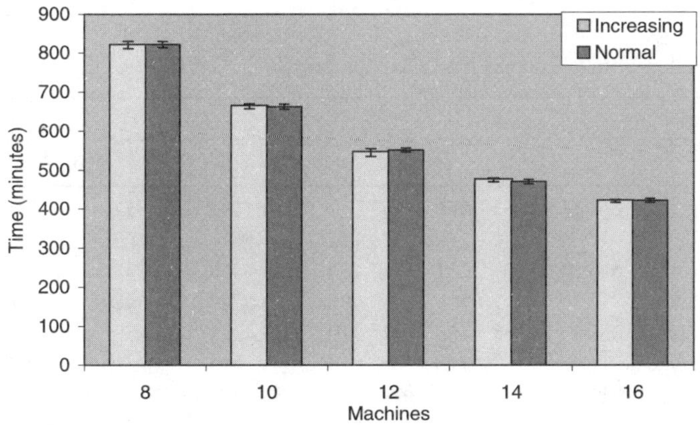

Fig. 10 Variation due to changing number of yard machines

Table 6 Solutions for the benchmark problem

	Approach						
	1	2	3	4	5	6	7
Average (min)	644.861	671.831	667.439	660.443	657.789	663.265	662.298
St. dev. (min)	17.535	20.979	20.986	4.871	8.607	6.306	6.285

Table 8 shows the mean and standard deviation of the seven approaches of solutions with varying numbers of allocated yard machines. Again, it is observed that the iterative searches provided more stable solutions (i.e. smaller standard deviations; so, it is more likely to converge to a similar solution for different replications) and were better than the CTM solutions. However, the CLM solutions were found to be better for all but eight-yard machines and in some cases, significantly so, and provided a lower bound to the solution value. The CTM fitness values are accepted as more accurate which is why the iterative searches concluded with a CTM solution.

7 Conclusion

This paper outlined a novel iterative search technique to solve an integrated model composed of two sub-models with dependent decision variables. An increasing algorithm, where the generations within each iteration increase, is compared with a nonincreasing approach. A genetic algorithm is also compared with a tabu search/ genetic algorithm hybrid. This iterative search was used to solve two models, CLM and CTM, using location and transfer feedbacks for successive iterations. The integrated iterative algorithm generally provided better solutions than those found using the only individual models, and the solutions were much more stable with less variation in the results.

Overall, the GA technique produced better results than the TS/GA hybrid, and in most cases, the nonincreasing algorithm performed better than the increasing

Table 7 Variation of levels across the seven approaches

Levels	Approach							
		1	2	3	4	5	6	7
2	Average	661.673	637.556	657.011	654.714	661.257	663.175	665.133
	St. dev.	16.729	22.228	20.125	6.561	3.562	5.177	5.412
3	Average	644.861	671.831	667.439	660.443	657.798	663.265	662.298
	St.dev.	17.535	20.979	20.986	4.871	8.607	6.306	6.285
4	Average	700.371	683.485	671.910	662.492	660.458	663.592	662.983
	St.dev.	16.370	20.750	21.913	6.185	8.450	9.047	8.016
5	Average	717.288	680.485	669.513	662.967	661.550	665.025	664.325
	St.dev.	23.342	21.788	20.501	6.013	8.776	9.404	5.376

Table 8 Variation of yard machine across the seven approaches

Machines	Approach							
		1	2	3	4	5	6	7
8	Average	824.927	935.581	856.730	811.900	812.300	820.842	822.700
	St.dev.	21.987	25.653	27.008	6.441	10.321	9.939	7.949
10	Average	644.861	671.831	667.439	660.443	657.798	663.265	662.298
	St.dev.	17.535	20.979	20.986	4.871	8.607	6.306	6.285
12	Average	531.597	604.799	553.661	545.958	544.425	545.033	551.583
	St.dev.	12.149	15.201	16.669	1.777	4.370	9.913	5.483
14	Average	449.314	513.106	463.975	468.049	464.135	475.357	470.924
	St.dev.	12.414	13.564	14.919	2.941	5.868	5.058	5.506
16	Average	378.618	435.559	394.712	413.533	418.283	421.275	422.750
	5.506	12.630	12.905	14.397	4.640	5.049	3.777	4.879

algorithm. It was found that reducing the maximum storage height resulted in a reduction in the turnaround time, although the nonincreasing algorithm performed worse for the two-level storage. A polynomial reduction in average throughput time resulted when the number of yard machines increased. Overall, it is recommended that the nonincreasing GA algorithm (approach 5) be used as it provided the best solutions for a wide range of infrastructure configurations.

References

Bish EK (2003) A multiple-crane-constrained scheduling problem in a container terminal. Eur J Oper Res 144:83–107

Kim KH, Kim KY (1999) Routing straddle carriers for the loading operation of containers using a beam search algorithm. Comput Ind Eng 36:109–136

Kim KH, Kim HB (2001) The optimal sizing of the storage space and handling facilities for import containers. Transp Res B 36:821–835

Kozan E (2000) Optimising container transfers at multimodal terminals. Math Comput Model 31:235–243

Kozan E, Preston P (1999) Genetic algorithms to schedule container transfers at multimodal terminals. Int Trans Oper Res 6:311–329

Preston P, Kozan E (2001a) A tabu search technique applied to scheduling container transfers. Transp Plan Technol 24(2):135–154

Preston P, Kozan E (2001b) An approach to determine storage locations of containers at seaport terminals. Comput Oper Res 28:983–995

Steenken D, Voss S, Stahlbock R (2004) Container terminal operation and operations research—a classification and literature review. OR Spectrum 26:3–49

Vis I, Koster R (2003) Transhipment of containers at a container terminal: an overview. Eur J Oper Res 147:1–16

Loo Hay Lee · Ek Peng Chew · Kok Choon Tan · Yongbin Han

An optimization model for storage yard management in transshipment hubs

Abstract This paper studies a yard storage allocation problem in a transshipment hub where there is a great number of loading and unloading activities. The primary challenge is to efficiently shift containers between the vessels and the storage area so that reshuffling and traffic congestion is minimized. In particular, to reduce reshuffling, a consignment strategy is used. This strategy groups unloaded containers according to their destination vessel. To reduce traffic congestion, a new workload balancing protocol is proposed. A mixed integer-programming model is then formulated to determine the minimum number of yard cranes to deploy and the location where unloaded containers should be stored. The model is solved using CPLEX. Due to the size and complexity of this model two heuristics are also developed. The first is a sequential method while the second is a column generation method. A bound is developed that allows the quality of the solution to be judged. Lastly, a numerical investigation is provided and demonstrates that the algorithms perform adequately on most cases considered.

Keywords Port operation · Storage allocation · Mixed-integer programming · Heuristic algorithm · Column generation

1 Introduction

Container traffic has been growing steadily and this trend is expected to continue (Yun and Choi 1999; Ryan 1998). A new generation of container vessels that have a greater carrying capacity and scarcity of the land will put an enormous pressure on port operators to develop effective container handling systems. High-density, automated container handling equipment is a potential candidate for improving the performance of container terminals and meeting the challenges of the future in

L. H. Lee · E. P. Chew · K. C. Tan · Y. Han (✉)
Department of Industrial and Systems Engineering, National University of Singapore,
10 Kent Ridge, 119260 Singapore, Singapore
E-mail: iseleelh@nus.edu.sg, isecep@nus.edu.sg, isetkc@nus.edu.sg, hanyongbin@nus.edu.sg

marine transportation. However, in order for these capital intensive equipments to function effectively, decision planning tool for integration and optimization becomes crucial.

A container terminal is the place where vessels dock on a berth and containers are loaded and unloaded. Based on the types of container handling operations, a container terminal can be roughly divided into two main areas, the quayside for berthing vessels and the storage yard for holding containers (as shown in Fig. 1). The quayside is made up of several berths for vessels to moor. The vessels moored at the berths are served by quay cranes (QCs) which load and unload containers. The storage yard is typically divided into several blocks where the containers are stored. Each container block is served by several yard cranes (YCs). The storage yard serves as an interface for loading (unloading) containers to (from) vessels to facilitate export (import) containers and to transship containers between vessels. To transport containers between the quayside and the storage yard, vehicles such as prime mover, or straddle carriers are used. A schematic diagram of the typical

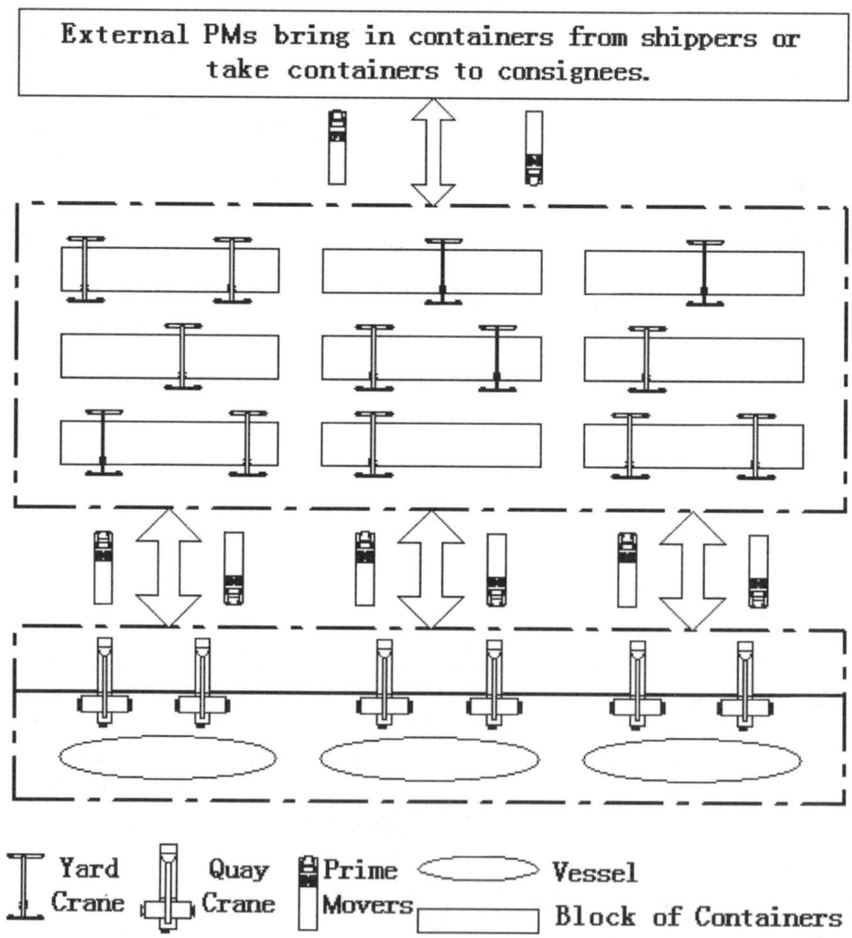

Fig. 1 A schematic diagram of a container terminal

processes in a container terminal is shown in Fig. 2 (Vis and De Koster 2003). The container activities can be categorized into three types: import, export, and transshipment activities. For export activities, the containers are brought in by shippers and will be stored at their designated locations in the storage yard. When it is time to load the containers, they are retrieved from the stored location and transported by vehicles to the quayside. The QCs then remove the containers from the vehicles and load them onto the vessels. The processes for import activities are similar but they are done in the reverse order. For transshipment activities, the processes are a little different. The containers will be stored in the storage yard after they are unloaded from the vessel and will be finally loaded onto other vessels. In this paper, our study is focused on the yard storage allocation problem in a transshipment hub where transshipment of containers is the major activity and the yard activity is intensive. This transshipment port uses prime mover as the main vehicle to transport the containers.

The storage yard plays a pivotal role in transshipment hubs. Most containers unloaded from one vessel are stored in the storage yard and are eventually loaded onto other vessels. Multi-level stacking of containers is a common practice when the volume of transshipment activities is intensive and the land is scarce. This can lead to high concentration of activities within a small area and may likely cause traffic congestion of prime movers. Another result of over stacking is unproductive reshuffles of containers. Traffic congestion and reshuffles can reduce the productivity of resources, which include prime movers, YCs, and QCs. Related works on the yard storage allocation problem is summarized as follows.

The efficiency of stacking depends greatly on the strategies of allocating storage space to arriving containers. Chen (1999) distinguishes several major factors that influence operational efficiency and cause unproductive container movements in terminal operations. Chung et al. (1988) propose the use of buffer space to increase the utilization of the material handling equipment and reduce the total container loading time. Kim and Kim (1999) propose a segregation strategy to allocate storage space for import containers. In Chen et al. (2000) the storage space allocation problem is examined with a time-space network with the objective of allocating containers to storage locations in advance. Taleb-Ibrahimi et al. (1993) describe handling and storage strategies for export containers and quantify their performance according to the amount of space and number of handling moves required. Kim (1997) proposes a methodology to estimate the expected number of reshuffles to pick up an arbitrary container and the total number of reshuffles to pick up all the containers in a bay for a given initial stacking configuration. Kim and Bae (1998) discuss how to reshuffle export containers in container terminals. Kim et al. (2000) propose a methodology to determine the storage location of an arriving export container by considering its weight. Kim and Park (2003) discuss how to allocate storage space for outbound containers that will arrive at a storage yard. Zhang et al. (2003) study the storage space allocation problem in the storage yard of terminals. Chen et al. (1995), Davies and Bischoff (1999), and Scheithauer (1999) study a strategy called consignment. This strategy attempts to store containers with the same destination, contents, and loading time together in some dedicated storage area. Crainic et al. (1993), Cheung and Chen (1998), and Shen and Khoong (1995) look into the yard storage allocation problem for empty containers.

From the literature, it can be seen that various problems associated with yard operations have been addressed. However, these papers do not sufficiently address

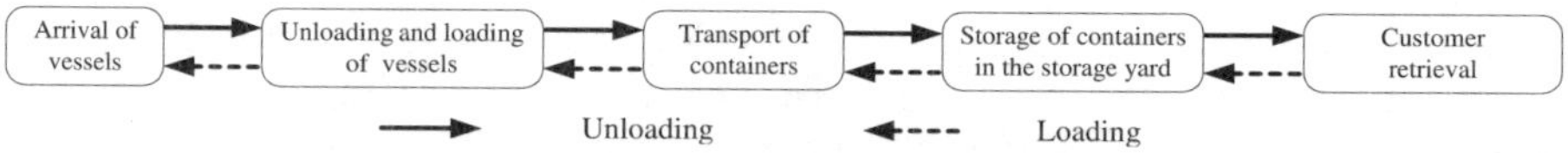

Fig. 2 A flow diagram demonstrating the interaction between container terminal processes

the particular needs of transshipment hubs, but more on the general terminals which emphasize on import and export activities. For transshipment hubs, the loading and unloading activities are both concentrated and need to be considered at the same time. This makes the planning problem much more challenging compared to port planning for general terminals where the loading and unloading activities can be considered independently by having different dedicated storage areas for import and export activities.

If the yard storage allocation problem is not handled properly at a transshipment yard, the problems of reshuffling and traffic congestion might arise. In the port that we study in this paper, the consignment strategy is used. This strategy stores export and transshipment containers with the same departing vessel together at the same designated storage areas. Import containers are not considered in this study as the port operator does not mix the storage area of the import containers with the export and transshipment containers. This strategy helps to reduce the reshuffling level to a negligible level. To handle the traffic congestion, a new workload balancing protocol is proposed and the details will be discussed in Problem description.

The rest of this paper is organized as follows. Problem description provides the detailed description of the problem. Model development describes the model development. Numerical experiments and the computational results are presented in Numerical experiments. Two heuristic algorithms are proposed and implemented in Heuristic algorithms. Conclusions and future research give conclusions and some future research topics.

2 Problem description

The port that we are studying handles a high volume of transshipment containers. One of its competitive edges is that the port is able to turn around the vessels within a short time. At the planning stage, this translates to the requirement for the port operator to allocate the incoming transshipment and export containers to the storage yard such that the traffic congestion can be kept at a minimal level.

To manage the yard allocation process more efficiently, the port operator organizes the storage yard into several blocks as shown in Fig. 3. The depth of each block is six containers and the length of the block is 40 containers. Every block is further divided into five subblocks, where the length of each sub-block is eight containers. The stacking height is five containers (which we call tier). The basic unit for the yard storage allocation process is at the subblock level, i.e., for the consignment strategy, we will assign the containers that are going to the same departing vessel to the same subblocks. There is an assigned lane for the movement of containers by prime movers (the "truck path") and a separate "passing" lane strictly to allow trucks to pass each other when required. The passing lane is narrow, and it is shared by yard cranes.

Traffic congestion may happen when too much workload needs to be handled within a small area at the same time. For example, if there are a lot of container movements in subblocks 7 and 12 (see Fig. 3), then there will be many prime movers waiting or moving nearby. This will cause traffic congestion. Similarly, if the workload in subblocks 6 and 7 is high, then the prime movers waiting at subblock 7 may block other prime movers from going to subblock 6 as they share the same path.

To ensure smooth flow of traffic, the port operator has imposed several restrictions during the planning stage. Among them are:

1. When a subblock is in the loading process, its neighboring subblocks should not have any loading or unloading activities.
2. There should not be two or more neighboring subblocks which are having high unloading activities.

To incorporate these restrictions into the planning model, we introduce a "high workload" rule/protocol and a vicinity matrix.

The protocol of high-low workload is to ensure that at any given time, many yard cranes will be highly utilized as the jobs are concentrated as they do not need to move around frequently to other subblocks to perform jobs. The ranges of high workload and low workload do not overlap. For example, the range of high workload is set between 50 and 100 containers per shift, while the low workload is set between 0 and 20 containers per shift by the port operator.

To capture the possible traffic congestion between subblocks, we use a vicinity matrix to represent the neighborhood structure between different sub-blocks. A sub-block is a neighbor if it is adjacent. Adjacency of subblocks inherently implies that trucks must use the same path. For example, subblock 7 is said to be a neighbor of subblocks 6 and 12. Subblock 7 is not a neighbor of subblock 2 even though they

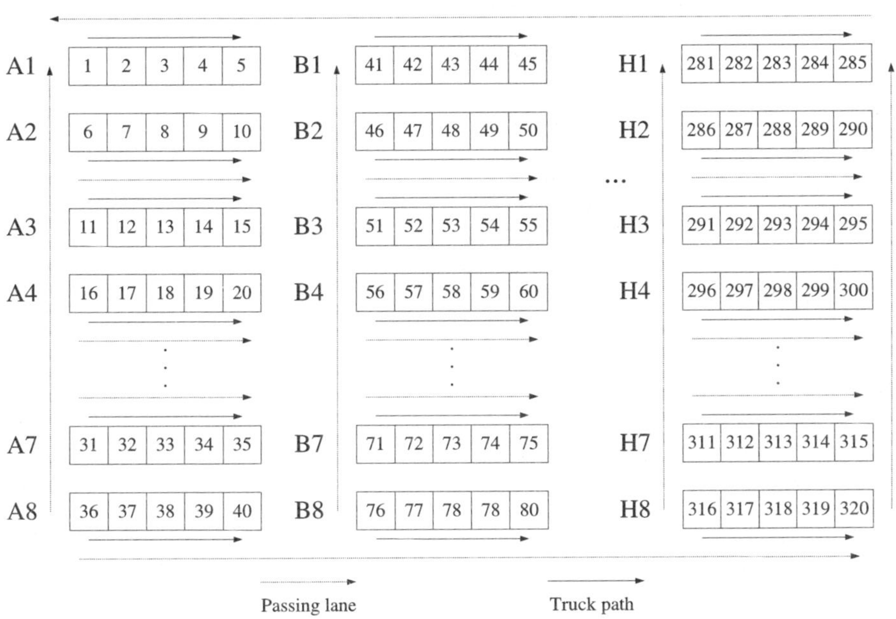

Fig. 3 A typical yard configuration

are back to back. In a vicinity matrix, a value of 1 means that the subblocks are neighbors to each other, and 0 means that they are not. For the layout shown in Fig. 3, the vicinity matrix is given in Table 1. As the vicinity matrix is symmetric, only the top right half is shown. The workload of a neighboring subblock should be low if the neighbor has been assigned a high workload.

For this problem, the port operator has pre-assigned a set of subblocks for each departing vessel based on their experience (this is also known as the yard template). Given this set of assigned subblocks, we must determine the minimum number of yard cranes to deploy and how many transshipment and export containers should be assigned to each subblock in each shift. The total loading activities in each subblock can be derived from the above decisions. (To simplify the discussion, in the subsequent section, we will refer to "containers" as both the transshipment and export containers, unless specified otherwise.) Currently, the port operator does not have any formal planning model to determine the allocation, and its decisions are based on intuition and past experiences. As a means to remedy this, a mixed integer programming model that incorporates the concepts discussed above is developed. The detailed model will be discussed in the next section.

3 Model development

In this section, the storage allocation problem is formulated as a mixed-integer linear programming model.

Table 1 Part of the vicinity matrix for the yard configuration shown in Fig. 2

$R_{ii'}$	1	2	3	4	5	6	7	8	9	10	11	12	13	14	15	16	17	18	19	20
1	0	1	0	0	0	0	0	0	0	0	0	0	0	0	0	0	0	0	0	0
2		0	1	0	0	0	0	0	0	0	0	0	0	0	0	0	0	0	0	0
3			0	1	0	0	0	0	0	0	0	0	0	0	0	0	0	0	0	0
4				0	1	0	0	0	0	0	0	0	0	0	0	0	0	0	0	0
5					0	0	0	0	0	0	0	0	0	0	0	0	0	0	0	0
6						0	1	0	0	0	1	0	0	0	0	0	0	0	0	0
7							0	1	0	0	0	1	0	0	0	0	0	0	0	0
8								0	1	0	0	0	1	0	0	0	0	0	0	0
9									0	1	0	0	0	1	0	0	0	0	0	0
10										0	0	0	0	0	1	0	0	0	0	0
11											0	1	0	0	0	0	0	0	0	0
12												0	1	0	0	0	0	0	0	0
13													0	1	0	0	0	0	0	0
14														0	1	0	0	0	0	0
15															0	0	0	0	0	0
16																0	1	0	0	0
17																	0	1	0	0
18																		0	1	0
19																			0	1
20																				0

3.1 Modeling assumptions

The following assumptions are made when developing the storage allocation model.

1. The time span of the model is 7 days with three planning periods in each day. Each planning period corresponds to an 8-h working shift.
2. We assume that the amount of containers arriving in every shift is given and will repeat weekly (this implies the planning period can be wrapped around). The actual number can vary, but for planning purposes, it is reasonable to assume that it is deterministic and an input to the model.
3. At any given time, if the subblock is during the loading activity, then, a dedicated yard crane will be assigned to that subblock as loading activities have higher priority.
4. Two different types of containers are handled in a container terminal. They are 20 ft containers and 40 ft containers. As one subblock consists of a few lanes, it is possible to have a mixture of different types of containers in one subblock. For simplicity, we assume they can be stored in the same subblocks. However, the model can easily be modified to handle the dedicated subblock for different types of containers.
5. All containers that arrive in a given shift will be stored in a subblock until they are loaded onto the departing vessel. The loading activities at any subblock have to be completed within two shifts because this is a requirement set by the port operator to reflect its current service level.
6. A yard crane assigned to a particular block should work until the end of the shift.

3.2 Notations

The model parameters are as follows:

I	The number of sub-blocks under consideration
J	The number of vessels under consideration in the planning horizon
K	The number of blocks under consideration
T	The number of shifts under consideration
N_i	The set of subblocks that are neighbors of subblock i, $1 \leq i \leq I$
V_j	The set of sub-blocks that are reserved for vessel j, $1 \leq j \leq J$
B_k	The set of subblocks that belong to block k, $1 \leq k \leq K$
WX_{jt}	The number of 20-ft containers of the departing vessel, j, which arrive in shift t. It is given and treated as input of the model, $1 \leq j \leq J$, $1 \leq t \leq T$
WY_{jt}	The number of 40-ft containers of the departing vessel, j, which arrive in shift t. It is given and treated as input of the model, $1 \leq j \leq J$, $1 \leq t \leq T$
CS	The capacity of each subblock in terms of TEUs, which is 240 (5 tiers×6 lanes×8 slots) in this model
CC	The capacity of each yard crane in terms of container moves per shift, which is 100 in this model
C_k	The maximum number of yard cranes allowed to reside in block k at any one time, $1 \leq k \leq K$

NL_{kt} The number of subblocks in loading process in block k in shift t, $1 \leq k \leq K$, $1 \leq t \leq T$

HL The lowest value that a high workload can take

HU The highest value that a high workload can take

LL The lowest value that a low workload can take

LU The highest value that a low workload can take

Subscript, i, is for subblock, j, for vessel, k, for block, t, for shift
The decision variables are as follows:

x_{it} The number of 20-ft containers that are allocated to subblock, i, for unloading in shift t, $1 \leq i \leq I$, $1 \leq t \leq T$

y_{it} The number of 40-ft containers that are allocated to subblock, i, for unloading in shift t, $1 \leq i \leq I$, $1 \leq t \leq T$

h_{it} =1 means that the total workload $(x_{it}+y_{it})$ that are allocated to subblock, i, for unloading in shift, t, is high, that is $H_l \leq x_{it}+y_{it} \leq H_u$, $1 \leq i \leq I$, $1 \leq t \leq T$
=0 means that the total workload $(x_{it}+y_{it})$ that are allocated to subblock, i, for unloading in shift t is low, that is $L_l \leq x_{it}+y_{it} \leq L_u$, $1 \leq i \leq I$, $1 \leq t \leq T$

d_{kt} The number of yard cranes allocated to block, k, for unloading in shift t, $1 \leq k \leq K$, $1 \leq t \leq T$

3.3 Model formulation

Managers in container terminals always attempt to reduce costs by efficiently utilizing resources, including berths, storage yards, quay cranes, yard cranes, prime movers, and human resources. In the proposed model, the total number of crane shifts required to handle all the workload should be minimized. It is not to determine the necessary number of yard cranes that should be available in the storage yard for the terminals. However, the operating cost should be reduced by putting less yard cranes in use. Each crane shift corresponds to one active yard crane in one shift. The model is formulated as follows.

$$(SAP)\text{Min } w = \sum_{k=1}^{K} \sum_{t=1}^{T} d_{kt} \tag{1}$$

Subject to:

$$\sum_{i \in V_j} x_{it} = WX_{jt} \text{ For all } 1 \leq j \leq J, 1 \leq t \leq T \tag{2}$$

$$\sum_{i \in V_j} y_{it} = WY_{jt} \text{ For all } 1 \leq j \leq J, 1 \leq t \leq T \tag{3}$$

$$\sum_{t=1}^{T}(x_{it}+2y_{it}) \leq CS \text{ For all } 1 \leq i \leq I \tag{4}$$

$$\sum_{t=1}^{T}(x_{it}+y_{it}) \leq 2\,CC \text{ For all } 1 \leq i \leq I \tag{5}$$

$$\sum_{i \in B_k}(x_{it}+y_{it}) \leq d_{kt}\ CC \text{ For all } 1 \leq k \leq K, 1 \leq t \leq T \tag{6}$$

$$HL+(LL-HL)(1-h_{it}) \leq x_{it}+y_{it} \leq LU+(HU-LU)h_{it} \text{ For all } 1 \leq i \leq I, 1 \leq t \leq T \tag{7}$$

$$x_{i't}=0, y_{i't}=0 \text{ For all } i' \in N_i, t \in L_i, 1 \leq i \leq I \tag{8}$$

$$\sum_{i' \in N_i \text{ or } i'=i} h_{i't} \leq 1 \text{ For all } 1 \leq i \leq I, 1 \leq t \leq T \tag{9}$$

$$d_{kt}+NL_{kt} \leq C_k \text{ For all } 1 \leq k \leq K, 1 \leq t \leq T \tag{10}$$

$$x_{it} \geq 0\ y_{it} \geq 0 \text{ For all } 1 \leq i \leq I, 1 \leq t \leq T \tag{11}$$

$$h_{it} \in \{0,1\} \text{ For all } 1 \leq i \leq I, \leq t \leq T \tag{12}$$

$$d_{kt} \text{ Integer For all } 1 \leq i \leq I, 1 \leq k \leq K, 1 \leq t \leq T \tag{13}$$

Constraints 2 and 3 ensure that all the workload unloaded from each vessel in each shift will be allocated to corresponding storage locations. Constraint 4 ensures the capacity restriction of subblocks. Constraint 5 ensures that the containers in each subblock should be loaded to the vessels in a certain time span (two shifts in the proposed model). Constraint 6 ensures that the yard cranes for unloading in each block can handle all the unloading workload in each shift.

A subblock with high unloading workload can share a yard crane with neighbors with low unloading workload. Also, several subblocks with low unloading workload can share a yard crane. To make full use of yard cranes, the workload allocated to each subblock in each shift should be either high or low. In this model, constraint 7 is used to ensure this restriction. Constraint 8 ensures that all the neighbors of a loading subblock cannot accept any containers arriving in that shift. Constraint 9 ensures that high unloading workload cannot be allocated to two subblocks that are neighbors to each other in the same shift.

As a result of the limitation of the length of the chassis trailer and due to safety consideration, each block can hold, at most, a certain number of yard cranes at any one time. One yard crane is required for each loading subblock, and hence, the number of loading subblocks is exactly equal to the number of yard cranes assigned to that block for loading. Constraint 10 ensures this restriction. Constraints 11, 12 and 13 are nonnegative and integer restrictions.

The model is not easy to solve because the MIP structure is poor and there are too many integer and binary variables. In the next section, the problem will be solved with CPLEX 8.1, a commercial software package.

4 Numerical experiments

In this section, the proposed storage allocation model (SAP) is first tested using two sets of input data for the simplified small-scale problem. Then it is used to solve a large-scale problem close to a moderate terminal.

4.1 Small-scale problem experiment

4.1.1 Input data

For the small-scale problem, there are eight blocks arrayed in four rows and two columns in the storage yard which is shown in Fig. 4. A vicinity matrix can be determined easily from the yard configuration.

It is assumed that there is exactly one vessel being loaded in each shift. The lowest and highest value that a high workload can take are 100 and 200, respectively, and those for a low workload are 0 and 100, respectively. The capacity of 1-yd crane is 200 container moves per shift. The maximum number of yard cranes that can reside in each block is two at any one time. Only five shifts are considered in the model. The parameters are drawn from the real practice with some minor changes to make the model feasible.

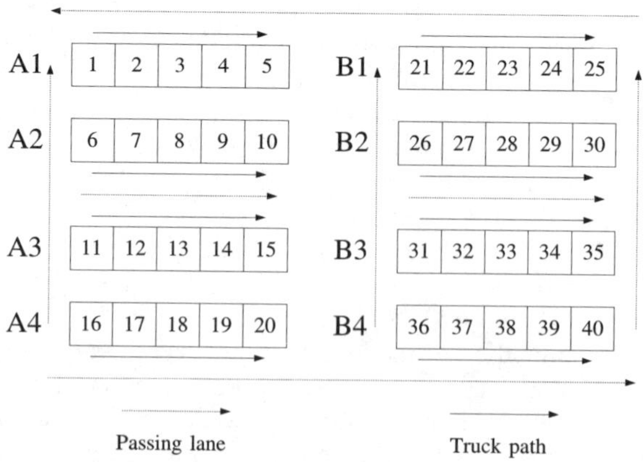

Fig. 4 Yard configuration for the small-scale problem

4.1.2 Implementation

The proposed model is implemented in C++ and run on a Pentium IV computer (CPU: 2.4 GHz, Memory: 512 M). The mixed-integer programming model is solved using CPLEX 8.1 with concert technology, a commercial software package with C++ optimization modeling library and interface.

The computational results are summarized in Tables 2 and 3. Two sets of input data are used to conduct the small-scale problem experiments. The utilization is defined as the ratio of the total storage space occupied by all the unloading containers to the total storage space in the storage yard.

As shown in Tables 2 and 3, for relatively low utilization scenarios (with utilization less than 0.45), the optimal solution can be obtained easily. For moderate utilization scenarios (with utilization between 0.45 and 0.60), it will take a longer time to solve as there are too many choices to allocate containers to their destination storage locations. In addition, too many choices can result in a very large branch and bound tree, which may cause the computer to run out of memory (for example, case 1 with utilization of 0.45). For high utilization scenarios (with utilization greater than 0.90 for case 1 and 0.70 for case 2), the problem is prone to be infeasible as the capacity constraints cannot be satisfied. The different results indicate that the proposed model is data dependent.

Table 2 Results of SAP for small-scale problem (case 1)

Utilization	Computation time (s)	Solution status	Objective Value (# of YCs)
0.05	1	Optimal	10
0.10	1	Optimal	10
0.15	1	Optimal	10
0.20	1	Optimal	10
0.25	1	Optimal	10
0.30	274	Optimal	15
0.35	340	Optimal	15
0.40	1	Optimal	15
0.45	34,196.03	Out of memory	17.56 (LB=20, gap=12.18%)
0.50	3,833	Optimal	20
0.55	110	Optimal	21
0.60	13,304	Optimal	25
0.65	114	Optimal	25
0.70	6	Optimal	25
0.75	11,602	Optimal	30
0.80	6	Optimal	30
0.85	328	Optimal	35
0.90	16	Optimal	35
0.95	1	Infeasible	–
1.00	1	Infeasible	–

Table 3 Results of SAP for small-scale problem (case 2)

Utilization	Computation time (s)	Solution status	Objective value (# of YCs)
0.05	1	Optimal	10
0.10	1	Optimal	10
0.15	1	Optimal	10
0.20	1	Optimal	11
0.25	1	Optimal	12
0.30	1	Optimal	13
0.35	2	Optimal	15
0.40	91	Optimal	17
0.45	1	Optimal	18
0.50	936	Optimal	21
0.55	643	Optimal	23
0.60	106	Optimal	23
0.65	36	Optimal	27
0.70	26	Optimal	28
0.75	1	Infeasible	–
0.80	1	Infeasible	–
0.85	1	Infeasible	–
0.90	1	Infeasible	–
0.95	1	Infeasible	–
1.00	1	Infeasible	–

4.2 Large-scale problem experiment

4.2.1 Input data

For the large-scale problem, there are 64 blocks arrayed in eight rows and eight columns. The layout of the yard configuration of the experiment problem is shown in Fig. 3. The time horizon is 7 days with three shifts in each day. The scale of the problem (around five million TEUs) is comparable to a section of the terminal that we are investigating.

The lowest and highest value that a high workload can take are 50 and 100, respectively, and those for low workload are 0 and 20, respectively. The capacity of 1-yd crane is 100 container moves per shift. The maximum number of yard cranes that can reside in each block is two at any one time. The parameters are fairly reflective on any terminal with high traffic intensity.

4.2.2 Implementation

The large-scale problem is also implemented in C++ and run on the same computer as that for the small-scale problems. The computational results are summarized in Table 4.

As shown in Table 4, the large-scale problem cannot be solved to optimality in a reasonable time and always terminates as a result of insufficient memory. The results just before running out of memory are presented. In addition, for scenarios with utilization between 0.3 and 0.35, it turns out to be infeasible. This is due to constraints 7, which restricts the workload allocated to each subblock to be either high or low. For some input data, it is impossible to satisfy these constraints.

For such a large-scale problem, the proposed MIP model (in total there are 7,392 integer variables and 24,370 constraints) is too complex to solve to optimality in a reasonable time. Therefore, heuristic algorithms should be developed to find a good enough solution to meet the requirement of port operators. To evaluate the performances of the heuristics, it is necessary to find a lower bound.

4.2.3 Finding a lower bound

One possible way to find a lower bound of this model is to solve each shift independently. Under this assumption, constraints 4 and 5 can be removed from the formulation. SPP is used to name the model to find a lower bound that is mentioned above. It is shown as follows.

$$(SPP)\ \text{Min}\ w = \sum_{k=1}^{K} d_{kt} \tag{14}$$

Table 4 Results of model SAP for the large-scale problem

Utilization	Computation time (s)	Solution status	Result details
0.05	30,346.80	Out of memory	104, gap=13.69% (lb=89.758)
0.10	24,640.89	Out of memory	149, gap=36.07% (lb=95.257)
0.15	26,680.84	Out of memory	188, gap=36.34% (lb=119.68)
0.20	25,721.92	Out of memory	217, gap=36.09% (lb=138.68)
0.25	64,013.97	Out of memory	221, gap=33.27% (lb=147.47)
0.30	1	Infeasible	–
$0.30<U<0.35$	1	Infeasible	–
0.35	1	Infeasible	–
0.40	11,060.56	Out of memory	285, gap=20.18% (lb=227.50)
0.45	11,037.89	Out of memory	315, gap=18.86% (lb=255.58)
0.50	13,642.97	Out of memory	352, gap=18.88% (lb=285.53)
0.55	16,114.41	Out of memory	364, gap=14.00% (lb=313.04)
0.60	25,609.03	Out of memory	396, gap=13.57% (lb=342.27)
0.65	23,034.78	Out of memory	415, gap=10.80% (lb=370.17)
0.70	26,500.50	Out of memory	453, gap=11.96% (lb=398.84)
0.75	58,407.64	Out of memory	486, gap=11.71% (lb=429.09)
0.80	220,047.20	Out of memory	524, gap=12.69% (lb=457.53)
0.85	196,890.58	Out of memory	557, gap=12.79% (lb=485.74)
0.90	554,958.83	Out of memory	605, gap=14.93% (lb=514.69)
0.95	684,439.48	Out of memory	623, gap=12.82% (lb=543.12)

Subject to:

$$\sum_{i \in V_j} x_{it} = WX_{jt} \text{ For all } 1 \le j \le J \tag{15}$$

$$\sum_{i \in V_j} y_{it} = WY_{jt} \text{ For all } 1 \le j \le J \tag{16}$$

$$\sum_{i \in B_k} (x_{it} + y_{it}) \le d_{kt} CC \text{ For all } 1 \le k \le K \tag{17}$$

$$H_l + (L_l - H_l)(1 - h_{it}) \le x_{it} + y_{it} \le L_u + (H_u - L_u) h_{it} \text{ For all } 1 \le i \le I \tag{18}$$

$$x_{i't} = 0, y_{i't} = 0 \text{ For all } i' \in N_i, t \in L_i, 1 \le i \le I \tag{19}$$

$$\sum_{i' \in N_i \text{ or } i'=i} h_{i't} \le 1 \text{ For all } 1 \le i \le I \tag{20}$$

$$d_{kt} + L_{kt} \le C_k \text{ For all } 1 \le k \le K \tag{21}$$

$$x_{it} \ge 0 \; y_{it} \ge 0 \text{ For all } 1 \le i \le I \tag{22}$$

$$h_{it} \in \{0, 1\} \text{ For all } 1 \le i \le I \tag{23}$$

$$d_{kt} \; Integer \text{ For all } 1 \le k \le K \tag{24}$$

The objective is to minimize the total number of yard cranes used in the current shift t. All constraints ensure the same restrictions as the original model, SAP, within one shift.

The SPP model is also implemented in C++ and run on the same computer as that for the original model, SAP. For comparison the same input data as those for both the small-scale problem and the large-scale problems are used. The computational results are presented in Tables 5, 6 and 7.

As shown in Table 5, model SPP for the first case small-scale problem gives good lower bounds for most scenarios. Most of them have the same objective value as the optimal solution. As model SPP for each shift can be solved efficiently, the lower bound for the original model SAP can be obtained within a short time. As shown in Table 6, for the second case small-scale problem, the lower bound for every scenario is exactly the same as the objective value of the optimal solution.

For the large-scale SAP model, the optimal solution cannot be obtained. Instead, only the best lower bound before running out of memory can be obtained. As shown in Table 7, the lower bound obtained from model SPP is always better than the best lower bound obtained from the model SAP before running out of memory.

5 Heuristic algorithms

The model is intractable when the size of the problem becomes large. In this section, two heuristic algorithms that may find a feasible solution close to the optimal solution in a reasonable time are proposed.

5.1 The sequential method

It can be seen from the procedure that a lower bound is found wherein the single shift model can be solved effectively. Inspired by this finding, the proposed model can be solved one shift at a time. This is called the sequential method (SQM). The difference between SPP and SQM is that in the sequential method, the linking constraints of Eqs. 4 and 5 will be considered so as to ensure that the solution is feasible. Specifically, a sequence of shifts should be picked first, and based on this sequence, the model is solved shift by shift. Note that after solving the model for each shift, the remaining capacity of each subblock should be updated. Therefore, constraints 4 and 5 can be modified as follows

$$x_{it} + 2y_{it} \leq CS - \sum\nolimits_{\tau \in \Gamma_t} (x_{i,\tau} + 2y_{i,\tau}) \text{ For all } 1 \leq i \leq I \tag{25}$$

$$x_{it} + y_{it} \leq CC - \sum\nolimits_{\tau \in \Gamma_t} (x_{i,\tau} + y_{i,\tau}) \text{ For all } 1 \leq i \leq I \tag{26}$$

where Γ_t is the set which consists of all those shifts that come before the current shift, t, in a given sequence. Constraints 25 and 26 ensure the capacity restriction of each subblock in terms of storage space and yard cranes, respectively. Hence, the SQM is the same as the procedure of SPP except that the two additional constraints, Eqs. 25 and 26 are added into SPP. The sequential method may be effective as it has much less decision variables and constraints.

It is also noted that the sequence of shifts chosen need not be in chronological order as the demand repeats weekly and the only linking constraints are the space

Table 5 Comparison of results of SAP and SPP for small-scale problem (case 1)

Utilization	Results of SAP	Lower bounds from SPP	
		Computation time (s)	Objective value
0.05	10	1	10
0.10	10	1	10
0.15	10	1	10
0.20	10	1	10
0.25	10	1	10
0.30	15	1	15
0.35	15	1	15
0.40	15	1	15
0.45	17.56 (lb=20, gap=12.18%)	1	20
0.50	20	1	20
0.55	21	1	20
0.60	25	1	25
0.65	25	1	25
0.70	25	1	25
0.75	30	1	30
0.80	30	1	30
0.85	35	1	35
0.90	35	1	35

Table 6 Comparison of results of SAP and SPP for small-scale problem (case 2)

Utilization	Results of SAP	Lower bounds from SPP	
		Computation time (s)	Objective value
0.05	10	1	10
0.10	10	1	10
0.15	10	1	10
0.20	11	1	11
0.25	12	1	12
0.30	13	1	13
0.35	15	1	15
0.40	17	1	17
0.45	18	1	18
0.50	21	1	21
0.55	23	1	23
0.60	23	1	23
0.65	26	1	26
0.70	28	1	28

restriction and crane capacity restriction. It does not matter which shift we choose first, as it will not violate any of these constraints. On the other hand, different solutions are obtained from different shift orderings which implies that the sequence is important. However as there are $T!$ different sequences that can be

Table 7 Comparison of results of SAP and SPP for large-scale problem

Utilization	Lower bounds from SAP before MR		Lower bounds from SPP	
	Lower bound	Computation time (s)	Lower bound	Computation time (s)
0.05	89.758	30,346.80	102	21
0.10	95.257	24,640.89	138	744
0.15	119.68	26,680.84	174	5,217
0.20	138.68	25,721.92	196	8,706
0.25	147.47	64,013.97	196	19,603
0.30	Not found	2	Not found	6,229
0.35	Not found	2	Not found	1,297
0.40	227.50	11,060.56	239	337
0.45	255.58	11,037.89	266	297
0.50	285.53	13,642.97	296	523
0.55	313.04	16,114.41	323	134
0.60	342.27	25,609.03	353	56
0.65	370.17	23,034.78	381	32
0.70	398.84	26,500.50	409	27
0.75	429.90	58,407.64	438	152
0.80	457.53	220,047.20	468	66
0.85	485.74	196,890.58	495	27
0.90	514.69	554,958.83	524	53
0.95	543.12	684,439.48	556	25

used, and it is time consuming to enumerate all of them, we propose to consider a subset of sequences which is as follows: select any shift in the planning horizon as the start shift of the sequence. Then starting from this shift, the sequence is formed by including the shifts according to the order of time. When the end of the planning horizon is reached, the sequence is wrapped around the beginning of the planning horizon, and it is continued until it reaches the shift immediately before the start shift.

5.2 The column generation method

Another heuristic algorithm is the column generation method (CGM). To be consistent, the same notations are used as those defined in the original model, SAP. In addition, some new notations are defined to represent the column coefficients and new decision variables.

In the column generation model, a column represents a storage allocation in a particular shift. The column coefficients represent the workload assigned to each subblock, the number of yard cranes required and the high-low pattern for that shift. They are listed below.

x_{itr} The number of 20-ft containers allocated to subblock, i, in shift, t, for column, r, $1 \le i \le I$, $1 \le t \le T$

y_{itr} The number of 40-ft containers allocated to subblock, i, in shift, t, for column, r, $1 \le i \le I$, $1 \le t \le T$

h_{itr} =1 if the workload allocated ($x_{itr}+y_{itr}$) to subblock, i, in shift, t, for column, r, is high, i.e., $H_l \le x_{itr}+y_{itr} \le H_u$, $1 \le i \le I$, $1 \le t \le T$
=0 if the workload allocated ($x_{itr}+y_{itr}$) to subblock, i, in shift, t, for column, r, is low, i.e., $L_l \le x_{itr}+y_{itr} \le L_u$, $1 \le i \le I$, $1 \le t \le T$

d_{ktr} The number of yard cranes allocated to block, k, in shift, t, for column, r, $1 \le k \le K$, $1 \le t \le T$

n_{tr} The total number of yard cranes required in shift, t, for column, r, $1 \le t \le T$

p_{tr} =1, if column, r, is selected for shift t
=0, otherwise.

In the column generation model, decision variable, p_{tr}, is used to represent whether column, r, is selected for shift t; it should be binary in the original MIP master problem, while it should be continuous between 0 and 1 in the relaxed master problem.

In the master problem, the total number of crane shifts required for the whole planning horizon under consideration should be minimized.

$$(CGM)\text{Min } w = \sum_{t=1}^{T} \sum_{r} n_{tr} p_{tr} \tag{27}$$

Subject to:

$$\sum_{r} p_{tr} = 1 \text{ For } 1 \le t \le T \tag{28}$$

$$\sum_{t \notin L_i} \sum_{r} (x_{itr} + 2y_{itr}) p_{tr} \leq CS \text{ For } 1 \leq i \leq I \tag{29}$$

$$\sum_{t \notin L_i} \sum_{r} (x_{itr} + y_{itr}) p_{tr} \leq 2\ CC \text{ For } 1 \leq i \leq I \tag{30}$$

$$p_{tr} \in \{0, 1\} \text{ For } 1 \leq t \leq T, \forall r \tag{31}$$

Constraint 28 ensures that one and only one column can be selected in each feasible solution for each shift. Constraint 29 and 30 ensure the capacity restriction of subblocks in terms of storage space and yard cranes. Constraint 31 is integer restrictions.

To solve the column generation model, the master problem should be feasible, and all the variables should be continuous, which are necessary to obtain the dual prices information. Therefore, in the relaxed master problem, p_{tr} is a continuous variable assuming a value between 0 and 1. A feasible solution to the original problem is added as the initial columns. This can ensure the feasibility of the relaxed master problem.

After obtaining the dual prices from the relaxed master problem, they can be used to price out the new columns. A new column with the most negative objective function coefficient in the master problem can be found from the pricing problem for each shift. After adding the new columns, the master problem can be solved again to obtain the updated dual prices.

The pricing problem is defined as follows:

$$\text{Min } z = n_{tr} - \pi_t - \sum_{i=1}^{I} (x_{itr} + 2y_{itr}) \sigma_i - \sum_{i=1}^{I} (x_{itr} + y_{itr}) \delta_i \tag{32}$$

where π_t, σ_i, and δ_i are dual prices for constraints 28, 29 and 30, respectively. The objective function of the pricing problem is to find the most negative objective function coefficient in the master problem.

Subject to:

$$\sum_{i \in V_j} x_{itr} = WX_{jt} \text{ For } 1 \leq j \leq J \tag{33}$$

$$\sum_{i \in V_j} y_{itr} = WY_{jt} \text{ For } 1 \leq j \leq J \tag{34}$$

$$\sum_{i \in B_k} (x_{itr} + y_{itr}) \leq d_{ktr} CC \text{ For } 1 \leq k \leq K \tag{35}$$

$$H_l + (L_l - H_l)(1 - h_{itr}) \le x_{itr} + y_{itr} \le L_u + (H_u - L_u)h_{itr} \text{ For } 1 \le i \le I \quad (36)$$

$$x_{i'tr} = 0, y_{i'tr} = 0 \text{ For } 1 \le i \le I, i' \in N_i, t \in L_i \quad (37)$$

$$\sum_{i' \in N_i \text{ or } i'=i} h_{i'tr} \le 1 \text{ For } 1 \le i \le I \quad (38)$$

$$d_{ktr} + L_{kt} \le C_k \text{ For } 1 \le k \le K \quad (39)$$

$$\sum_{k=1}^{K} d_{ktr} = n_{tr} \quad (40)$$

$$x_{itr} \ge 0 \; y_{itr} \ge 0 \text{ For } 1 \le i \le I, 1 \le t \le T, \forall r \quad (41)$$

$$h_{itr} = 0 \text{ or } 1 \text{ For } 1 \le i \le I, 1 \le t \le T, \forall r \quad (42)$$

$$n_{tr} \ge 0 \text{ Integer For } 1 \le t \le T, \forall r \quad (43)$$

All constraints ensure the same restrictions as the original model, SAP, within one shift. Constraints 4 and 5 are removed from the formulation as only one shift is considered at one time.

Table 8 Results of heuristic algorithms for small-scale problem (case 1)

Utilization	Results of SAP	Results of SQM	Results of CGM
0.05	10	10	10
0.10	10	10	10
0.15	10	10	10
0.20	10	10	10
0.25	10	10	10
0.30	15	15	15
0.35	15	15	15
0.40	15	15	15
0.45	17.56 (gap=12.18%)	20	20
0.50	20	20	20
0.55	21	21	21
0.60	25	25	25
0.65	25	26	26
0.70	25	Not Found	38
0.75	30	Not Found	40
0.80	30	Not Found	40
0.85	35	Not Found	40
0.90	35	Not Found	40
>=0.95	Infeasible	Infeasible	Infeasible

To accelerate the column generation procedure, the results obtained from model SPP and SQM are treated as alternative columns. In conjunction with the initial feasible solution, the column generation method may improve the quality of the results obtained from the sequential method.

5.3 Implementation

The sequential method and the column generation method are both implemented in C++ and run on the same computer as that for the original model, SAP. They use the same input data as those for the original model, SAP. The computational results are presented in Tables 8, 9 and 10.

Tables 8 and 9 present the results of the sequential method and the column generation method for small-scale problems. For relatively low utilization scenarios, the obtained feasible solutions are quite close to the optimal solutions. For some scenarios, the objective value of the feasible solution obtained from SQM is exactly the same as that of the optimal solution. For relatively high utilization scenarios, the sequential method may not find any feasible solution. It is because the sequential method is a greedy algorithm and cannot ensure feasibility.

For relatively low utilization scenarios, the column generation method can yield near-optimal or optimal solutions as the results of model SQM are considered as alternative columns. For relatively high utilization scenarios, there may be a big gap between the solutions obtained and the lower bounds.

Table 10 compares the results of the sequential method and the column generation method for large-scale problem. For most relatively low and moderate utilization scenarios (with utilization up to 0.80 except between 0.3 and 0.35), the sequential method can give near-optimal or exact the optimal solution. This is also true for the column generation method that uses the results of the sequential method as alternative columns. However, for very high utilization scenarios (with

Table 9 Results of heuristic algorithms for small-scale problem (case 2)

Utilization	Results of SAP	Results of SQM	Results of CGM
0.05	10	10	10
0.10	10	10	10
0.15	10	10	10
0.20	11	11	11
0.25	12	12	12
0.30	13	13	13
0.35	15	15	15
0.40	17	17	17
0.45	18	18	18
0.50	21	21	21
0.55	23	23	23
0.60	23	24	24
0.65	26	Not Found	40
0.70	28	Not Found	40
>=0.75	Infeasible	Infeasible	Infeasible

Table 10 Results of heuristic algorithms for large-scale problem

Utilization	Lower bounds from SPP	Results of SQM	Results of CGM
0.05	102	102	102
0.10	138	138	138
0.15	174	174	174
0.20	196	196	196
0.25	196	197	197
0.30	Infeasible	Infeasible	Infeasible
0.35	Infeasible	Infeasible	Infeasible
0.40	239	239	239
0.45	266	266	266
0.50	296	296	296
0.55	323	323	323
0.60	353	353	353
0.65	381	381	381
0.70	409	409	409
0.75	438	438	438
0.80	468	472	468
0.85	495	Not Found	2,048
0.90	524	Not Found	2,048
0.95	556	Not Found	2,048
1.00	584	Not Found	2,048

utilization greater than 0.8), the sequential method does not work well. Instead, the column generation method can give a feasible solution.

In the experiments conducted, the column generation method gives a better solution than that of the sequential method only for those problems wherein the sequential method can get a feasible answer. This means that the column generation model actually doesn't improve the solution quality effectively. Even though the relaxed master problem can be solved to optimality easily, the columns generated cannot improve the quality of solution to the MIP version master problem. For each single shift, there are too many optimal solutions as there are too many possible allocations with the same number of cranes. Once the master problem finds some linear combination of the columns that can satisfy the linking constraints, the column generation procedure will stop right away. Consequently, the solutions for different shifts will not be explored enough. We can use some meta-heuristics to generate more columns for a future research topic.

6 Conclusions and future research

In this paper, we study an actual problem faced by a leading transshipment port operator. Currently, the port operator uses consignment strategy and attempt to assign containers to different subblocks with the objective of preventing traffic congestion. However, they do not have any formal planning tool and its decisions are based on intuition and past experiences. Hence, we develop a tool that is able to provide a holistic and systematic way to address the problem which takes into consideration the port operator's actual requirements. The model is based on a MIP formulation and is particularly useful for transshipment hubs where transshipment of containers is the major activity, and the yard activity is intensive. Although the model under some scenarios cannot be solved to optimality by the commercial software package, we have developed two heuristics based on the MIP model to provide good results. While the heuristic algorithms cannot guarantee an optimal solution, we have developed a bound which is useful in quantifying the quality of these solutions. So far this is the first paper to address the yard allocation problem with consignment strategy and vicinity matrix for a transshipment port.

In the current model, we assume that the yard template is given, i.e., the set of subblocks which is assigned to a certain departing vessel is known. Given this information, we determine where we should allocate the containers in each shift. A future research direction is to look at how to design the yard template and integrate this to the current yard storage allocation model. Another possible future research topic is to identify good ways to solve the MIP model efficiently, which include the integration of meta-heuristics with the MIP model, a different decomposition method which can exploit the structure of the MIP model.

References

Chen CS, Lee SM, Shen QS (1995) An analytical model for the container loading problem. Eur J Oper Res 80(1):68–76

Chen CY, Chao SL, Hsieh TW (2000) A time-space network model for the space resource allocation problem in container marine transportation. Paper presented at the 17th international symposium on mathematical programming, 2000, Atlanta, USA

Chen T (1999) Yard operations in the container terminal—a study in the "unproductive moves". Marit Policy Manage 26(1):27–38

Cheung RK, Chen CY (1998) A two-stage stochastic network model and solution methods for the dynamic empty container allocation problem. Transp Sci 32(2):142–162

Chung YG, Randhawa SU, Mcdowell ED (1988) A simulation analysis for a transtainer-based container handling facility. Comput Ind Eng 14(2):113–115

Crainic TG, Gendreau M, Dejax P (1993) Dynamic and stochastic models for the allocation of empty containers. Oper Res 41:102–126

Davies AP, Bischoff EE (1999) Weight distribution considerations in container loading. Eur J Oper Res 114: 509–527

Kim KH (1997) Evaluation of the number of reshuffles in storage yards. Comput Ind Eng 32 (4):701–711

Kim KH, Bae JW (1998) Re-marshaling export containers in port container terminals. Comput Ind Eng 35:655–658

Kim KH, Kim HB (1999) Segregating space allocation models for container inventories in port container terminals. Int J Prod Econ 59:415–423

Kim KH, Park KT (2003) A note on a dynamic space-allocation method for outbound containers. Eur J Oper Res 148(1):92–101

Kim KH, Park YM, Ryu KR (2000) Deriving decision rules to locate export containers in storage yards. Eur J Oper Res 124:89–101

Ryan NK (1998) The future of maritime facility designs and operations. In: Simulation Conference 1998, pp 1223–1227

Scheithauer G (1999) LP-based bounds for the container and multi-container loading problem. Int Trans Oper Res 6(2):199–213
Shen WS, Khoong CM (1995) A DSS for empty container distribution planning. Decis Support Syst 15:75–82
Taleb-Ibrahimi M, De Castilho B, Daganzo CF (1993) Storage space vs handling work in container terminals. Transp Res B 27:13–32
Vis IFA, De Koster R (2003) Transshipment of containers at a container terminal: an overview. Eur J Oper Res 147(1):1–16
Yun WY, Choi YS (1999) A simulation model for container-terminal operation analysis using an object-oriented approach. Int J Prod Econ 59(1–3):221–230
Zhang C, Liu J, Wan Y-w, Murty KG, Linn RJ (2003) Storage space allocation in container terminals. Transp Res B 37:883–903

Rommert Dekker · Patrick Voogd · Eelco van Asperen

Advanced methods for container stacking

Abstract In this paper, we study stacking policies for containers at an automated container terminal. It is motivated by the increasing pressure on terminal performance put forward by the increase in the size of container ships. We consider several variants of category stacking, where containers can be exchanged during the loading process. The categories facilitate both stacking and online optimization of stowage. We also consider workload variations for the stacking cranes.

Keywords Container stacking · Marine terminals · Simulation · Container rehandling

1 Introduction

World trade, especially the Asia–US and Asia–Europe trade, has developed rapidly over the last decades. As a result, container traffic has increased at a high rate as well. Ocean carriers have responded by ordering more and much larger ships. For example, the recently built *PONL Mondriaan* can carry up to 8,450 TEU, whereas Maersk/Sealand's largest ships are considered to be equally large or larger. The consequence of having larger container ships is that terminal activities become more a bottleneck and its productivity has to go up. This was already acknowledged in the FAMAS research project started in The Netherlands in 1999 (Celen et al. 1999). In this paper, we will report on explorative research concerning container-stacking policies at an automated terminal. Before a detailed discussion, we will first give an overview on container activities.

R. Dekker (✉) · P. Voogd · E. van Asperen
Econometric Institute, Erasmus University Rotterdam, Burg. Oudlaan 50,
3062 PA Rotterdam, The Netherlands
E-mail: rdekker@few.eur.nl

2 Container operations and trends

Several reviews on container handling have been published (Meersmans and Dekker 2001; Steenken et al. 2004; Vis and de Koster 2003). The overview below is based on them as well as on own experience with terminal studies. Although marine container terminals vary all over the world, they have a number of similarities. Ocean-going ships moor at a berth where quay cranes unload and load containers from the ship. Containers that have been unloaded are then transported to the main stack where they are positioned through cranes or straddle carriers. Containers can again be loaded in sea ships. Alternatively, they can be further transported on land through truck, train, or barge. In those cases, the container is moved from the stack to a rail or barge terminal or it is directly positioned on a truck, which has entered the terminal. Most terminals are manually operated; a few terminals use semiautomated equipment such as automatic guided vehicles (AGV), to transport containers, and automatic stacking cranes (ASC), to stack containers. These are ECT in Rotterdam, CTA in Hamburg, and Thamesport in London. In this paper, we will focus on these automatic terminals such as the Delta Dedicated Terminals at ECT's Maasvlakte complex in Rotterdam.

2.1 Implications of larger ships

Large ships are more expensive to buy and to operate than small ships. As a ship's port time can be considered as nonproductive, a large ship's port time is more costly per hour than a small ship's time. Larger ships, however, take more time to unload and load due to the larger amount of cargo. This is a kind of paradox, which puts a limit to the size of ships, as pointed out in (Cullinane and Khanna 2000). The port time consists of port entry and departure time, (un)mooring time, preparation time, and the actual loading/unloading time. Larger ships are therefore likely to make fewer and larger calls than small ships to reduce unproductive time. For example, the *PONL Mondriaan* loaded and unloaded some 4,000 TEU in one port. This will put much more stress on the terminal logistics and stack.

2.2 Structure of stacking strategies

Several decision horizons can be identified in stacking, viz. strategic/design for the long-term, tactical for the medium-term, operational for the short-term, and real-time for the direct operations. Strategic decisions concern the choice of equipment, the size of the terminal in general, and the stack in particular. Automated stacks have less flexibility and apply more costly equipment than manually operated stacks; hence, the design is very important. Tactical stacking decisions concern capacity decisions on months to year. In manually operated stacks, there are more tactical decision freedoms than in automated stacks, viz. layout of the stack, number of cranes employed. Decisions on a tactical level include the use of operation strategies, such as using a prestack or the application of stack reorganizations (also called remarshalling) at those moments where no ships need to be served. Operational decision making concerns the reservation of space for ships, the decision to store a container at a particular location, the allocation of equipment to jobs, etc. Finally, the real-time phase is mainly relevant for automated equipment, as it concerns speed control and collision avoidance of equipment. These are mostly technical decisions taken by control systems. In this paper, we mainly address strategic and operational decisions. The way the latter are carried out is captured in a stacking strategy. The main objectives of a stacking strategy are

- Efficient use of storage space
- Efficient and timely transportation from quay to stack and further destination and vice versa
- Avoidance of unproductive moves

The second objective implies, e.g., that an export container should be stacked close to the ship with which it will sail and that its retrieval time should be short. A stack with a maximum height of one container would be optimal for the third objective. This would however lead to an inefficient ground use and long travel times, so it is rarely applied in practice (apart from some stacks on wheels in the US). Accordingly, one has to decide whether a container should be stacked on another one or not.

A main input for a stacking strategy is the information available on a container. This is usually its type (size, reefer, dangerous goods), modality and date/time of departure. Unfortunately, this information may change or not be completely known upon arrival.

There are several types of stacking strategies. In category stacking, one defines categories and stacks containers of the same category on top of each other. In the residence time strategy, one stacks a container on others if its departure time is earlier than that of all containers, which will be below it.

Steenken et al. (2004) distinguish storage planning and scattered stacking. In storage planning, space in specific areas of the stack is reserved before the ship's arrival. In scattered stacking, yard areas are not assigned to a ship's arrival but to a berthing place. The stacking position is then determined in real-time and containers are stochastically distributed over the area. Scattered stacking results in higher yard utilization and a significant reduction in the number of reshuffles. The category stacking employed in this paper is a form of scattered stacking.

Some containers (e.g., reefers) require special locations because they need to be supplied with electricity. The determination of the stack capacity is a major design problem of a terminal, as the physical space required for the stack is often restricted and expensive. Stacking high may be advocated, but the expected number of reshuffles increases sharply with the stacking height. We define a reshuffle as an unproductive move of a container, which is required to access another container that is stored beneath it (this implies that reshuffles occur only when removing containers from the stack).

Quite often, stacks are separated into import and export parts. Import containers are those containers that arrive in large container ships from overseas and continue their destination through inland transport. These arrivals are somewhat predictable. The departure of import containers, however, is likely to be in an unpredictable order, so they cannot be stacked that high. Export containers that arrive via land transport may arrive somewhat randomly, but their departure is usually connected to a ship; hence, they can be stacked in a much better way.

2.3 Loading or stowage plan

Every ship which is loaded at a terminal has a stowage plan. According to Steenken et al. (2004), it is made in two steps. First the shipping line makes a rough plan based on categories, which is sent to the terminal. Later, somewhat before the arrival of the ship, a more detailed plan is made by the terminal planner who fills the categories in with detailed containers. The stowage plan specifies which container will be loaded at which location in the ship. As containers vary in size and weight, the load distribution is essential for the ship's stability. Heavy containers should be stored as low as possible. The stowage plan, however, also directly influences the ease of unloading the containers and, hence, containers of the same destination should be loaded on top of each other or on top of containers destined for ports further away. Apart from these restrictions, there are also containers with dangerous goods, which should be stored preferably below decks, reefers that have special positions, etc. Advanced software is used to perform offline optimization of the stowage plan also to avoid reshuffles as much as possible. Although the stowage plan fixes the load order per quay crane, it does not fix the exact order in which the containers leave the stack as the crane loading cycles are quite stochastic and a difference in progress between cranes may occur. Therefore, this software does not take the actual operations of the loading into account (Steenken et al. 2004). Online stowage planning does take the details of yard operations into account and will be employed in this paper; it is not yet in use at container terminals.

The stacking problem can be considered to be more difficult than the stowage planning as there can be uncertainty about which container will be needed before another. For import containers, this uncertainty exists because trucks arrive more or less randomly to pick up a specific container.

3 Stacking research

3.1 Literature overview

Little has been published in scientific literature on stacking problems. A main reason may be that the practical problems are quite complex and do not easily allow for analytical results, which are relevant for practice. Steenken et al. (2004) gave a high-level overview of stacking both in theory and in practice.

Stacking problems can be dealt with in two ways: simplified analytical calculations or detailed simulation studies. The first gives insight into the relationships between the various parameters on a more abstract level. The second can go in much more detail, with the negative side effect that it is time-consuming and only few people really understand its ins and outs. No comprehensive stacking theory exists today, and a good stack design not only depends on local space conditions but also on the information characteristics of the ingoing and outgoing flow of containers, which may vary from place to place. Examples of both approaches are given below.

Sculli and Hui (1988) were among the first to develop yardsticks for the relation between stacking height, utilization (or storage space needed), and reshuffles by applying a comprehensive simulation study. Taleb-Ibrahimi et al. (1993) discussed this relation for export containers both at a long-term scale, as well as operationally. They discussed dynamic strategies that store early-arriving containers in a rough pile until a certain date, after which all containers for a ship are put in a dedicated storage area (usually close to the berthing place of the ship). The procedures developed calculate the storage space needed as function of the stacking height. De Castilho and Daganzo (1993) continue these studies with the stacking of import containers. They consider two strategies: one that keeps stacks of the same size vs one that segregates the containers on arrival time. A slightly more detailed discussion resulting in tables and yardsticks (looking at stacking blocks with bays of similarly sized containers served by gantry cranes), both analytically and by simulation, was given by Kim (1997). Kim and Kim (1998) extended these studies by also taking the number of stacking cranes into account. They developed a simple cost model for optimizing this number using analytical approximations for the various performance measures.

In case the stowage plan is available some time before the sailing, the containers in the export stack may be remarshalled. This results in an "ideal" stack and, thus, less handling work during the loading operation of the vessel. Kim and Bae (1998) describe a two-stage approach to minimize the number of containers to be moved and to do so in the shortest possible traveling distance. Segregating space allocation strategies of import containers was studied by Kim and Kim (1999). In segregation strategies, stacking newly arrived containers on top of containers that arrived earlier is not allowed. Spaces are thus allocated for each arriving vessel. They study cases with constant, cyclic, and varying arrivals of vessels.

An empirical statistical analysis of the actual performance at a Taiwanese container terminal was provided by Chen et al. (2000). The number of reshuffles (Chen et al. use the term "shift moves") was related to the storage density, the volume of containers loaded, and the volume of containers discharged both for stacking crane blocks and straddle carrier blocks.

Decision rules using weight groups for locating export containers were derived and validated through dynamic programming by Kim et al. (2000). Weight is a useful criterion as heavy containers are usually stored deep in a ship.

Fig. 1 Overview of ECT's DDE terminal

Stacking policies for automated container terminals are investigated by Duinkerken et al. (2001), who use a detailed simulation model that not only models the stack, but also the quay transport in an automated container terminal. They also apply categories, but in a much more simplified way than we do in this paper. All in all, a comprehensive analysis of how stacking should be done at an operational level is still lacking; hence, this paper will deal with it.

3.2 Selection of research object

In this paper, we will investigate a container terminal with an automated stack as it is envisaged that future developments will move into that direction. A picture and logical layout of such a stack are given in Figs. 1 and 2. (Notice a slight difference in Figs. 1 and 2 with regard to the reefer platforms; in this paper, we will follow Fig. 2).

Figure 2 gives the general layout of the stack. On each lane, we assume one automatic stacking crane. Transfer points are located on both the sea- and landsides. The lanes are perpendicular to the seaside, where jumbo (very large) and deep-sea ships, as well as short-sea/feeders, are loaded and unloaded. The

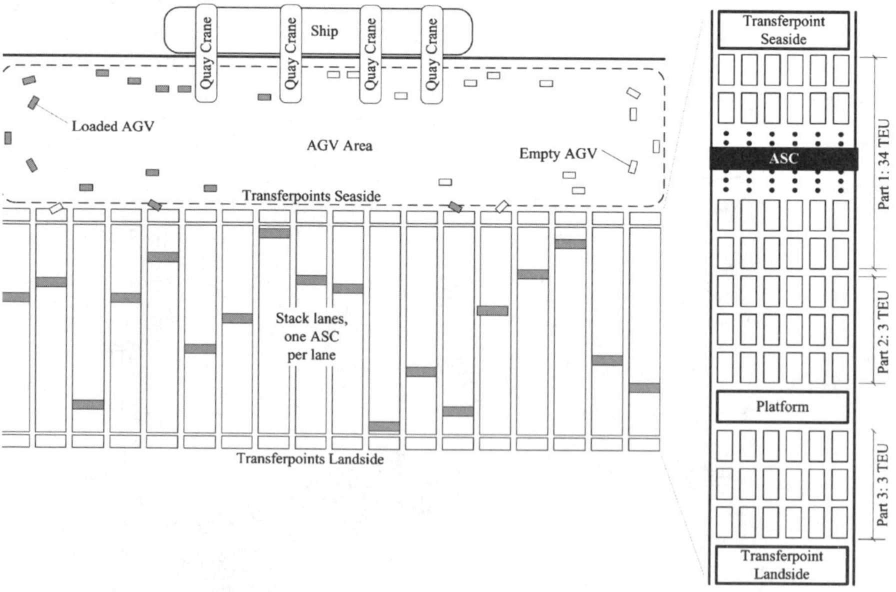

Fig. 2 Schematic overview of the stack

containers are placed with their long side parallel to the direction of the lane. The transfer point on the landside is used for rail, truck, and barge. All lanes have the same length, width, and height, expressed in terms of locations. These locations are slightly larger than one TEU to allow for some space between the containers, which will be used to pick up and put down containers. We define a pile as zero or more containers stacked on top of each other. We will refer to a pile with zero containers (the ground location is empty) as an empty pile; a pile that is stacked to the maximum stacking height is called a full pile.

Every lane is partitioned in three parts. Part 1 starts at the seaside and is used for nonreefer containers. There are two parts adjacent to a special platform for reefer containers on the landside. Part 2 is closest to the landside, has a length of three locations, and is used for reefer containers only. Part 3 is located between part 1 and the platform; it has a length of three locations and is used for reefers and reshuffles of other containers. The reefers are stored directly adjacent to the platform between parts 2 and 3. The platform is 10 ft deep and supplies the reefers with electricity.

Automatic guided vehicles (AGV) transport the containers from the transfer points on the seaside to the quay cranes and vice versa. In the stacking algorithm, it is assumed that the number of AGV is sufficient to handle all transport to and from these transfer points in time.

The base case is a stack with 27 lanes, length 40, width 6, and height 3. This implies that the theoretical stack capacity is 19,440 TEU. We do not make a distinction between an export and import stack. This is partly because separate space is already reserved for reefers and partly because it would cause inflexibility. The difference between export and import is implicitly incorporated in the analysis, because we will introduce different categories of containers, which are stacked together and the import/export property is part of these categories.

3.3 Simulation program

A large simulation program was developed in the "MUST" language (Upward Systems 1994). This is a Turbo Pascal add-on, which allows easy programming in Turbo Pascal, while using a number of modules from the package. It was also extensively used within ECT. It is fast, memory can be managed well, and complex algorithms can easily be written and incorporated. Two separate programs have been developed: a generator program and an evaluator program.

The generator program creates entry and departure times of some 175,000 containers covering a period of 15 weeks of operation. The first 3 weeks are used as a warm-up period to fill up the stack. The output was written to a file, which was used as input for the evaluator program where different stacking procedures could be tested. The generation of the containers was tied to the modalities with which they would arrive or depart. Several types of ships were considered, viz. deep-sea ships and 8,000 TEU large jumbo ships, the latter arriving once a week with a call size of about 3,000 containers. From a high-level modal split matrix, we developed cyclic ship schedules, as well as detailed arrivals of all other modalities. The matrix in Table 1 illustrates the flow between the different modes for a 3-week period.

We also developed detailed ship loading plans that specify the locations of individual containers and detailed crane sequences for loading and unloading. The call size of the jumbo ship was set at some 3,000 containers. We assumed a

Table 1 Modal split matrix

From	To						
	Jumbo	Deep-sea	Short-sea/feeder	Truck	Rail	Barge	Total
Jumbo	0	2,332	3,630	407	965	1,405	8,739
Deep-sea	2,691	1,344	1,466	270	389	568	6,728
Short-sea/feeder	3,870	2,000	0	967	1,876	2,735	11,448
Truck	438	368	967	0	0	0	1,773
Rail	1,047	540	1,877	0	0	0	3,464
Barge	1,524	788	2,736	0	0	0	5,048
Total	9,570	7,372	10,676	1,644	3,230	4,708	37,200

50:40:10% ratio between 20, 40, and 45 ft containers, which gives a TEU container ratio of about 1.5. This means that the jumbo ship loads some 4,500 TEU. We have also modeled other transport modalities: short-sea/feeder, rail, truck, and barge, with the daily fluctuations in truck arrivals and a stationary pattern with fluctuations for all the other modes. An average container residence time of 3.7 days was used, in line with information available at ECT. This implies an average utilization of 50% of the base stack configuration. Detailed information about the generator program and its output is available in (Voogd et al. 1999).

The evaluator program performs a deterministic simulation of an experimental setting, based on the stochastic output of the generator program. The output of the generator program contains exact departure times for all containers. The evaluator program uses these times to trigger events and adds a small perturbation for use in the stacking algorithms. These perturbations are used to model the information uncertainty that occurs in practice. AGV routing was not modeled in the simulation program. We took a constant time depending on the quay crane and ASC lane where the container came from or had to go to.

This experimental setup enables accurate evaluation of various stacking algorithms; the generator program provides the same scenario for each experiment. Any change in the results is due to the stacking algorithm selected for the experiment and to the minor perturbations. This way of experimental setup, however, does not facilitate different demand scenarios, as for each scenario, a quite detailed arrival modeling needs to be constructed, which is a difficult scientific problem on its own.

3.4 Stacking algorithms

A stacking algorithm describes the way in which containers are handled, both in case of moves into and out of the stack as well as in case of reshuffles. For containers leaving the stack, we have no options unless the containers are exchangeable with others. In that case (see also below), there might be other containers of the same category for the same ship (or other modality), which can be retrieved in a better way.

The main part of a stacking algorithm decides where to put a new container or a reshuffled container. In this paper, we investigate two main concepts, viz. random stacking and category stacking. In random stacking, there is no preference for particular places, and it is used to evenly spread containers over the stack. In category stacking, we define categories of containers on the basis of the loading plan. Containers in the same category may be exchanged freely. In category stacking, one tries to exploit this property as much as possible. We supplement these concepts with decision rules for specific cases.

A stacking algorithm is also influenced by the information available at the moment of stacking. If the departure time of a container is known at stacking time, then we can stack the container on top of a pile of containers with a later departure time. This does, however, require a sufficiently large stack to allow the creation and maintenance of these "ordered" piles.

3.4.1 Common rules

There are some basic rules for all stacking algorithms in this paper:

- Twenty-foot containers occupy one TEU location in the stack, 40-ft containers occupy two locations, and 45-ft containers occupy three locations.
- Containers of different sizes cannot be stacked on top of each other.
- Containers have to be stacked precisely on top of each other (no overhang and a container can be on top of just one container).
- Containers can only be stacked in the direction along the lane, not transverse.
- Reefer containers are not placed on top of normal containers, or vice versa.

Reefer containers have a special requirement: the need for a power connection. This limits the locations available for stacking these containers. Thus, we have implemented the same stacking algorithm for reefers in the first five experiments. The only locations with power connections are directly adjacent to the platform. Thus, the number of locations available to reefers is limited to twice the lane width (once for each side of the platform). The stacking algorithm for reefers selects a random, nonfull pile within the special reefer section of the stack. If the pile is empty, the container is only stacked there when no more than three of these six reefer positions are occupied. This helps to make sure that all reefer reshuffles can be carried out. Otherwise this could cause a problem, because there are very few possibilities for the container to be reshuffled to. If the pile is not full, the reefer can be stacked if they are containers (reefers) of the same size. Whenever no suitable location is found in 5,000 random choices of a lane, the aim is changed to the reefer locations on the other side of the platform. This way of stacking probably causes low occupancy in parts 2 and 3 of the stack.

3.4.2 Random stacking

This algorithm is used as a benchmark. Suppose a 20-ft container has to be stacked. The program uses random search to find a pile that is not full. If the pile is empty or if the containers in this pile are also 20 ft, an acceptable position has been found, and the container can be stacked in this pile. If the pile consists of containers of a different size, then the container cannot be stacked here. The program then determines a new random position by choosing at random a new lane, row, and position until a location is found where the container can be stacked. Forty and 45 ft containers are handled in the same way, but in those cases, the algorithm searches for either an existing nonempty pile of the same size or for a sufficient number of adjacent empty piles (two for 40-ft containers, three for 45-ft containers).

For reshuffles, the program searches all piles in the lane except for the reefer positions on the landside of the platform. The container is reshuffled to one of the possible piles closest to the original pile.

3.4.3 Category stacking

This algorithm is based on defining categories of containers. These are defined through the export modality and, in case of a ship, the place of a container in it. We assume that for certain categories (especially those defined for jumbo and deep-sea, but not for trucks), containers are exchangeable in the loading plan or in the actual loading, if they are either in the same or different piles. The algorithm keeps track of a variable for every combination of lane, ship, and category. This variable indicates how many piles of containers exist, within that lane, with only containers of that specific ship/category combination and an empty top position. The variable is used to facilitate the search for a good location (note that searching over 19,000 locations upon each of the 175,000 container entries is very time-consuming in the simulation).

Now, suppose a new container has to be stacked. The first step is to determine if there is a pile that is not full and only with containers of that same category and for the same ship. All lanes are checked for such a pile; to spread the load evenly across the lanes, we start the search at a random lane. Using the variable described above, a zero indicates that no such pile exists, whereas a positive value means one or more of those piles exist.

When the variable indicates that one or more of those piles exist, the program starts searching, randomly within that lane, for one of those piles. When found, the container is stacked on top of that pile. If this creates a full pile, the variable associated with the current ship/category combination is decreased with one for that lane.

When no such pile can be found in the current lane, i.e., the variable has value zero for that ship/category combination in that lane, the aim shifts to the next lane. If value of the variable equals zero for that ship/category combination for all lanes, the container is stacked using random stacking (see description above).

3.5 Performance measures

Below we discuss appropriate performance measures of stacking policies.

3.5.1 Reshuffles and reshuffle occasions

There are two performance measures concerning the reshuffles. First of all, we define a reshuffle occasion as one or more reshuffle operations required to retrieve a container from the stack. We measure the reshuffle occasions as a percentage of containers that leave the stack. The total number of reshuffles is also counted (again as a percentage of the total number of containers leaving the stack). These measures are calculated separately for import and export containers. An export reshuffle is a container (export or import) that is reshuffled because the export container needed is under that container (so it is not necessarily an export container that is reshuffled). It seems obvious that a situation with many reshuffles or reshuffle occasions is undesirable, for reshuffling takes a valuable amount of time.

3.5.2 No positions available

We may not always find an empty location in the stack, especially considering the randomness in positioning a container when it enters the stack. This will most likely concern the 45-ft containers, because they require three adjacent empty locations and they form a minority in comparison to the 20- and 40-ft containers. Therefore, there will be few piles with 45-ft containers. Although the maximum utilization is always less than 100%, we may not find an empty location for a 40- or 45-ft container. We assume that there is an emergency stack for these containers and leave them out of consideration, as they would otherwise cause a deadlock in the program. The aspect does imply that the real capacity is much lower than the physical capacity, which is also a known practical fact. We may also encounter this problem when reshuffling a container; if we cannot find an empty location in the same lane, we move these containers to the emergency stack. A small number of reshuffles and reshuffle occasions indicate a better performance. Larger numbers indicate that the current stack size might be too small to be used with the current algorithm.

3.5.3 Workload of the automatic stacking cranes

A third group of performance measures deals with the workloads of the ASC. These workloads are determined every quarter of an hour as the proportion of time the ASC are busy. The design of the simulation program allows ASC workloads to exceed the capacity, i.e., workloads of more than 900 s per quarter. Since the focus of this research is on the stacking algorithms and not on ASC scheduling, we have chosen to allow these overloads and consider the frequency and gravity of these occasions as one of the criteria for the performance of an algorithm. Details about ASC technical performance can be found in (Voogd et al. 1999).

A move is handled at the same moment in time as specified in the container files, even when the ASC is not ready at that moment. Every move starts, when not already in the right position, with shifting the ASC from the previous position to the position for picking up the container (transfer point for containers that enter the stack) and ends at the position where the container is put down (transfer point for containers that leave the stack).

To give an indication of traveling times for ASC, the times are calculated for going from one of the transfer points to the first container position, to the twentieth position, and to the last (fortieth) position (all positions relative to the transfer point; see Table 2). The implementation code contains a precise model of the ASC movements, including maximum speed and acceleration along the three axes (longitudinal, lateral, and vertical).

The difference between the traveling times to the twentieth position is incurred by the reefer platform.

The workloads for all ASC are written to a file at the end of each quarter. The maximum and average workloads are determined, given as percentages of one quarter. An average workload of 50% therefore means that, on average, an ASC is busy half of the time, which is 450 s per quarter. Concerning the actual scheduling of an ASC, a workload of 80% is already pretty high. This is why the proportions of ASC quarters, with the ASC working more than 80, 90, 100, 110, and 120% of the time, are measured.

3.5.4 Occupation

The degree of occupation is measured for the ground locations. For this purpose, at the end of each quarter, the number of ground locations in use is recorded. The maximum and average numbers are calculated separately for the three parts of the stack. The overall occupation of the stack depends only on the size of the stack, as the number of containers that will be handled during the simulation is constant for all experiments. The occupation is 51% for the first three experiments and 47% for the other experiments; this is low, but a consequence of the large call sizes of the jumbo ships.

For the ground locations, we expect a larger number of reshuffles when few ground locations are occupied. The average height of the nonempty piles is higher, which increases the possibility of reshuffles. If, on the other hand, almost all ground locations are covered, then we expect a negative influence on the number of reshuffles and new containers that cannot be stacked in the regular stack.

4 Features of the stacking algorithms

In this paper, we explore the use of categories for the stacking of containers. For each experiment, we will indicate for which categories containers are considered to be exchangeable. Here, we define exchangeable to mean that a different container from the same category may be substituted when a container is requested for loading. The categories defined for large containerships are typically exchangeable. All containers to be picked up by trucks also form a category, but these containers are not exchangeable. To facilitate the exchange operationally, we stack containers of the same category in the same pile as much as possible, but exchange is also possible for containers of different lanes.

The definition of the categories is based on the weight class, destination, and type of container (the same criteria are mentioned in (Steenken et al. 2004)). Thus, only the export modality is a feature in the definition of the categories; the import modality is not taken into consideration. Using the data from ECT, we defined

Table 2 Typical ASC travel times

Transfer point	First position (s)	Twentieth position (s)	Last position (s)
Seaside	9.2	45.3	79.8
Landside	9.2	47.8	79.8

some 45 different categories for jumbo ships and 90 categories for deep-sea ships. Containers destined for short-sea/feeder, truck, rail, and barge transport will be allocated to a single category for each mode, even though they cannot be exchanged in operation. As we will see in the experiments, it is not wise to stack them in the same pile.

In addition to categorization, we have implemented several other features for the stacking algorithm.

4.1 Preference for ground locations

We use a preference for ground locations to decrease the possibility of spoiling a uniform pile, i.e., a pile with containers that all belong to the same category. The implementation of this feature tries to avoid stacking a container of a different category onto an existing uniform pile. This causes a preference for stacking on empty piles and for stacking on multiform piles. It will reduce the number of empty piles and may cause problems for stacking or reshuffling (45-ft) containers.

4.2 Workload control

The workload control feature associates a workload variable with each lane. We defined the workload variable as the percentage of time of the current quarter that the ASC for the lane was busy. When the workload variable exceeds a specified threshold, the lane is skipped in the search for a stacking position.

4.3 Alternative algorithm for reefers

Reefer containers can be stacked in just a small part of the stack. Therefore, our initial experiments exhibited some problems with reefer reshuffles. For every reefer reshuffle there are only up to five possible new positions (within the same lane). When stacking these containers at random, a lot of containers could not be reshuffled (within the same lane). The number of reefer reshuffles however was substantial. We therefore introduced category stacking for reefers with a modification to avoid the creation of full piles. In this way, we aim to leave a sufficient number of feasible empty positions for reefers.

4.4 Use empty pile closest to departure transfer point

When an empty pile has been selected for a container and multiple empty piles are available in the same lane, the algorithm will select the pile that is closest to the point where the container will leave the stack. The aim is to lower the ASC workloads during ship loading. The ASC will have to travel a shorter distance to get to the container, which decreases the time needed to unstack a container. Furthermore, it is expected that this feature will also increase the number of nonempty piles. This is due to the fact that we will now use the empty pile directly adjacent to an existing nonempty pile, leaving no space (TEU position) open. The proportion of “unusable” empty piles will then be lower. Using more ground locations is also thought to decrease the number of reshuffles. We will explain this feature with the following example.

Consider the case where we have to stack four 45-ft containers with a maximum stack height of three containers. Furthermore, suppose that all ground locations are occupied except for the last six TEU positions in front of a transfer point and that all piles of 45-ft containers are full. In this case, random stacking might put the first container on the second, third, and fourth TEU location instead of the first, second, and third TEU location. The second and third containers will then be stacked on top of the first container. Even with three TEU ground locations available, the fourth container cannot be stacked in this lane: the locations that are available are not adjacent. With the new rule, the first container will be stacked upon the first three empty TEU locations, leaving the other locations open for one (or more) of the other three containers.

At first, we will use this feature for all modalities. A variation of this feature is designed to reduce the ASC travel time (and thus the workload) when unloading jumbo or deep-sea ships: containers destined for the landside are not subjected to this rule. When there are no jumbo or deep-sea ships present at the quays, the ASC workloads are lower and the additional travel time does not pose a problem. This feature will probably have a negative effect on the average distance to travel for export containers (because the import containers can use positions close to the seaside). It will also decrease the effect described above concerning the use of ground locations.

4.5 Combine parts 1 and 3 of the stack

Initial experiments showed a low use of the locations in the third part of the stack. We therefore decided to use this part of the stack for both regular containers and reefers (for reshuffles and new containers). The reefers can still be stacked onto the last (one, two, or three) piles of the second part of the stack. We expect this feature to generate a better use of ground locations and thus reduce the number of reshuffles. An obvious disadvantage of this feature is that the number of available positions for reefers is reduced.

4.6 Exchanging containers from different lanes

Categories can also be used to select a container from a different lane in the loading operation. We can use this to avoid overloading an ASC. This feature is therefore triggered if the ASC in a selected lane is too busy. The algorithm scans all lanes of the stack for a lane that contains a container of the required category and an ASC that has a workload below the predefined limit.

4.7 Using the expected departure (residence) time of the containers

The expected departure time can be used to store containers that will leave shortly on top of containers that will stay in the stack for a longer period. This feature is used whenever a container has to be stacked and there is no nonfull pile of that category. The container will then be stacked on top of a container for which the expected departure time is later than the expected departure time of the incoming container.

The expected departure time for jumbo and deep-sea containers is approximated by the middle of the time interval during which the ship lies alongside the quay. For the other modalities, the average dwell time of a container is approximately half a week: the expected departure time is therefore approximated by adding 3.5 days to the time of arrival. Note that this option does not use detailed information. It can also be applied if no information on the departure time is available.

4.8 Choosing the ASC that has the lowest workload

The lowest ASC workload feature can be used for both incoming and outgoing containers. For incoming containers, creating uniform piles takes precedence over the lowest workload. Thus, a container will be stacked on top of a uniform pile of the same category even if the ASC for that lane is very busy. If there are uniform piles in multiple lanes, then the lane with the lowest ASC workload will be selected. For outgoing containers, we select the lane with the lowest ASC workload from lanes in which containers from the target category are stored.

5 Experiments

The following data applies to all experiments. The stack has 27 lanes for experiments A0 to C and 29 lanes for all other experiments. A lane is 40 TEU long, 6 TEU wide, and the maximum stacking height is three containers. Categories and exchanges are possible for jumbo, deep-sea, as well as for rail and barge; temporary substacks are used for rail and barge to loosen the loading order restrictions when leaving the main stack. Category stacking is applied for all modalities, except where stated differently. The experiments are listed in Appendix A and the numerical results from the experiments are in Appendix B. We will now describe the experiments and analyze the results.

5.1 Base Case

A null experiment (A0) uses random stacking without the possibility to exchange containers of the same category for the same (jumbo or deep-sea) ship. The number of reshuffles in case of random stacking is high (89%). Although it is hard to validate such stacking programs, the number is not considered unrealistic by terminal operations people.

Experiment A considers category stacking for all modalities without any of the additional features. This yields much better results than random stacking: the percentage of reshuffles drops from 89 to 46%.

In experiment B, short-sea/feeder and truck containers are not stacked as categories, because these containers are not exchangeable. The percentage of reshuffles for these containers is reduced significantly (short-sea/feeder from 112 to

82%; truck from 104 to 84%), while the percentage of reshuffles for all other modalities has increased. The average use of ground locations rises from 65 to 70%.

5.2 Preference for ground positions

Experiment C extends experiment B with a preference for ground locations as discussed in "Features of the stacking algorithms" section. This has a pretty large effect, mainly on the number of reshuffles and reshuffle occasions. On aggregate, those percentages are approximately half of the percentages when using no preference. The percentages of reshuffles are shown in Fig. 3.

The workloads of the ASC are also influenced by this preference, although the effects are moderate.

As expected, the number of empty piles drops especially in part 1 of the stack. This causes an increase in the number of containers that cannot be stacked. The probability that containers cannot be stacked or cannot be reshuffled is higher when there are fewer empty piles; on the other hand, the percentages of reshuffles and reshuffle occasions are lower.

In this case, almost one out of every 1,000 containers cannot be stacked, which is a very high proportion. One way to reduce this number is to increase the size of the stack. Therefore, we added two lanes (29 instead of 27) to create experiment D (this configuration of the stack will be used for all other experiments). As can be expected, this decreases the number of reshuffles, as well as the average workloads and the proportion of containers that cannot be stacked. Finally, it also reduces the use of ground locations a little.

5.3 Workload control

In experiment E, we add a workload control variable for each lane. A container is not stacked into a certain lane when the workload of the ASC in the current quarter exceeds 80%. This workload control variable is only used when (un)stacking regular (nonreefer) containers.

The workload control variables do not affect the reshuffles. The aim of this feature is to reduce the number of busy or very busy ASC quarters. The most significant

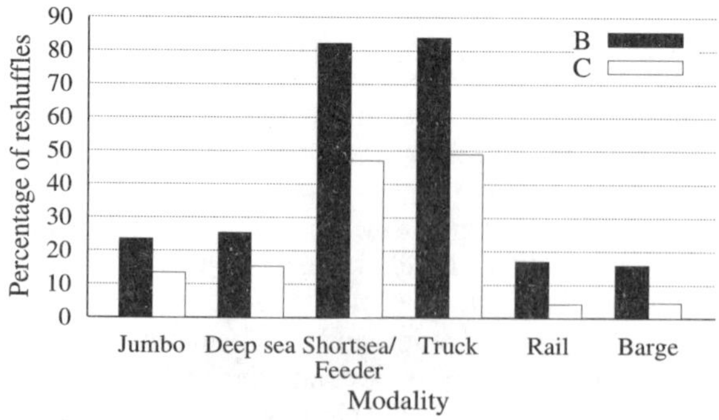

Fig. 3 The effect of preference for ground positions on reshuffles

impact can be observed in the percentage of ASC quarters with a workload over 100% during jumbo operations: this percentage drops from 11.8 to 8.3%.

5.4 Improved reefer stacking

The next experiment (F) adds the modified category stacking policy for reefers to the setup of experiment E. Experiment G adds workload control for reefer containers; the limit is set to 80%.

This seems to have a few positive effects. The overall number of reshuffles and reshuffle occasions are reduced (from 19.0 to 16.0% and from 13.4 to 11.4%, respectively). There are no more reshuffles for reefer containers (this cannot be deduced directly from the table). In addition, it is now possible to find a position for all new containers and reshuffles. Finally, the use of ground locations in the third part is much lower when using category stacking for reefer containers.

Adding a workload control variable for reefers (experiment G) has a (small) positive effect (it reduces the proportions of busy ASC quarters a little).

5.5 Use ground position closest to transfer point for unstacking

In experiment H, whenever a container is to be stacked on an empty pile in a lane, we select the pile that is closest to the transfer point where the container will leave the stack. The result is an increase of approximately 2% in the use of ground locations in part 1 of the stack (both average and maximum). The overall percentage of reshuffle occasions decreases from 11.4 to 10.4%; the percentage of reshuffles drops from 16.2 to 14.8%. The percentage of quarters with a high workload is lower during jumbo handling (7.8 vs 6.3% quarters with a workload over 100%). This is also true during deep-sea handling and overall.

5.6 Combine parts 1 and 3 of the stack

Experiment I was motivated by an observed low use of ground locations in the third part of the stack. Thus, experiment I extends experiment H with the option to stack

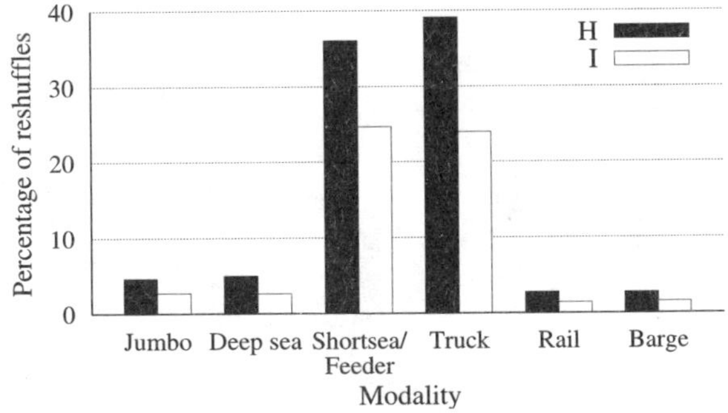

Fig. 4 The effect of combining parts 1 and 3 of the stack on reshuffles

regular containers in the third part of the stack. The average and maximum use of ground locations increase and lead to a clear reduction in reshuffles and reshuffle occasions (Fig. 4). Compared to experiment H, there are a significant number of reshuffles for which no position could be found (15 per 100,000 containers). The maximum workload for jumbo containers rises from 220 to 279%.

The total percentage of reshuffles decreased from 14.8 to 9.7. The total percentage of reshuffle occasions dropped from 10.4 to 6.9. The percentage of busy ASC quarters has decreased from 6.3 to 5.6 for jumbo containers.

5.7 Exchanging containers from different lanes

To study the influence of this feature, we have defined experiments J and K.

J Experiment I modified to exclude import containers from the closest transfer point rule.
K Experiment J, with the added possibility of exchanges between different lanes.

The exchange candidate has to be on top of its pile. The algorithm looks for exchange candidates whenever the workload of the ASC for the original container exceeds 80%.

Experiment J does not yield favorable results in comparison to experiment I: the percentages of reshuffles and reshuffle occasions are higher. Adding the exchange from different lanes feature in experiment K causes the percentage of reshuffles to drop from 9.9 to 9.5. The primary purpose of adding this feature was to obtain lower proportions of ASC quarters with high workloads. Figure 5 below illustrates the overall percentages of high ASC workloads: the percentage of busy quarters is reduced significantly. We have explored several additional ways to implement this feature but the results are similar. From these experiments, we conclude that adding the possibility of exchanging containers from the same category within different lanes has a positive effect. It reduces the number of reshuffles and reshuffle occasions, as well as the proportion of high ASC workloads.

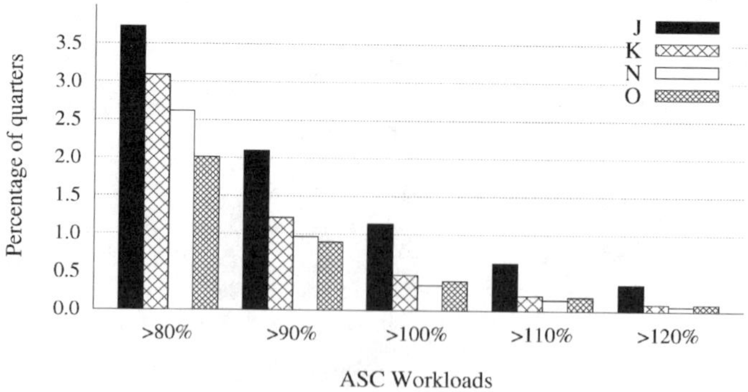

Fig. 5 The effect of using ASC workload on the percentage of busy quarters

5.8 Using the expected departure time of the containers

In practice, it is often difficult to obtain a reliable indication of the departure time. Therefore, we use the expected departure time as a relative measure to create an ordering for the containers. For this feature, we have to define a boundary value that controls whether a container can be stacked on top of another one. When we make this restriction too loose, a lot of containers will be stacked on a container that will leave earlier, which causes a reshuffle. If, on the other hand, the restriction is too tight, we will make less use of the opportunity to use the expected departure times of the containers.

To get some insight into the effects of adding a rule based on the expected departure times of the containers, we can compare the results of the experiments K, L, and M. Experiment K makes no use of this rule; experiments L and M extend experiment K with the expected departure time feature. For experiment L, the value of the boundary is 3 h after the expected departure time of the container already in the stack. Experiment M sets the boundary to the expected departure time of the container that is currently on top of the selected pile.

This feature was designed to lower the number of reshuffles. The percentages of reshuffles and reshuffle occasions are lowest for experiment L (8.8 and 6.2). For experiment M, these percentages (9.6 and 7.0) are even higher than for experiment L (9.5 and 6.8). The restriction on the expected departure times may be too tight for experiment M. The differences between these experiments concerning the high ASC workloads are small. Furthermore, using the departure times of the containers leads to a somewhat higher use of ground locations.

5.9 Choosing the ASC that has the lowest workload

We have designed two experiments to determine the effects of starting in the lane for which the ASC has the lowest workload when stacking or unstacking. Experiment K is used for comparison.

N Algorithm K with the ASC workload feature implemented for incoming containers for which multiple uniform piles in different lanes have been found.
O Same as experiment N, with lowest ASC workload feature implemented for outgoing, regular (non-reefer) containers.

The percentages of reshuffles and reshuffle occasions increase when adding this feature. However, the feature was designed to improve the workloads, so Fig. 5 shows the percentages of high workloads for these experiments.

As we can see, the percentage of high workloads has indeed decreased by starting in the lane where the ASC has the lowest workload. We have also experimented with the lowest ASC workload rule for reefer containers and a lower maximum stacking height (two) for truck containers as an extension of experiment O: these experiments yielded no additional benefits.

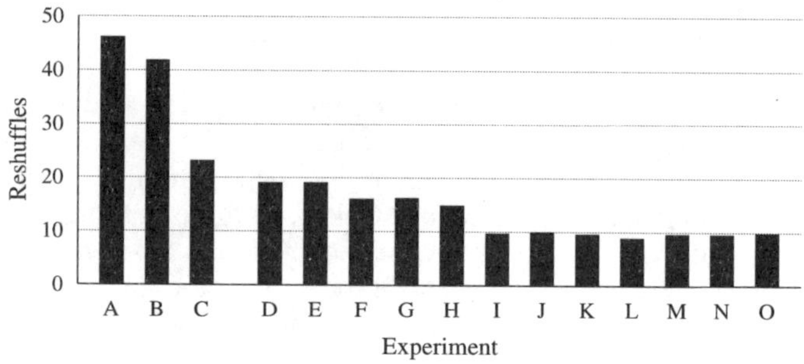

Fig. 6 Overall percentage of reshuffles

6 Comparison of all scenarios

In this section, we will focus on the overall results rather than compare individual experiments. Again, we will visualize some of the results in graphs.

First of all, Fig. 6 indicates that the percentage of reshuffles can be significantly reduced. For our benchmark, this was 46.1%; for experiment L, it is just 8.8%. That is less than 20% of the initial percentage. The graph also shows that a number of other experiments have a similar percentage of reshuffles.

Maybe, the most important performance measure is the proportion of busy ASC quarters. Figure 7 shows for all experiments the percentage of ASC quarters with the ASC working more than possible. We have decreased this value a lot. In the benchmark case, this is equal to 3.8%. The best result is obtained using experiment N (0.3%), but there are several experiments with similar performance (in terms of this percentage).

For some experiments, there are (relative to the numbers for other experiments) a lot of containers that cannot be stacked (either new containers or reshuffles; see Fig. 8). This is a highly undesirable effect. Note that, because we just took these containers out of the stack or we did not stack them at all, this also positively biases the results.

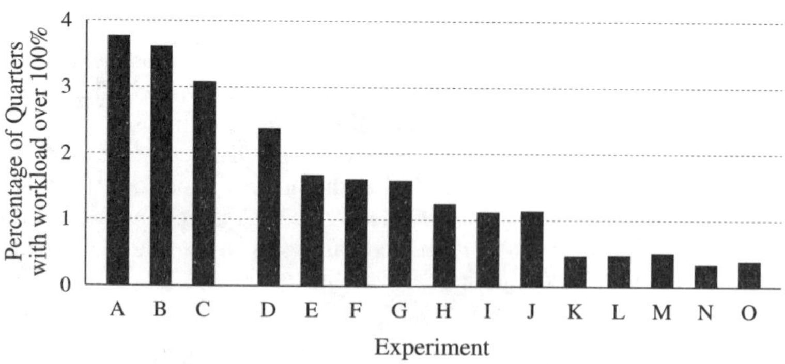

Fig. 7 Percentage of ASC quarters with a workload over 100%

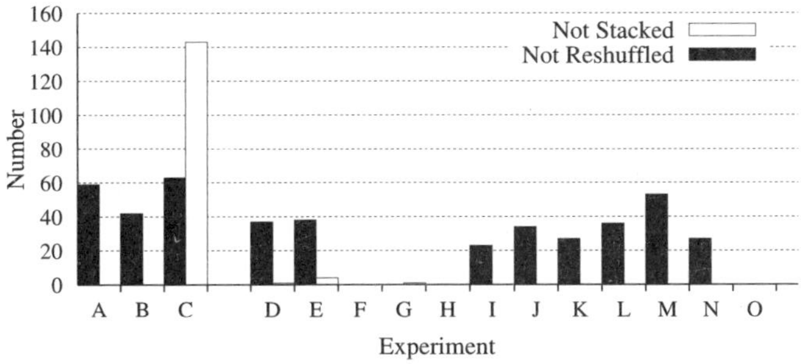

Fig. 8 Containers that cannot be stacked or shuffled

7 Conclusions

In this paper, we have investigated a number of policies for stacking containers in a yard by means of simulation. The following conclusions can be derived from the experiments.

Loading and unloading operations for jumbo containers ships creates workloads that exceed the capacity of the set of ASC (27 to 29 in total). The average workload over time is well below 100%, but the workload during the handling of a jumbo ship is very high with many short-term bottlenecks. This means that the stack configuration is not able to follow the quay crane production.

Category stacking yields much better results than random stacking. Allowing exchanges for containers for the same category jumbo or deep-sea ship further improves the results. The number of actual reshuffles and the number of reshuffle occasions can be reduced by adding a preference for ground locations. This also reduces the ASC workloads. However, there is the possibility of creating a higher proportion of nonstackable containers due to the reduced number of empty piles; this feature requires careful implementation.

Treating containers for short-sea/feeder, rail, truck and barge as categories to be stacked together seems to have no large effect on the whole. Although it reduces the number of reshuffle occasions for jumbo and deep-sea, this number increases for the other modalities. Using fewer piles, on average, for the same containers leads to a higher number of reshuffles. The effect of stacking piles with only truck containers up to a height of two is negligible.

The peaks in ASC workloads can be reduced by adding a workload control variable, as well as stacking on piles close to the transfer point where the containers are to leave the stack. Finally, the possibility of exchanging containers of the same category within different lanes decreases the proportion of high workloads as well.

The definition of the categories is based on parameters used in stowage planning. This allows online optimization in which we can avoid suboptimal yard operations that might be caused by a predefined (offline) stowage plan.

Overall, we conclude that detailed simulation experiments of the stacking operations can drastically improve the stacking performance and is thus essential for constructing automated container terminals.

Acknowledgements The authors would like to thank A. Nagel, F. J. A. M. Nooijen, and R. Th. van der Ham (ECT) for assistance during the research. The authors also thank the referees for useful comments, which helped to improve the paper.

Appendix A

Experiments

A0 This is a reference experiment that uses random stacking without exchanges.
A This experiment considers category stacking for all modalities.
B Category stacking without exchangeability for short-sea/feeder and truck containers.
C Same as the previous experiment, with an added preference for ground locations in the random part of the algorithm.
D Experiment C with 29 lanes instead of 27.
E Experiment D, with the workload control variable set to 80% rather than 88.9%.
F Experiment E with alternative reefer stacking policy (no workload control variable for reefer containers though).
G Experiment F, with reefer containers also subject to the workload control feature with a limit of 80%.
H Experiment G with the closest transfer point feature. This feature selects an empty pile closest to the transfer point at which the container will leave the stack.
I This setup is based on experiment H: We allow the stacking of regular containers in the third part of the stack (this part is usually reserved for reefer containers).
J In this modification of experiment I, we exclude import containers from the closest transfer point rule.
K Experiment I, with the option of exchanges between different lanes. Exchanges are considered whenever the ASC workload of the selected lane exceeds 80%. Feasible exchange locations are limited to the top containers of each pile and are located using a random search approach.
L Same as experiment K, with the expected departure time rule: a container can only be stacked on top of other containers if the new container has an expected departure time less than 3 h after the expected departure time of the current topmost container in the pile.
M Experiment L, but the expected departure time of the new container must be before or equal to the expected departure time of the topmost container of the pile.
N Experiment K with the ASC workload feature for incoming containers, for which multiple uniform piles in different lanes have been located.
O Experiment N, with the addition of the ASC workload feature for outgoing, regular (i.e., nonreefer) containers.

Appendix B

Numerical results of the experiments

Experiment	A0	A	B	C	D	E	F	G	H	I	J	k	L	M	N	O
Reshuffle occasions																
Total	60.9	31.0	31.0	16.1	13.3	13.4	11.4	11.4	10.4	6.9	7.1	6.8	6.2	7.0	6.8	6.9
Jumbo	67.8	12.4	18.5	10.2	8.5	8.3	4.3	4.1	3.6	2.1	2.3	2.1	1.8	2.4	2.0	1.4
Deep-sea	55.9	12.3	19.5	11.4	9.4	9.5	5.0	4.8	4.1	2.2	2.5	2.2	2.4	2.8	2.3	2.3
Short-sea/feeder	62.8	71.5	58.6	31.5	26.1	26.6	26.3	26.6	24.5	17.3	17.5	17.2	15.2	16.9	17.1	17.9
Export	55.9	35.1	34.2	18.7	15.5	15.6	12.9	12.9	11.8	8.0	8.2	7.9	7.1	8.1	7.9	8.0
Truck	–	68.0	58.4	32.3	26.7	26.4	26.5	26.6	26.0	16.2	16.3	14.9	12.5	14.9	16.1	16.1
Rail	–	9.5	14.7	3.8	3.0	3.2	2.7	3.1	2.5	1.3	1.2	1.1	1.5	1.5	1.3	1.0
Barge	–	8.8	13.8	3.9	2.8	2.9	2.8	2.9	2.4	1.4	1.4	1.1	1.5	1.6	1.2	1.3
Import	–	19.2	21.8	8.7	7.0	7.1	6.9	7.0	6.5	3.9	3.9	3.5	3.4	3.8	3.8	3.7
Reshuffles performed																
Total	89.3	46.1	41.8	23.0	19.0	19.0	16.0	16.2	14.8	9.7	9.9	9.5	8.8	9.6	9.5	9.8
Jumbo	99.7	15.7	23.5	13.4	11.3	10.8	5.5	5.3	4.7	2.8	3.0	2.8	2.0	2.8	2.6	1.8
Deep-sea	81.8	16.4	25.4	15.3	13.0	13.1	6.1	5.8	5.0	2.7	3.0	2.6	2.8	3.1	2.7	2.7
Short-sea/feeder	92.1	112.2	82.2	47.0	38.5	39.0	38.3	39.3	36.0	24.7	25.1	24.7	23.0	24.1	24.5	26.1
Export	81.9	52.9	46.6	26.8	22.2	22.3	18.3	18.5	16.8	11.2	11.5	11.2	10.3	11.1	11.0	11.4
Truck	–	103.5	83.9	49.0	40.2	39.2	39.4	39.5	39.1	24.0	24.3	22.6	20.1	22.7	24.2	24.8
Rail	–	10.8	17.0	4.3	3.4	3.7	3.0	3.5	2.8	1.4	1.4	1.2	1.6	1.6	1.4	1.1
Barge	–	9.8	15.9	4.5	3.2	3.4	3.2	3.3	2.7	1.5	5.4	1.2	1.6	1.6	1.4	1.3
Import	–	26.2	27.9	12.1	9.6	9.6	9.4	9.6	9.0	5.3	9.9	4.9	4.8	5.2	5.3	5.3
No position (per 100,000)																
For new container	0	0	0	96	1	3	0	1	0	0	0	0	0	0	0	–
For reshuffle	74	40	28	42	25	25	0	0	0	15	23	18	24	36	18	–
Ground locations: maximum																
Overall	84.5	75.3	79.9	91.1	89.2	89.1	88.7	87.8	89.4	95.2	94.7	95.7	96.4	96.5	95.1	95.2
Part 1 of the stack	89.6	80.8	86.3	98.7	97.7	97.5	97.5	97.4	99.4	98.4	97.8	98.9	99.6	99.8	98.4	98.3
Part 2 of the stack	50.6	50.2	51.4	48.6	48.1	46.7	58.2	58.2	58.2	58.2	58.2	58.2	57.9	57.9	58.2	58.2
Part 3 of the stack	68.9	50.4	50.6	53.5	42.2	40.2	22.4	22.2	14.4	–	–	–	–	–	–	–
Ground locations: average																
Overall	77.4	64.7	70.3	81.3	79.0	79.0	78.3	78.2	79.4	83.9	83.5	84.0	85.1	85.3	83.9	84.3
Part 1 of the stack	83.5	69.6	76.5	89.0	87.5	87.4	87.4	87.4	89.3	87.6	87.1	87.7	88.9	89.0	87.5	88.0
Part 2 of the stack	38.7	38.9	38.6	39.2	37.5	38.1	38.7	38.7	38.7	38.7	38.7	38.6	38.7	38.6	38.6	38.7
Part 3 of the stack	47.1	36.0	31.4	36.5	24.5	24.8	13.6	13.3	8.8	–	–	–	–	–	–	–
Workload ASC: overall																
Maximum (%)	301.9	312.6	289.4	270.4	302.5	260.8	246.5	287.8	220.2	278.9	258.9	259.1	193.6	238.1	214.0	367.3
Average (%)	31.2	27.0	26.8	25.4	23.3	23.4	23.3	23.3	22.5	22.3	22.3	22.3	22.3	22.4	22.1	22.0
Percentage >80%	10.3	7.3	7.1	6.1	5.1	4.7	4.6	4.6	3.9	3.7	3.7	3.1	3.1	3.2	2.6	2.0
Percentage >90%	7.7	5.3	5.1	4.4	3.5	2.8	2.7	2.7	2.2	2.1	2.1	1.2	1.2	1.3	1.0	0.9
Percentage >100%	5.7	3.8	3.6	3.1	2.4	1.7	1.6	1.6	1.2	1.1	1.1	0.5	0.5	0.5	0.3	0.4
Percentage >110%	4.1	2.7	2.5	2.2	1.6	1.0	1.0	1.0	0.7	0.6	0.6	0.2	0.2	0.2	0.1	0.2
Percentage >120%	3.0	1.9	1.7	1.5	1.1	0.6	0.6	0.6	0.4	0.3	0.4	0.1	0.1	0.1	0.1	0.1
Workload ASC: jumbo																
Maximum (%)	–	312.6	289.4	270.4	302.5	260.8	246.5	225.3	220.2	278.9	258.9	259.1	193.6	218.0	214.0	367.3
Average (%)	–	59.5	60.4	57.6	53.2	53.2	53.0	52.9	50.0	49.6	49.8	49.5	49.4	49.8	49.4	49.0
Percentage >80%	–	28.4	28.6	25.9	22.2	21.3	21.1	20.7	17.7	16.8	17.3	14.5	14.6	15.3	12.7	9.5
Percentage >90%	–	22.3	22.0	19.8	16.5	13.3	13.1	12.8	10.5	9.8	10.4	5.9	6.0	6.3	4.8	4.4
Percentage >100%	–	17.1	16.5	14.7	11.8	8.3	8.0	7.8	6.3	5.6	5.9	2.3	2.3	2.6	1.7	1.9

Percentage >110%	–	13.0	12.3	10.8	8.4	5.2	5.0	5.0	3.7	3.2	3.3	1.0	1.0	1.1	0.7	0.9
Percentage >120%	–	9.6	8.8	7.8	5.9	3.1	3.1	3.0	2.1	1.8	1.9	0.5	0.4	0.5	0.3	0.5
Workload ASC: deep-sea																
Maximum (%)	–	266.3	242.7	239.8	226.5	190.9	199.0	287.8	206.5	204.3	197.9	176.6	159.6	238.1	213.4	205.8
Average (%)	–	39.7	40.1	38.3	35.2	35.3	35.1	35.0	33.2	33.0	33.1	33.0	33.1	33.2	32.9	32.5
Percentage >80%	–	11.8	11.8	10.1	8.1	7.3	7.0	7.0	5.7	5.4	5.4	4.2	4.4	4.3	3.3	2.8
Percentage >90%	–	8.0	8.2	6.8	5.1	4.0	4.0	4.0	2.9	2.9	2.8	1.7	1.8	1.6	1.2	1.2
Percentage >100%	–	5.4	5.4	4.4	3.2	2.2	2.2	2.2	1.6	1.5	1.4	0.6	0.7	0.6	0.4	0.5
Percentage >110%	–	3.5	3.5	2.9	2.0	1.3	1.2	1.2	0.8	0.7	0.7	0.2	0.3	0.3	0.2	0.2
Percentage >120%	–	2.3	2.2	1.8	1.1	0.7	0.7	0.7	0.4	0.4	0.4	0.1	0.1	0.1	0.1	0.1

References

de Castilho B, Daganzo CF (1993) Handling strategies for import containers at marine terminals. Transp Res B 27:151–166

Celen HP, Slegtenhorst RJW, Van der Ham RTh, Nagel A, Van den Berg J, De Vos Burchart R, Evers JJM, Lindeijer DG, Dekker R, Meersmans PJM, De Koster MJM, Van der Meer R, Carlebur AFC, Nooijen FJAM (1999) FAMAS–NewCon: phase 1: starting points, Phase 2: architecture integrating information system, CTT publicatiereeks 32 (in Dutch)

Chen T, Lin K, Yuang YC (2000) Empirical studies on yard operations part 2: quantifying unproductive moves undertaken in quay transfer operations. Marit Policy Manage 27:191–207

Cullinane K, Khanna M (2000) Economies of scale in large containerships: optimal size and geographical implications. J Transp Geogr 8:181–195

Duinkerken MB, Evers JJM, Ottjes JA (2001) A simulation model for integrating quay transport and stacking policies in automated terminals. In: Proceedings of the 15th European Simulation Multiconference (ESM2001), SCS, Prague

Kim KH (1997) Evaluation of the number of rehandles in container yards. Comput Ind Eng 32:701–711

Kim KH, Bae JW (1998) Re-marshalling export containers in port container terminals. Comput Ind Eng 35:655–658

Kim KY, Kim KH (1998) The optimal determination of the space requirement and the number of transfer cranes for import containers. Comput Ind Eng 35:427–430

Kim KH, Kim HB (1999) Segregating space allocation models for container inventories in port container terminals. Int J Prod Econ 59:415–423

Kim KH, Park YM, Ryu KR (2000) Deriving decision rules to locate export containers in container yards. Eur J Oper Res 124:89–101

Meersmans PJM, Dekker R (2001) Operations research supports container handling. Report Econometric Institute EI/2001-22, Erasmus University Rotterdam

Sculli D, Hui CF (1988) Three-dimensional stacking of containers. Omega 16:585–594

Steenken D, Voß S, Stahlbock R (2004) Container terminal operation and operations research—a classification and literature review. OR Spectrum 26:3–49

Taleb-Ibrahimi M, De Castilho B, Daganzo CF (1993) Storage space vs handling in container terminals. Transport Res B 27:13–32

Upward Systems (1994) Must simulation software: user and reference manual. Delft, The Netherlands

Vis IFA, de Koster R (2003) Transshipment of containers at a container terminal: an overview. Eur J Oper Res 147:1–16

Voogd P, Dekker R, Meersmans PJM (1999) FAMAS–Newcon: a generator program for stacking in the reference case. Report Econometric Institute EI-9943/A

Martin Grunow · Hans-Otto Günther · Matthias Lehmann

Strategies for dispatching AGVs at automated seaport container terminals

Abstract Control of logistics operations at container terminals is an extremely complex task, especially if automated guided vehicles (AGVs) are employed. In AGV dispatching, the stochastic nature of the handling systems must be taken into account. For instance, handling times of quay and stacking cranes as well as release times of transportation orders are not exactly known in advance. We present a simulation study of AGV dispatching strategies in a seaport container terminal, where AGVs can be used in single or dual-carrier mode. The latter allows transporting two small-sized (20 ft) or one large-sized (40 ft) container at a time, while in single-mode only one container is loaded onto the AGV irrespective of the size of the container. In our investigation, a typical on-line dispatching strategy adopted from flexible manufacturing systems is compared with a more sophisticated, pattern-based off-line heuristic. The performance of the dispatching strategies is evaluated using a scalable simulation model. The design of the experimental study reflects conditions which are typical of a real automated terminal environment. Major experimental factors are the size of the terminal and the degree of stochastic variations. Results of the simulation study reveal that the pattern-based off-line heuristic proposed by the authors clearly outperforms its on-line counterpart. For the most realistic scenario investigated, a deviation from a lower bound of less than 5% is achieved when the dual-load capability of the AGVs is utilized.

Keywords AGV dispatching · Container terminals · On-line and off-line control · Simulation

---

M. Grunow
Department of Manufacturing Engineering and Management, Technical University of Denmark, Building 423, 2800 Kgs., Lyngby, Denmark
E-mail: grunow@ipl.dtu.dk

H.-O. Günther (✉) · M. Lehmann
Department of Production Management, Technical University Berlin, Wilmersdorfer Str. 148, 10585 Berlin, Germany
E-mail: Hans-Otto.guenther@tu-berlin.de, Matthias.Lehmann@tu-berlin.de

1 Introduction

Driven by the trend towards globalization of the economy, world trade volumes have increased dramatically during the last decade. Today, maritime cargo transportation has become the predominant transportation mode in international trade. For instance, 78.7% of the USA foreign trade in 2001 was accomplished by maritime cargo transportation (cf. BTS 2004). At the same time, the number of container terminals worldwide increased considerably. Their major function is to serve as multi-modal interfaces between sea and land transport.

To cope with increased transportation volumes and to benefit from the economies of scale, ship owners have constantly increased the capacity of their deep-sea container vessels, recently culminating in the 10,000 TEU (20-ft equivalent units) container ship generation. Operators of seaport container terminals have primarily responded to this development by increasing their terminals in size and making use of more efficient transportation and handling equipment. There are, however, a great number of existing terminals which have reached their limits for further expansion. Hence, new automated container terminals are constructed worldwide. These terminals are better suited to serve the huge, modern deep-sea container vessels and to employ improved logistics equipment.

One direction for improving the overall productivity of a container terminal and to reduce the berthing times of vessels is to enhance the degree of automation of the handling and transportation equipment. Hence, manually operated cranes have been replaced by automated ones and AGVs are used instead of manually driven carts. An example of the AGV application in the Container Terminal Altenwerder (CTA) in Hamburg, Germany is given in Fig. 1. Nevertheless, for transportation between different terminals at one location, as is the case in the city of Busan (Korea), conventional trucks are still the primary mode of transportation (cf. Koo et al. 2004a). For intra-terminal operation, dual-load AGVs represent a recent development in transportation technology. Such vehicles offer the advantage of being able to transport two 20-ft containers or one 40-ft container at a time. Another recent development is represented by so-called automated lifting vehicles (ALVs) which, in contrast to AGVs, are capable of lifting a container from the ground by itself (cf. Vis and Harika 2004; Yang et al. 2004). The only container

Fig. 1 AGV employed at the Container Terminal Altenwerder, Hamburg, Germany (source: http://www.hhla.de/de/Geschaeftsfelder/index.jsp, visited on August 22, 2005)

terminal employing ALV systems so far is the port of Brisbane, where they have been introduced for commercial use in December 2005.

As a container terminal represents a complex system with various interrelated components, computerized logistics control systems have recently gained considerably higher attention. The use of automated equipment in turn requires much more sophisticated control strategies to exploit the capabilities of advanced automated equipment (cf. Günther and Kim 2004; Steenken et al. 2004). For instance, in automated container terminals, dual-load AGVs are still operated in single-carrier mode, mainly because adequate dispatching strategies, which allow for the efficient use of their enhanced transportation capacity, are missing. The dispatching problem for dual-load carriers is obviously considerably more complex than for single-load carriers.

In the academic literature, the AGV dispatching problem arising in seaport container terminals has been widely neglected. Two exceptions are the papers by Bae and Kim (2000) and Koo et al. (2004b). Their investigations, however, consider selected issues related to dispatching of single-load carriers. Another noticeable exception is the paper by Kim and Bae (2004). They develop an efficient look-ahead heuristic for dispatching single-load AGVs. In a numerical investigation, it is shown that their heuristic outperforms conventional dispatching rules. A problem similar to AGV dispatching is the yard trailer routing problem investigated by Nishimura et al. (2005). They consider man-driven multi-load trailers and develop a genetic-algorithm-based dispatching approach. In a simulation study, it is shown that a dynamic routing strategy, i.e., one which does not assign a vehicle to a specific crane, is superior to a static routing strategy with dedicated crane-vehicle assignments. However, because of the excessive computational requirements, their approach is barely applicable in a real-time dispatching strategy. Finally, Grunow et al. (2004) developed a heuristic for dispatching dual-load AGVs. For two idealized seaport container terminal configurations, they compared their approach against a benchmark solution obtained from an MILP model application.

The main contributions of this paper are:

- Contrary to most papers in the academic literature, the multi-load capability of the AGVs is taken into account.
- The approach developed in Grunow et al. (2004) is extended. In an off-line approach, we now assign all transportation orders in the planning horizon to AGVs. This bears the following benefits:
  - The new conception is less myopic.
  - During the assignment of transportation orders to AGVs, availability restrictions do not have to be taken into account.
  - Vehicle-initiated dispatching is no longer required. Hence, the corresponding triggering events do not have to be monitored.
  - The latter two improvements lead to a significant reduction of the information system complexity.
- A comprehensive simulation model has been developed. Contrary to previous studies on vehicle dispatching, this model reflects conditions which are typical of a real automated terminal environment. The simulation model is used to compare the off-line to the on-line approaches as they are applied in the harbour

practice and to analyze the suitability of the approaches with regard to real time requirements. Furthermore, a lower bound was derived as an additional performance measure.

A specific issue of considerable importance in decentralized control of complex logistics systems is the handling of deadlock situations. Various strategies can be pursued to handle deadlocks arising between different resources in the terminal configuration. Related procedures for application in real-time control of AGV systems at automated container terminals are presented in a companion paper (see Lehmann et al. 2006, for a review).

This paper is organized as follows. In Section 2, the AGV dispatching problem is explained in greater detail. Next, on-line and off-line dispatching modes are discussed in Section 3. This is followed by the detailed presentation of related dispatching strategies (see Section 4). Results of a simulation study are presented in Section 5. Finally, conclusions are drawn and directions of future research are highlighted.

2 AGV dispatching

A typical seaport container terminal is divided into a berthing, an AGV, and a storage area. Figure 2 illustrates the layout of one of the latest highly automated seaport container terminals. The berthing area is equipped with quay cranes for the loading and unloading of vessels. When a vessel arrives at the port, it has already been determined at which position the vessel is berthed and which quay cranes will be working on the vessel (cf. Guan and Cheung 2004). The unloading sequence of the containers is equally known in advance for each vessel (cf. Kim et al. 2004). Thus, detailed schedules for the quay cranes can be derived from the given unloading sequence (cf. Park and Kim 2003). At the same time, the final destination in the storage area is determined for each container. The storage area is divided into blocks each of which is serviced by one or more stacking cranes. After unloading a container, the stacking cranes at the affected block are scheduled to meet the estimated arrival time of the container. The transport of the containers from the berthing area to the storage yard is realized by dual-load AGVs (for a general framework of scheduling operations in container terminals, see Hartmann 2004a).

In the container terminal considered, AGVs are operated in single-load carrier mode but shall be used as dual-load carriers in the future. The particular difficulty of AGV dispatching in a highly automated container terminal is that AGV pick-up and drop-off times for each container have to coincide with the schedules of the quay and stacking cranes to avoid idle times of this equipment and to guarantee short berthing of the vessels. The operations necessary to load a vessel are similar.

AGV dispatching usually consists of three sub-problems, namely, assigning AGVs to transportation orders, routing the AGVs, and traffic control. Algorithms for routing and traffic control are generally already included in the control software provided by the AGV manufacturer. Thus, only the assignment problem is investigated in this paper. In contrast to applications of AGVs in manufacturing systems, rigid pick-up and drop-off time constraints have to be considered, which

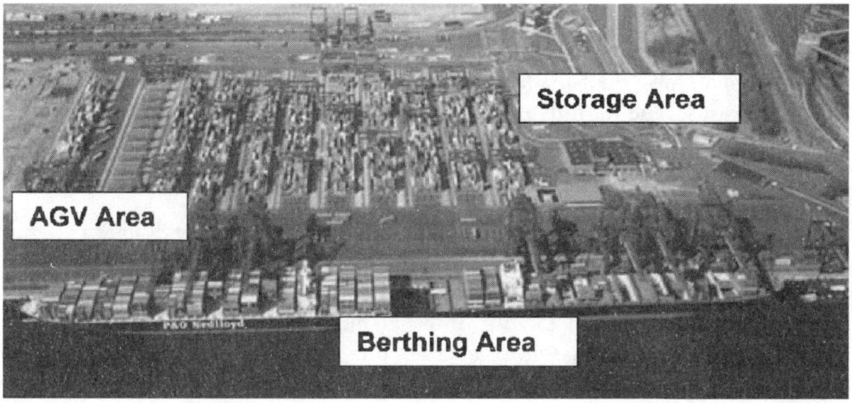

Fig. 2 Layout of the Container Terminal Altenwerder, Hamburg, Germany (source: http://www.hhla.de/de/Geschaeftsfelder/HHLA_Container/Altenwerder_(CTA)/Daten_und_Fakten.jsp, visited on August 22, 2005)

significantly increase the problem complexity. In addition, the lack of buffers requires an exact synchronization of the operations.

In the case of single-load carriers (cf. Bish et al. 2005), AGV dispatching can be reduced to an m:n assignment problem with the objective of minimizing the costs associated with not meeting target times imposed by the quay cranes' schedule. (Note that quay crane waiting times directly affect the vessels' turnover time and, thus, the productivity of the container terminal.) The corresponding linear optimization model can be solved rather efficiently due to its pure binary nature. However, in the case of multi-load carriers, the assignment problem is significantly more complex. In addition to the basic order-vehicle assignment, the various pick-up and drop-off operations have to be sequenced for each AGV.

Throughout the paper, we make the following basic assumptions:

- Each AGV is capable of carrying one 40-ft container or two 20-ft containers at a time.
- All AGVs in the fleet are identical in their function, loading capacity, speed, etc.
- AGVs are not pooled, i.e., they operate independently from each other and are not dedicated to a specific quay or stacking crane.
- AGV travel times are assumed to be deterministic. In particular, effects of congestion among AGVs on the guide path are neglected.
- Transportation of special-purpose containers, e.g., reefer or hazardous goods containers, is not considered.

3 On-line and off-line dispatching mode

Scheduling in dynamic application environments has been an active research area in recent years. Much work has been carried out to compare on-line and off-line scheduling strategies and to find out which of them is more suitable. However, a general answer to this question will always depend on the specific application environment. While for master production planning in manufacturing systems a predictive approach might be adequate, in short-term scheduling, for instance, plant managers often prefer to initiate only the next operation in an on-line manner.

When dispatching dual-load vehicles in seaport container terminals, the choice is not so obvious. The high degree of stochasticity seems to favor myopic on-line strategies, whereas predictive plans constructed by off-line strategies promise to exploit the optimization potential resulting from the combination of different transportation orders into a joint tour.

On-line dispatching is usually seen as appropriate in a highly dynamic planning environment where only limited information about future events is available. In the case of container terminals, the stochastic nature of the handling system is due to internal as well as external factors. Internal factors are, for instance, short-term decisions of quay crane operators to alter the sequence of handling operations, while external factors include weather conditions, the unknown state of a container, or congestion in the AGV traffic system. Because of these uncertainties, decisions must be made without complete knowledge of the future events. One option to deal with the stochastic nature of the logistics system is to employ on-line dispatching. According to this dispatching mode, a decision is made when needed and immediately executed (cf. Fiat and Woeginger 1998; Sgall 1998). In this case, no predictive plan is generated. The schedule rather results from a sequence of on-line decisions, which are made one at a time as the system status changes (cf. Sabuncuoglu and Bayiz 2000). While the application of these rules is simple, their inherent myopic and greedy nature may sacrifice their performance.

Off-line dispatching requires decisions to be made simultaneously for all transportation orders occurring within a short-term look-ahead period. Thus, a predictive schedule is constructed. However, due to the uncertainty of the future events, the schedule may have to be revised when significant deviations occur, e.g., late arrival of AGVs, breakdown of equipment, or delays in performing the loading and unloading tasks. This type of planning approach is therefore also termed *reactive planning* (cf. Sabuncuoglu and Bayiz 2000).

Depending on the factors which trigger rescheduling, the following policies can be distinguished:

- *Periodic rescheduling* takes place after predefined time intervals using rolling time horizons (cf. Church and Uzsoy 1992).
- *Event-driven rescheduling* is carried out on significant deviations from the current schedule. But also specific events, such as arrival of a new job, may cause rescheduling (cf. Smith 1994; Vieira et al. 2003).
- In *hybrid rescheduling*, a combination of the above policies is applied (cf. Church and Uzsoy 1992).

In this paper, we consider an event-based logic of the logistics control systems. Thus, decisions are triggered by certain events, e.g., the completion of a transportation order, or when the development of the logistics system deviates from its predicted behavior, e.g., loading or unloading operations take significantly longer than expected. A typical on-line dispatching strategy, adopted from flexible manufacturing systems, is compared with a more sophisticated off-line heuristic developed by the authors.

4 Dispatching strategies

4.1 Characteristics of dispatching strategies

As stated above, the vehicle-dispatching problem at hand consists of assigning transportation orders to AGVs and of determining the sequence of transportation orders assigned to each vehicle. In the case of dual-load AGVs, which allow up to two 20-ft containers to be loaded on one vehicle at the same time, also the individual pick-up and drop-off operations of each order have to be sequenced. Once the assignment and sequencing decisions have been made, the corresponding pick-up and drop-off times can be derived in a straightforward manner for single-load as well as for dual-load carriers.

As scheduling in a dynamic environment is usually accomplished by solving a sequence of static problems, it has to be decided when a new static problem should be solved. Within the paradigm of event-driven dispatching, certain triggering events have to be identified. In on-line dispatching, no predictive schedule is constructed, only a local decision on how to deal with an immediately upcoming event (e.g., a new transportation order) is made. The advantage of this approach is that AGVs are dispatched according to events which are relatively certain. On the other hand, the myopic nature of this approach may lead to low-quality solutions. As an alternative, we therefore construct a predictive schedule using the entire information for a predefined look-ahead period. However, due to the stochastic nature of the logistics processes in a seaport container terminal, this early uncertain information is bound to change during the execution of the schedule and may therefore require an adaptation of the schedule.

4.2 On-line dispatching strategies

In on-line dispatching, triggering events are generated when a new transportation order is released (*transportation-order-initiated dispatching*) or an AGV becomes available (*vehicle-initiated dispatching*). A transportation order is released once the execution of the previous order has begun. In the case of a discharge order, the order is released when the quay crane starts the lifting operation of the predecessor container from the vessel. The next charge order of a quay crane is similarly released once the pick-up operation of the predecessor container has been started at the storage area. The transportation order is then assigned to an AGV if one is available. Otherwise, it is kept in the set of unassigned orders.

However, the concept of vehicle availability has to be further specified. There are two different views on when a vehicle should be considered available. From a physical point of view, a vehicle is available when it is unloaded, i.e., no container is placed on its loading platform. This concept, however, is rather myopic and not suited for most planning decisions, as information about the logical status of the vehicle (i.e., its actual schedule) is neglected. We therefore determine the availability of an AGV based on its status *after* completing the current trip. A single-load AGV is considered available during its trip to the drop-off location. Dual-load vehicles are fully available during the trip to their last drop-off location and partially available during the trip to the first pick-up location of a 20-ft container or to the first drop-off location. Of course, both types of vehicles are considered available when parked idle at some dwell point in the guide path. Vehicle-initiated dispatching is performed when a vehicle becomes fully available. In this case, a job is selected from the set of unassigned orders.

If the AGVs are used as single-load carriers, one can build upon the dispatching rules known from manufacturing and warehouse applications. In these environments, basic rules are used for the dispatching of single-load carriers or to find an initial assignment for multi-load carriers (see, e.g., Egbelu and Tanchoco 1984; Hwang and Kim 1998; Klein and Kim 1996; Lim et al. 2003; de Koster et al. 2004; Le-Anh and de Koster 2005, and recently Lee and Srisawat 2006). These approaches are typically restricted to a one-to-many assignment. Accordingly, either one out of the feasible vehicles is assigned to a transportation order or from a set of unassigned transportation orders one is assigned to an available vehicle.

Certainly the most popular representative for the first case, transportation-order-initiated dispatching, is the *nearest-vehicle* (NV) rule which assigns the vehicle located the closest to the pick-up location of a transportation order whenever a new transportation order is initiated. This rule, however, may discriminate vehicles that are very far from *any* active quay or stacking crane and may, thus, lead to a rather disproportionate use of the available vehicles. One way to avoid this drawback is to apply the *least-utilized-vehicle* (LUV) rule instead. This rule aims at balancing the vehicles' workload by preferring less utilized vehicles for actual assignment. The utilization of a vehicle is measured by counting the transportation orders completed so far and those already assigned to the vehicle. Vehicle-initiated dispatching normally resorts to the *first-come-first-served* (FCFS) strategy, which is applied to prioritize waiting transportation orders. Another adequate dispatching strategy is the *shortest-travel-time* (STT) rule, which is the vehicle-initiated counterpart of the NV rule. By this rule, transportation orders are chosen according to the distance the vehicle would have to cover to service them.

After some initial experiments, we decided to define an on-line strategy, which consists of a transportation-order-initiated and a vehicle-initiated component. Based on our numerical experience, we combined the nearest-vehicle rule and the first-come-first-served rule.

To utilize the full loading capacity of the AGVs, they must be operated in dual-load carrier mode which requires more elaborate dispatching rules. The rules used for multi-load carriers in manufacturing systems cannot as easily be employed for the problem at hand. Rules used in manufacturing systems normally select additional orders if they can be reached en route to the drop-off location of an already loaded shipment. A common criterion for such rules is the deviation from the route that has been scheduled so far. These rules therefore clearly require information about the routing of the AGVs. These are easy to obtain for a manufacturing system with its limited guide path network, where the actual routes between the pairs of working stations are highly predictable. In a seaport container terminal, however, the guide path is of much larger size and the grid structure of the network allows far greater routing flexibility. The actual route may, in fact, depend on the current traffic situation in the network. Furthermore, the fleet size of 50–100 AGVs operated in container terminals is significantly larger compared to manufacturing systems. Hence, due to the increased traffic volume, the routing complexity in container terminals is considerably higher. Finally, in the case of container terminals, routing and traffic control routines are often provided by the AGV manufacturer and encapsulated in the vehicles' traffic control software.

Therefore, they have to be considered a black box for vehicle dispatching. It is mainly for these reasons that the rules from manufacturing systems could not be adopted for the dispatching problem at hand.

Hence, we propose the following extended on-line dispatching rule for dual-load AGVs in container port terminals: (1) The pick-up of a second container should always take place after that of a container already assigned to the AGV. As a result, no deviation from the actual trip could occur. (2) The sequence of the drop-off operations of two loaded containers is determined by the *nearest-destination rule*, prioritizing the container with the nearest drop-off location. In Fig. 3, the pseudo-code for the on-line dispatching rule is given. The procedure is executed each time an AGV has finished the last operation in its schedule or a new order is released.

4.3 Off-line dispatching strategies

For off-line dispatching, we consider two types of triggering events, i.e., the completion of a quay crane operation and exceeding a delay threshold. Off-line approaches, in contrast to their on-line counterparts, generate a predictive schedule. In our case, a schedule for the next *t* transportation orders at each of the quay cranes is created. Hence, the schedule covers *t*· (number of quay cranes) transportation orders. Once a quay crane operation has been completed, a new transportation order must be included in the schedule to maintain the look-ahead horizon of *t* transportation orders at this quay crane. The transportation orders are then assigned to the AGVs. However, in the turbulent environment of container ports, deviations between the generated schedule and the actual AGV trips are bound to occur.

Triggering events: „AGV finished last operation in schedule“, “new order released”.

1. ***IF*** (triggeringEvent = „AGV finished last operation in schedule“
   ***THEN* go to step 2**
   ***ELSE* go to step 3**.

2. ***(vehicle initiated dispatching)***
   IF (no unassigned order left)
   THEN park AGV and **exit** procedure
   ELSE assign AGV to the order with the earliest start time (FCFS-rule), return schedule and **exit** procedure.

3. ***(transportation order initiated dispatching)***
   IF (released order comprises 40 ft container OR AGV type is “single-load”)
   ***THEN* go to step 3.1**
   ***ELSE* go to step 3.2**.

3.1 ***IF*** (no fully available AGV left)
   ***THEN* exit** procedure and include transportation order in set of unassigned orders
   ELSE assign order to the nearest AGV (NV-rule), return schedule and **exit** procedure.

3.2 ***IF*** (neither fully nor partially available AGV left)
   ***THEN* exit** procedure and include transportation order in set of unassigned orders
   ELSE assign order to the nearest AGV (NV-rule), determine sequence of the drop-off operations of the two orders assigned to this AGV according to the nearest-destination-rule, return schedule and **exit** procedure.

Fig. 3 Pseudo-code for the on-line dispatching rule

Hence, the second type of triggering events is introduced, which prompts the creation of a new schedule if the deviation exceeds a given threshold level.

There are two categories of transportation order assignments: temporary and fixed assignments. An order is temporarily assigned if, during a future dispatching request, the assignment can be broken up and the order can be assigned to another AGV, while this is not feasible in the case of a fixed assignment. Fixed assignments obviously decrease the possibility (and therefore the optimization potential) of reactive scheduling and should be used carefully. In our off-line approach, fixed assignments are only used for the actually performed order of each vehicle.

As an off-line dispatching strategy, a pattern-based heuristic has been developed by the authors. In the sequel, only a sketch of the heuristic procedure will be given (for details, the reader is referred to Grunow et al. 2004). This is followed by some extensions of the basic version of the heuristic. In the pattern-based heuristic, an m:n assignment of vehicles to transportation orders is determined by iteratively solving an m:1 assignment problem. The transportation orders in the planning horizon are considered one by one as they are released by the overall logistics control system. For each transportation order in this sequence, the possible assignment to each (partially or fully) available vehicle is evaluated. Furthermore, for each possible assignment to a partially available vehicle, different assignment patterns are tested, reflecting the feasible sequences of pick-up and drop-off operations of the new transportation order and the one that has already been assigned to the same vehicle in a previous step.

In our heuristic, we allow for patterns where pick-up and drop-off operations of the new order are sequenced after those of the already assigned order [assignment pattern "aann", read *a*ssigned (pick-up)–*a*ssigned (drop-off)–*n*ew (pick-up)–*n*ew (drop-off)], in between them ("anna") or alternating ("anan"). Similar sequences can be generated starting with the pick-up of the new transportation order ("nnaa", "naan", and "nana"). In Fig. 4, all possible assignment patterns for 20-ft containers are shown. Pick-up and drop-off operations are indicated by an arrow pointing upwards or downwards, respectively. From the six theoretically possible assign-

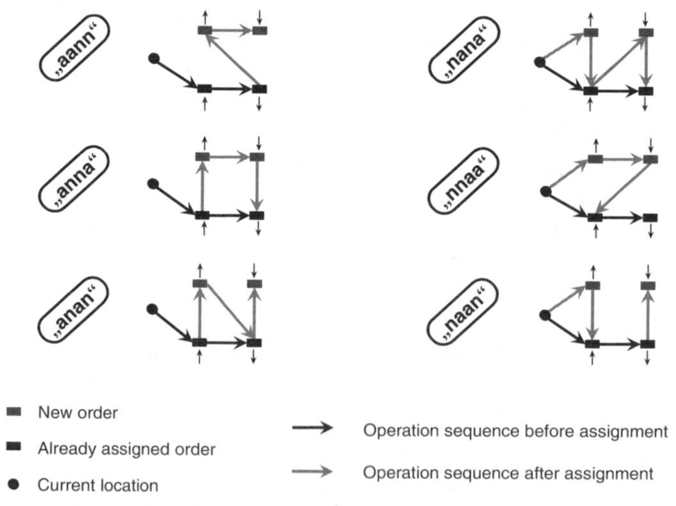

Fig. 4 Possible assignment patterns

ment patterns, only three are considered in our pattern-based heuristic. At the time the dispatching request is initiated, the vehicles might already be on their way to the service point of the next operation. Thus, to avoid re-routing of a vehicle's mission and to prevent that an already assigned transportation order is infinitely delayed, assignment patterns "nnaa", "nana", and "naan" are not considered here.

It should be noted that, in the basic version of this procedure (cf. Grunow et al. 2004), at most two transportation orders (of 20-ft containers) can be assigned to a vehicle before it becomes unavailable. As the schedule of each AGV could thus comprise at most four operations (two pick-up and two drop-off operations), the possibility of constructing extended tours is not given. Especially in an off-line strategy, AGV schedules comprising more than two transportation orders may be advantageous. Hence, a natural extension of the basic pattern-based heuristic is to allow more than two transportation orders to be assigned to each AGV.

However, an increased number of transportation orders leads to an exponential growth of combinations of pick-up and drop-off operations, resulting in a prohibitive runtime requirement for extended schedules. Therefore, in the extended pattern-based heuristic, we restrict the number of feasible pick-up–drop-off patterns in such a way that each transportation order can be interlocked with at most one other transportation order (in the case of two 20-ft containers), i.e., pick-up and drop-off operations p1 and d1 of the first order may only be interlocked with pick-up and drop-off operations p2 and d2 of the second order but never with the corresponding operations p3 and d3 of a third order, unless the drop-off operation d1 of the first transportation order has been completed. Another reason for this restriction is a practical one. The more transportation orders are interlocked, the more orders are obviously affected, if a specific order cannot be performed in time. As a result, extensive delays for a great number of orders could occur. Despite these restrictions, the proposed approach is able to create extended schedules, taking more advantage of the capabilities of the off-line dispatching strategy. Note that a transportation order for a 40-ft container can obviously not be interlocked with other transportation orders.

In the *extended pattern-based heuristic*, instead of identifying the status of an AGV as simply fully available, partially available, or unavailable, each AGV shows only two conditions, depending on the last order in its current schedule. If the pick-up and drop-off operation of the last order refer to a 20-ft container and they are scheduled successively, the vehicle is labeled as "S | pd", meaning that its schedule consists of some sequence of operations "S" followed by the *p*ick-up and *d*rop-off operation of the last order in the current sequence (for a 20-ft container). A new (20-ft container) order can now be appended to the current operation sequence of the AGV according to the best of the three assignment patterns "aann", "anan", or "anna" shown in Fig. 4. If, on the other hand, pick-up and drop-off operations of the last order in the current sequence of the AGV are interlocked with those of another transportation order (e.g., "p1-p2-d1-d2" or "p1-p2-d2-d1"), the label of the vehicle is set to "S". This label indicates that the schedule of an AGV consists of a sequence of operations, where the last transportation order is interlocked with some other order. The label is also set to "S" if the last order was a 40-ft container. In such a case, a new order can only be assigned to that AGV according to the pattern "aann", i.e., appending pick-up and drop-off operation of the new order at the end of the current schedule.

In Fig. 5, the possible transitions between condition "S" and condition "S | pd" are shown. There are two arcs leaving from "S". They indicate that assignment pattern "aann" is the only feasible one in condition "S". If a transportation order for a 40-ft container is appended to the schedule, the label remains at "S". If the new transportation order is for a 20-ft container, the status of the AGV changes to "S | pd" On the other hand, should the AGV be in condition "S | pd", then assignment pattern "aann" for the inclusion of a 20-ft container transportation order maintains the initial condition, while patterns "anan" and "anna" and the appendage of a transportation order for a 40-ft container convert the AGV's condition into "S".

The feasible options for generating a chain of transportation orders for a single AGV are illustrated in Fig. 6. (For the sake of clarity, only 20-ft containers are considered. Note that an order for a 40-ft container can only be appended at the end of a schedule, always leading to condition "S".) At first, pick-up and drop-off operations p1 and d1 of the first order are assigned to the vehicle. The resulting condition of the vehicle is "S | pd" (node 1). The second order with operations p2 and d2 can be appended by use of any of the assignment patterns, "aann", "anan", or "anna" leading to nodes 2, 3, and 4. In node 2, the two orders are executed successively, i.e., the corresponding handling operations are not interlocked and the condition of the vehicle is identified as "S | pd". Thus, any of the assignment patterns can be used to append order 3 with operations p3 and d3 leading to nodes 5, 6, and 7. If, however, patterns "anan" or "anna" are selected, the handling operations of the two orders are interlocked and the vehicle changes to condition "S" (nodes 3 and 4). Hence, pattern "aann" is the only feasible to append the third order leading to nodes 8 and 9, respectively. For each of the nodes 5 to 9, the condition of the vehicle is identified as "S" or "S | pd" and the next order is appended to the existing chain.

Regardless of which assignment option is used, after evaluating all feasible assignments, the one with the lowest cost (e.g., waiting time of the quay crane) is selected. The vehicles' availability (or condition) is updated and a new iteration is initiated for the next transportation order, now considering the modified condition for each AGV. The heuristic terminates if all transportation orders in the planning horizon are assigned to a vehicle. In Fig. 7, the pseudo-code for the pattern-based heuristic is given. The procedure is executed each time the delay of an order exceeds a predefined threshold level or a new order is released.

The pattern-based heuristic could also be used as an on-line strategy. However, it only reveals its full potential in the off-line mode and has therefore been exclusively used as an off-line approach in our investigations.

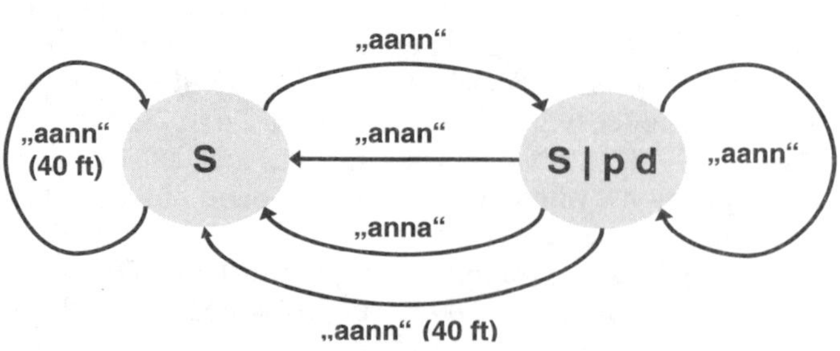

Fig. 5 Feasible transitions between the conditions of an AGV schedule

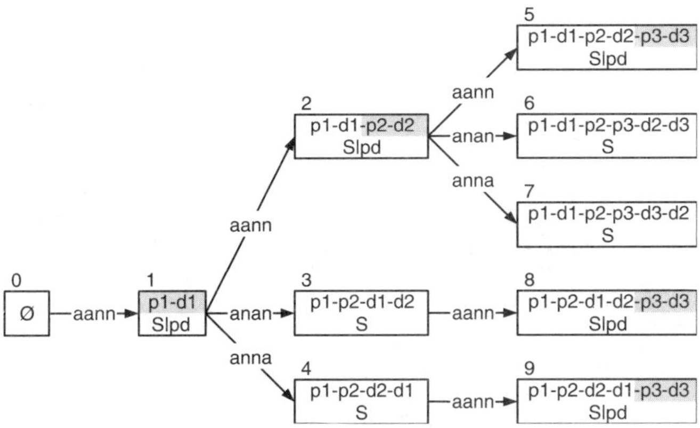

Fig. 6 Generation of chains of transportation orders for a single AGV (20-ft containers; *grey-shaded* areas show non-interlocked pick-up and drop-off operations at the end of an operation chain)

A mixed-integer linear programming (MILP) model, which shows the clearest off-line character, could alternatively be applied. In this approach, each vehicle is assigned a sequence of transportation orders for the given time window. A detailed description of the MILP model formulation of the dispatching problem for the case of dual-load carriers can be found in Grunow et al. (2004). However, for problem instances of realistic size, runtimes of the MILP model often exceed 1 min, which is not acceptable for the problem at hand. Thus, this approach is not considered in our numerical investigations.

The main problem when using MILP models (solved by use of standard optimization software, such as ILOG OPL studio) for real-time applications is that the resulting models are hardly scalable in terms of runtime requirement. After a certain amount of time, the optimal solution is achieved, but one cannot expect, for instance, to get a solution "half as good" in half the time.

5 Simulation study

5.1 Design of the simulation model

To evaluate the effectiveness of the various dispatching approaches, a comprehensive simulation study has been conducted. A discrete event-based simulation model has been developed using the eM-Plant 6.0 simulation system. For modeling a real logistics system through simulation, a major issue in the design of the simulation model refers to the definition of the system boundaries. We decided to build up the simulated system around an AGV guide path and a fleet of vehicles which transport 20- or 40-ft containers between quay cranes located at the berth side and automated stacking cranes which operate at the different storage blocks arranged at the opposite side of the guide path. Thus, sub-systems not included in the simulation model are, for instance, the stowage and berth planning for vessels, the storage planning for containers inside the storage blocks, the interface to the hinterland, and the traffic control of the AGV system.

Triggering events: „delay of order exceeds threshold level", "new order released".

1. Break all order-AGV-assignments apart from fixed ones (next operation, loaded containers).
   Move all unassigned orders to the set ORDERPOOL. Determine status of AGVs ("S" or "S|pd").
2. ***IF*** (ORDERPOOL = ∅)
   THEN return (new) schedule
   ***ELSE* go to step 3.**
3. Set bestDelay:=∞. Define currOrder as the order in ORDERPOOL with the earliest start time. Move all AGVs to the set AGVPOOL.
4. Define currAGV as any element of AGVPOOL. AGVPOOL:= AGVPOOL\{currAGV}.
5. Evaluate pattern "aann" for the combination currOrder-currAGV.
   IF (currDelay < bestDelay)
   THEN
   bestDelay:=currDelay, bestAGV:=currAGV.
   IF (currOrder is a 20 ft container) ***THEN*** nextStatus:= "S|pd" ***ELSE*** nextStatus:= "S".
6. ***IF*** (status of currAGV is "S|pd" AND currOrder is a 20 ft container AND AGV type is "dual-load")
   ***THEN* go to step 7**
   ***ELSE* go to step 9**.
7. Evaluate pattern "anan" for the combination currOrder-currAGV.
   IF (currDelay < bestDelay)
   THEN bestDelay:=currDelay, bestAGV:=currAGV, nextStatus:= "S".
8. Evaluate pattern "anna" for the combination currOrder-currAGV.
   IF (currDelay < bestDelay)
   THEN bestDelay:=currDelay, bestAGV:=currAGV, nextStatus:= "S".
9. ***IF*** (AGVPOOL = ∅)
   ***THEN* go to step 10**
   ***ELSE* go to step 4**.
10. Assign bestAGV to currOrder. Update status of bestAGV:=nextStatus.
    ORDERPOOL:= ORDERPOOL\{currOrder}. **Go to step 2**.

Fig. 7 Pseudo-code for the pattern-based heuristic

To simulate automated container terminals of a different size, a basic module was defined which constitutes the building block of a flexible terminal configuration (see Fig. 8). Hence, by combining various modules, a larger terminal configuration can be generated. The basic module consists of four elements: (1) the AGV guide path laid out as a four-lane uni-directional loop, (2) a fleet of AGVs, (3) a single quay crane, and (4) two storage blocks equipped with two automated stacking cranes each. However, we do not include the detailed operations of the stacking cranes into the simulation model. In optional modules, one or two of the storage blocks or the quay crane are omitted. Thus, an arbitrary combination of quay cranes and storage blocks can be simulated. This design is used to generate a terminal configuration with a series of storage blocks concentrated in the center of the storage yard. AGVs are not dedicated to a single module but can freely commute in all modules. To generate a specific terminal configuration, only five parameters are required:

1. the number of quay cranes
2. the number of storage blocks
3. the number of AGVs
4. the AGV travel time between two quay or stacking cranes
5. the AGV travel time between the storage area and the berth side

As an example, Fig. 9 displays a medium-sized terminal configuration with 10 quay cranes, 30 storage blocks, and AGV travel times of 20 and 10 s between two quay cranes and between two storage blocks, respectively. The trip from the storage area to the berth side or vice versa requires 10 s. All cranes in the system are linked by a uni-directional mesh-type guide path in which only the traversals between the quayside and the storage yard show a bi-directional orientation.

Moreover, our simulation model is based on the following major assumptions:

- The loading and unloading sequence of containers is known for each vessel.
- Travel times of vehicles are assumed to be deterministic.
- The cycle times of a quay crane and stacking crane are generated according to the empirical distributions observed by Vis and Harika (2004) (for details, see the next sub-section).
- One of the two stacking cranes at the storage blocks is used for loading and unloading AGVs, while the other serves at the interface to the hinterland, e.g., at docking stations for trucks or at the railway link.
- Transportation orders are generated according to the working cycle of the quay cranes. For each quay crane, the storage block from where an export container is to be picked up and to where an import container is to be delivered is randomly selected.

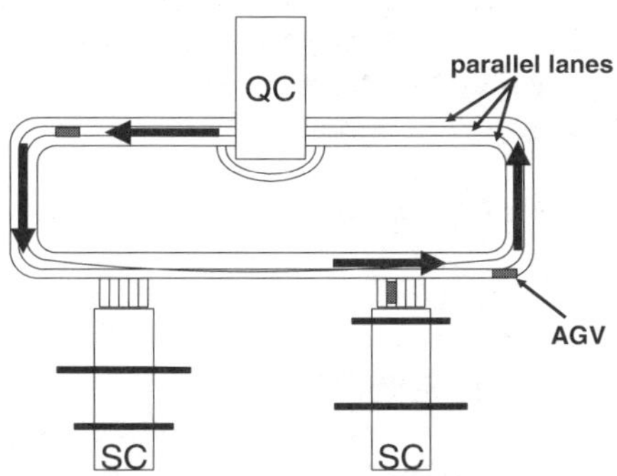

Fig. 8 Basic module of a terminal configuration. *QC*, quay crane; *SC*, stacking crane

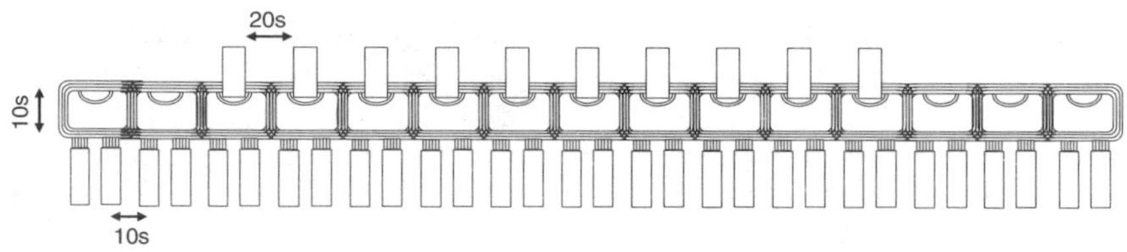

Fig. 9 Medium-sized terminal configuration generated from basic modules

5.2 Experimental scenarios

The scenarios investigated in our study reflect realistic terminal environments and consider stochastic variations in the timing and processing of loading and unloading operations of containers. To evaluate the performance of the dispatching strategies in extreme situations, a very high workload was simulated. It was therefore assumed that no quay crane is running short of jobs—either loading or unloading jobs—during the simulation. In real life, each quay crane is generally idle for some time while waiting for the next vessel to berth. However, berth planning for vessels is outside the defined system boundaries and therefore not considered.

Throughout the numerical experiments, the degree of stochasticity is varied so that the relative performance of the various dispatching methods can be assessed. Each scenario is characterized by the number of quay and stacking cranes in the terminal configuration and the stochastic variations of the handling time per container. All the detailed data required to feed the simulation model (e.g., container and equipment attributes) were generated according to the guidelines of Hartmann (2004b), which were derived from the simulation project of a modern automated container terminal. We specifically generated a number of scenarios by varying the following experimental factors:

- Small, medium, and large terminal configurations were generated consisting of 5, 10, and 15 quay cranes as well as 15, 30, and 45 storage blocks, respectively.
- The average travel distance of a transportation order increases with the terminal size. Hence, the appropriate AGV fleet size had to be determined experimentally. We use 32, 72, and 120 AGVs in the small, medium, and large terminal configuration, respectively.
- Different degrees of stochasticity were simulated by considering the cycle times of the quay cranes and stacking cranes as random values which are determined according to the empirical distributions observed by Vis and Harika (2004). We distinguish four degrees of stochasticity:
  - Deterministic: The cycle times were set to the mean values of the Vis/Harika distributions.
  - Low: The Vis/Harika distributions are compressed such that the structures (including the mean value) are maintained but their variances are reduced to half of the original value.
  - Normal: Use of the original Vis/Harika empirical distributions.
  - High: The Vis/Harika distributions are expanded such that the structures (including the mean value) are maintained but their variances are doubled.

This distribution refers, for instance, to a scenario in which adverse weather conditions prevail.

- Vehicles were operated alternatively as single- and dual-load carriers. In the latter case, the capability of the vehicles of loading one 40-ft or two 20-ft containers at a time was utilized.
- The share of 40-ft containers was set to 50%, which is a realistic value for common terminals.

For each scenario, simulation experiments were repeated ten times with different randomly generated input data. Based on the minimal cycle time for the quay cranes, 1,000, 2,000, and 3,000 transportation orders were generated for the small, medium, and large configuration scenario, respectively. For each data set, the following two approaches were tested once for the single-load mode ("SLC") and once for the dual-load mode ("MLC"):

1. On-line dispatching using the combination of the basic rules "nearest-vehicle/ first-come-first-served (NV/FCFS)"
2. Off-line dispatching using the extended pattern-based heuristic ("Pattern")

While the on-line approach only uses information about the next transportation order of each quay crane or storage block, for the off-line heuristic a look-ahead window of four transportation orders per quay crane was used. All these transportation orders are considered by the pattern-based heuristic for the generation of the actual predictive schedule. Reassignment of all operations scheduled during the last dispatching request is allowed apart from the one to which the vehicle is currently en route (to avoid deviations) and apart from the drop-off of already picked up containers (which clearly must be done by the AGV currently transporting the container). Apart from a new transportation order, the off-line heuristic is also triggered once a significant deviation between the schedule and the system status occurs. In our experiments, we use the delay of a pick-up or drop-off operation as an indicator. The corresponding threshold value is set to 60 s.

5.3 Numerical results

As minimizing turnover time of the vessels is the most important performance criterion for AGV dispatching, the different approaches are assessed with respect to the overall processing time required to complete all transportation orders. We compare the simulation results to a lower bound which is calculated individually for each problem instance considering the sequence of the transportation orders at each of the quay cranes and the derived sequences at the individual storage blocks. For each operation, the actual handling times are taken into account, which were generated according to the distribution of the corresponding degree of stochasticity. Based on this information, an un-capacitated project scheduling problem is formulated in which the precedence constraints refer to the sequences at the quay and stacking cranes as well as the pick-up and drop-off operations of each transportation order. The problem was modeled in ILOG OPL-Studio and solved using the commercial standard solver CPLEX 8.1 to generate lower bounds for all problem instances. The complete model formulation is defined by:

Index sets and parameters	
I^+	Set of pick-up operations
$i^+ \in I^+$	Pick-up operation for container i
I^-	Set of drop-off operations
$i^- \in I^-$	Drop-off operation for container i
$I = I^+ \cup I^-$	Set of all operations
$\widetilde{\Pi}$	Set of ordered pairs $(i1, i2)$ of operations with $i1$ being the direct predecessor of $i2$ in the actual schedule of a crane (as realized in the simulation)
$s \in S$	Locations in the guide path (quay or stacking cranes)
$\widetilde{t}_i^h$	Actual duration of operation i (as realized in the simulation)
s_i	Location at which operation i takes place
$d(s_i, s_k)$	Travel time between s_i and s_k
Decision variables	
t_i^{start}	Earliest start time of operation i
t_i^{end}	Earliest finish time of operation i
F	Makespan
Model formulation	
$Min\, F$	
subject to	
$F \geq t_i^{end}$	$\forall i \in I$
$t_i^{start} + \widetilde{t}_i^h = t_i^{end}$	$\forall i \in I$
$t_{i1}^{end} \leq t_{i2}^{start}$	$\forall (i1, i2) \in \widetilde{\Pi}$
$t_{+}^{end} + d(i^+, i^-) \leq t_{i^-}^{start}$	$\forall (i^+ i^-)$
$t_i^{start}, t_i^{end} \geq 0$	$\forall i \in I$
$F \geq 0$	

The first two constraints define the makespan and the earliest start and finish time of each operation. The next two constraints assure that the precedence constraints imposed by the crane schedules and the relation between pick-up and drop-off operation of the same order are met. The minimum makespan referred to in the objective function gives the lower bound.

In our experiments, the lower bounds for the overall processing time assume values of 6 h and more. This compares to an average operation time of the quay cranes of 3 h (=average cycle time of 1:05 · 200 transportation orders per quay crane). This gap can only be closed if the selection of the storage block and the sequencing of the operations at the stacking cranes are improved. However, these decisions are beyond the scope of this paper which focuses on AGV dispatching. In compliance with the common container terminal practice, the AGV system is regarded as a subordinate service system which is operated on the basis of fixed target data supplied by the quay and stacking cranes.

By examining the detailed simulation results of preliminary numerical tests, we detected deadlock situations, which hampered the system performance. We therefore developed specific deadlock handling strategies and included them in the simulation model used for the numerical investigation. A detailed presentation of these deadlock handling strategies can be found in the companion paper by Lehmann et al. (2006).

The main research questions addressed in our numerical investigation are the following:

- How does the degree of stochasticity affect the performance of the dispatching strategies?
- Does the size of the terminal configuration have a major impact on the relative performance of the dispatching strategies?
- How do the on-line ("NV/FCFS") and the off-line ("Pattern") strategy perform against each other?
- Can the system performance of the terminal be improved by utilizing the multi-load capability ("MLC") of the AGVs compared to the single-load carrier ("SLC") mode?

Figure 10 shows the final results of the simulation experiments in comparison to the lower bound. As a general result, we found that the performance of the investigated dispatching strategies shows similar characteristics for the three investigated terminal configurations. In all cases, the extended pattern-based heuristic clearly outperforms the on-line ("NV/FCFS") heuristic. Its overall processing time deviates from the lower bound between 4.5 and 14% for the small, 2.5 and 10% for the medium, and 1.5 and 7.5% for the large terminal configuration. The superior performance for the larger configurations is mainly due to the fact that the scheduling frequency increases, as—due to the increased number of quay cranes—a larger number of transportation orders is considered in the planning horizon. On average, every 4.3 s, an additional transportation order triggers a dispatching request in the large terminal configuration, while the average time between dispatching requests triggered by transportation orders is 13 s for the small terminal configuration. In particular, the result for the most realistic scenario, i.e., the large terminal configuration with a normal degree of stochasticity, seems to indicate that the developed approach is appropriate. In this case, a deviation from the lower bound of less than 5% was observed.

The tested on-line heuristics show a deviation from the lower bound of between 17 and 33%. These results clearly demonstrate that the on-line approach, in contrast to the pattern-based heuristic, is unable to exploit the optimization potential which results from coordinated dispatching of the entire AGV fleet over a limited time horizon.

The general effect of an increasing degree of stochasticity is identical for all approaches. It impairs the performance of the heuristics. However, the performance reduction for the pattern-based heuristic is far less than expected. No convergence of the off-line and on-line heuristics can be observed especially. The off-line character of the pattern-based heuristic is apparently not very distinctive. This is probably due to the high scheduling frequency and to additional triggering of dispatching requests once the delay threshold has been exceeded.

The version of the pattern-based heuristic, which utilizes the multi-load capability of the AGVs, achieves better results than the pattern heuristic, which treats the AGVs as single-load carriers. This result is highly relevant for terminal operators being currently reluctant to actually use the additional dispatching flexibility and dual-load capability provided by the vehicle technology.

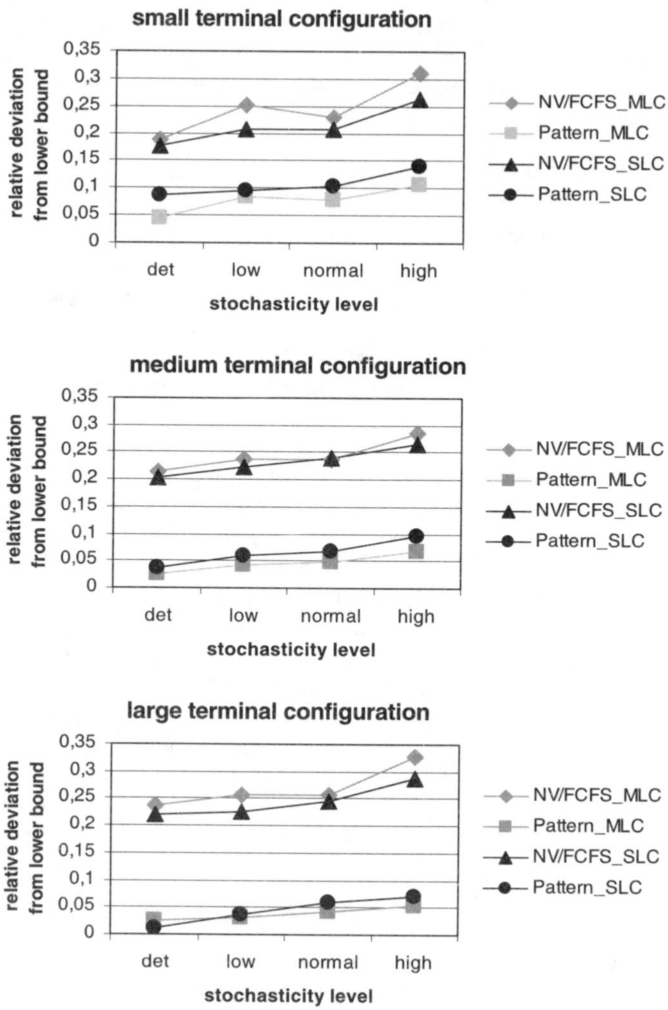

Fig. 10 Performance of dispatching heuristics for different sizes of terminal configurations and different degrees of stochasticity

As can be seen from our experimental results, the additional scheduling complexity does result in improved system performance. In our simulation study, the advantage of the multi-load heuristic shrinks for the larger terminal configurations (in one case, it even becomes negative). This is mainly due to the random assignment of storage locations employed in our simulation experiments. As a result, the average distance between storage blocks from where a container is retrieved or to where it is to be delivered increases with enhanced terminal size. Hence, combining transportation orders becomes less appealing. This effect can also be seen from Fig. 11, which shows that the share of dual-loads decreases with the terminal size. A less simplistic storage block assignment method would definitely help to further realize the benefits of the multi-load capability of AGVs in a large-terminal configuration.

Figure 11 also shows that—contrary to the pattern-based heuristic—the share of 20-ft containers which is transported in dual-load mode increases with the size of the terminal configuration for the on-line heuristic ("NV/FCFS_MLC"). At any given point in time, when a dispatching request is triggered, the probability increases that an AGV is underway which is only partially loaded and to which a second 20-ft container can be assigned. However, the choice of partially loaded AGVs is rather limited. This is particularly true for small-terminal configurations. If one or more AGVs are available, an assignment must still be made. The determined dual-load assignment may therefore result in an AGV tour which leads to large delays. Due to the myopic nature of the on-line heuristic, this decision is not revised at a later stage. This also explains the poor performance of the on-line heuristic for the multi-load mode which is even worse than its performance for the single-mode case (cf. Fig. 10).

In contrast, the pattern-based heuristic also considers AGVs which will become available in the future. A larger choice of AGVs (in fact the entire fleet) is considered for the assignment. This results in AGV schedules of a higher quality. Moreover, the pattern-based heuristic revises assignments, if the AGV is not already on its way to execute the corresponding transportation order.

Figure 12 shows the average number of dispatching requests per order for the large scenario. (The figures for the smaller scenarios are very similar.) For the deterministic scenario, there are about 1.6 requests per order for the on-line heuristic and about 1.4 requests for the pattern-based heuristic. For each order, one request is initiated by a transportation order. For the on-line approach, additional 0.6 vehicle-initiated requests occur. For the pattern-based approach, additional 0.4 requests are triggered by delays. In contrast to the on-line heuristic for which the number of vehicle-initiated dispatching requests is rather insensitive to the stochasticity level, the number of delay-triggered dispatching requests increases significantly for the pattern-based heuristic. In this case, the average number of requests per order reaches 2.5 for the highly stochastic scenario. By adapting the scheduling frequency to the requirements of the problem environment, the pattern-based heuristic thus shows a considerable flexibility.

For the pattern-based heuristic, the average number of dispatching requests per order decreases with the scenario size (see Fig. 13). The frequency of transportation-order-initiated dispatching requests is higher than for the larger

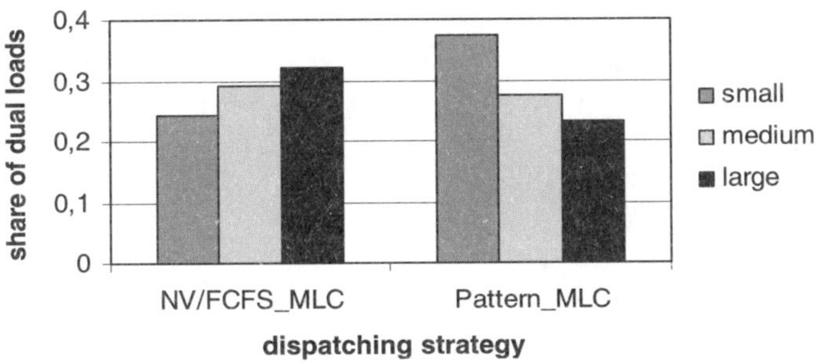

Fig. 11 Share of 20-ft containers transported as dual loads (depending on the size of the terminal configuration)

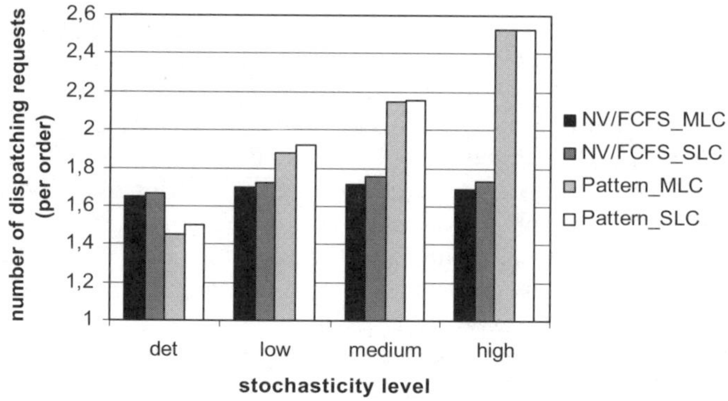

Fig. 12 Average number of dispatching requests per order for the large scenario

scenarios. Therefore, less additional delay-triggered dispatching requests are needed.

The high scheduling frequency required for the pattern-based heuristic obviously poses high demand on the computation time of the heuristics. However, even for the large-sized terminal configurations, which represent scenarios of practical size, the average runtime of each approach was less than 1 s per dispatching request even though the procedures were executed in the interpreter mode of the used simulator eM-Plant. Thus, the approaches perfectly meet the requirements of a real-time application environment of seaport container terminals in practice.

6 Summary and outlook

The main contribution of this paper is the development of rule-based strategies for the AGV dispatching problem in seaport container terminals and their evaluation by use of a scalable event-driven simulation model which allows to model terminal configurations of practical size. The generation of realistic experimental scenarios including the determination of adequate values of relevant model parameters and

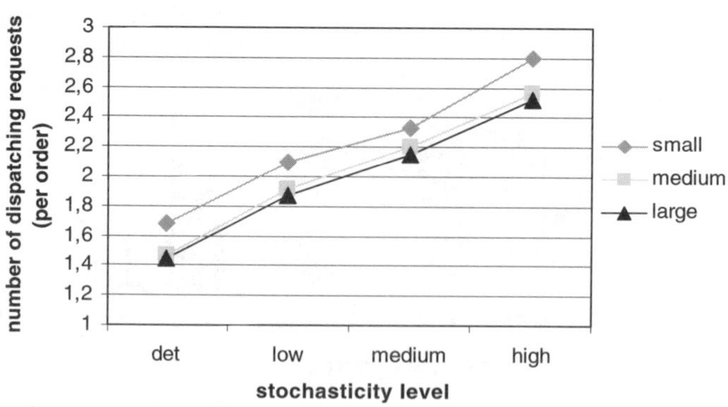

Fig. 13 Dual-load carrier dispatching requests for the pattern-based heuristic

the definition of the system boundaries of the simulation model are described in detail. After a brief discussion of the AGV dispatching problem at hand, two principle approaches—on-line and off-line dispatching—are introduced. A basic on-line dispatching rule and a more sophisticated off-line heuristic are developed and compared in a detailed simulation study considering stochastic handling times, various terminal configurations, and operation modes of the AGVs. In particular, single- and dual-load carrier modes are evaluated.

As a result of the numerical investigation, it could be shown that all dispatching approaches were rather insensitive to the size of the terminal configuration. Moreover, runtimes of less than 1 s meet the requirements of real-time dispatching systems. The off-line heuristic clearly outperformed their on-line counterpart. It shows a relative deviation from a lower bound (which was derived on the basis of given quay and stacking crane sequences) of less than 5% for the most realistic case of a large-terminal configuration and a degree of stochasticity, which corresponds to results observed in empirical studies. The use of the dual-load capability plays a key role in obtaining this result. It is employed for more than 30% of all 20-ft containers.

One way to close the remaining gap is to implement improvement heuristics such as neighborhood search, which further improve the solution found by the pattern-based heuristic, until the result has to be transmitted to the logistics control system. The authors plan to implement this approach in their future research work. Detailed investigations of common neighborhoods for the problem at hand can be found in Nanry and Barnes (2000) and Nanry (1998).

Even without these potential enhancements, the proposed method for AGV dispatching derives close-to-optimum solutions for the case of an AGV system which is regarded as a service system for quay and stacking cranes. However, the lower bound derived in this paper indicates that the quay and stacking crane sequences have a large impact on the utilization of these resources. Our future research will thus aim at integrating the decisions on storage block assignment and sequencing the operations of stacking cranes.

References

Bae JW, Kim KH (2000) A pooled dispatching strategy for automated guided vehicles in port container terminals. Int J Manag Sci 6:47–67

Bish EK, Chen FY, Leong YT, Nelson BL, Ng JWC, Simchi-Levi D (2005) Dispatching vehicles in a mega container terminal. OR Spectrum 27:491–506

BTS (2004) Bureau of transportation statistics. Department of Transportation, USA (http://www.bts.gov/, visited on 27.5.2004)

Church LK, Uzsoy R (1992) Analysis of periodic and event-driven rescheduling policies in dynamic shops. Int J Comput Integr Manuf 5:153–163

de Koster R(M)BM, Le-Anh T, van der Meer JR (2004) Testing and classifying vehicle dispatching rules in three real-world settings. J Oper Manag 22:369–386

Egbelu PJ, Tanchoco JMA (1984) Characterization of automatic guided vehicle dispatching rules. Int J Prod Res 22:359–374

Fiat A, Woeginger GJ (eds) (1998) Online-algorithms: the state of the art. Springer, Berlin Heidelberg New York

Grunow M, Günther H-O, Lehmann M (2004) Dispatching multi-load AGVs in highly automated seaport container terminals. OR Spectrum 26:211–235

Guan Y, Cheung RK (2004) The berth allocation problem: models and solution methods. OR Spectrum 26:75–92

Günther H-O, Kim KH (eds) (2004) Container terminals and automated transport systems. Springer Berlin Heidelberg New York

Hartmann S (2004a) A general framework for scheduling equipment and manpower at container terminals. OR Spectrum 26:51–74

Hartmann S (2004b) Generating scenarios for simulation and optimization of container terminal logistics. OR Spectrum 26:171–192

Hwang H, Kim SH (1998) Development of dispatching rules for automated guided vehicle systems. J Manuf Syst 17:137–143

Kim KH, Bae JW (2004) A look-ahead dispatching method for automated guided vehicles in automated port container terminals. Transp Sci 38:224–234

Kim KH, Kang JS, Ryu KR (2004) A beam search algorithm for the load sequencing of outbound containers in port container terminals. OR Spectrum 26:93–116

Klein CM, Kim J (1996) AGV dispatching. Int J Prod Res 34:95–110

Koo PH, Lee WS, Jang DW (2004a) Fleet sizing and vehicle routing for container transportation in a static environment. OR Spectrum 26:193–209

Koo PH, Lee WS, Koh SH (2004b) Vehicle dispatching for container transportation in seaport container terminals. Proceedings of the 7th international conference on computers and industrial engineering, Jeju, Korea

Le-Anh T, de Koster MBM (2005) On-line dispatching rules for vehicle based internal transport systems. Int J Prod Res 43:1711–1728

Lee J, Srisawat T (2006) Effect of manufacturing system constructs on pick-up and drop-off strategies of multiple-load AGVs. Int J Prod Res 44:653–673

Lehmann M, Grunow M, Günther H-O (2006) Deadlock handling for real-time control of dual-load AGVs at automated container terminals. OR Spectrum (in press)

Lim JK, Kim KH, Yoshimoto K, Lee JH, Takahashi T (2003) A dispatching method for automated guided vehicles by using a bidding concept. OR Spectrum 25:25–44

Nanry WP (1998) Solving the precedence constrained vehicle routing problem with time windows using the reactive tabu search metastrategy. PhD thesis

Nanry WP, Barnes JW (2000) Solving the pickup and delivery problem with time windows using reactive tabu search. Transp Res B Methodol 34:107–121

Nishimura E, Imai A, Papadimitriou S (2005) Yard trailer routing at a maritime container terminal. Transp Res Part E Logist Trans Rev 41:53–76

Park Y-M, Kim KH (2003) A scheduling method for berth and quay cranes. OR Spectrum 1 (25):1–23

Sabuncuoglu I, Bayiz M (2000) Analysis of reactive scheduling problems in a job shop environment. Eur J Oper Res 126:567–586

Sgall J (1998) On-line scheduling—a survey. In: Fiat A, Woeginger GJ (eds) Online algorithms: the state of the art, lecture notes in computer science, vol 1442. Springer, Berlin Heidelberg New York, pp 196–231

Smith S (1994) OPIS, a methodology and architecture for reactive scheduling. In: Zweben M, Fox MS (eds) Intelligent scheduling. Morgan Kaufmann, San Francisco, pp 29–66

Steenken D, Voβ S, Stahlbock R (2004) Container terminal operation and operations research—a classification and literature review. OR Spectrum 26:1–49

Vieira GE, Herrmann JW, Lin E (2003) Rescheduling manufacturing systems: a framework of strategies, policies and methods. J Sched 6:39–62

Vis IFA, Harika I (2004) Comparison of vehicle types at an automated container terminal. OR Spectrum 26:117–143

Yang CH, Choi YS, Ha TY (2004) Simulation-based performance evaluation of transport vehicles at automated container terminals. OR Spectrum 26:149–170

Ebru K. Bish · Frank Y. Chen · Yin Thin Leong · Barry L. Nelson · Jonathan Wing Cheong Ng · David Simchi-Levi

Dispatching vehicles in a mega container terminal

Abstract We consider a container terminal discharging and uploading containers to and from ships. The discharged containers are stored at prespecified storage locations in the terminal yard. Containers are moved between the ship area and the yard using a fleet of vehicles, each of which can carry one container at a time. The problem is to dispatch vehicles to the containers so as to minimize the total time it takes to serve a ship, which is the total time it takes to discharge all containers from the ship and upload new containers onto the ship. We develop easily implementable heuristic algorithms and identify both the absolute and asymptotic worst-case performance ratios of these heuristics. In simple settings, most of these algorithms are optimal, while in more general settings, we show, through numerical experiments, that these algorithms obtain near-optimal results for the dispatching problem.

Keywords Port terminal operations · Vehicle dispatching · Heuristics

Research was supported in part by the Port of Singapore Authority (PSA).

Ebru K. Bish
Virginia Polytechnic Institute and State University, Department of Industrial and Systems Engg., Blacksburg, VA 24061-0118, USA

Frank Y. Chen (✉)
The Chinese University of Hong Kong, Department of System Engg. and Engg. Mgmt., NT, Hong Kong, China, E-mail: yhchen@se.cuhk.edu.hk

Yin Thin Leong
Port of Singapore Authority (PSA), Singapore, Singapore

Barry L. Nelson
Northwestern University, Department of Industrial Engg. and Management Sciences, Evanston, IL, USA

Jonathan Wing Cheong Ng
University of Hong Kong, Department of Industrial and Manufacturing Systems Engg., Hong Kong, China

David Simchi-Levi
Massachusetts Institute of Technology, Department of Civil and Environmental Engg., Cambridge, MA, USA

1 Introduction and motivation

In the last few years we have seen the breakdown of many trade barriers and the globalization of trade. These developments have increased the importance of logistics and transportation, and in particular, the importance of marine transportation systems. These systems include a network of terminals around the globe that allow manufacturers and shippers to deliver goods quickly to their customers. These terminals serve as hubs for the transshipment of containers from ship to ship or to other modes of transportation, e.g., rail and trucks. In this paper we analyze a container terminal, where the majority of the terminal operations consists of ship-to-ship transshipments.

In today's competitive market place, a speedy transshipment of containers to and from ships is important to both the carrier, since it provides significant operational efficiencies, and to the terminal, which can handle a large number of ships per day. Unfortunately, in many regions around the globe, the terminals are now working at, or close to, capacity and there is significant pressure from the business sectors to increase terminal throughput and, in particular, to *decrease ship turn-around time at the terminal*. In most cases, this requires the development of methodology and tools which will allow the efficient coordination of activities within the terminal area. In this paper we consider one aspect of the terminal operation, which is to dispatch vehicles to containers in the terminal. This research was motivated by our industry partner, who operates a major container terminal. In what follows, we describe the operations of this container terminal, noting here that most container terminals operate in a similar way.

When a ship arrives at the terminal, containers are first discharged from the ship onto vehicles by *quay cranes*; the vehicles then transport the containers to various storage locations in the *yard area*. Typically, after most, or all, containers have been discharged from the ship, other containers are uploaded onto the ship. These containers are carried by vehicles from the yard to the *ship area*, and are loaded onto to ship by the quay cranes. Thus, two types of cranes exist in the terminal: *quay cranes*, which are used to load and unload containers to and from the ship, and *yard cranes*, which are used to load and unload containers at the terminal yard storage area.

Most containers handled by the terminal are standard size (Twenty-foot-equivalent unit (TEU)) containers. Due to the large container sizes, a crane would unload a container only onto a vehicle; unloading to the ground would require additional crane operation to lift the container from the ground and load onto a vehicle, and therefore, is not desirable. Thus, a vehicle needs to be available by the crane throughout the loading and unloading operations. This constraint will be further discussed in Section 3.

The terminal typically handles a small number of ships at a time, and each ship is served by a number of quay cranes. A few hours before the arrival of an incoming ship, the terminal receives detailed information about its contents; i.e., containers that are to be discharged into the yard, as well as a list of containers currently in the yard that should be uploaded onto the ship. This information allows the terminal dispatchers to generate the so-called *crane job sequence*: For each quay crane serving the ship, a detailed sequence specifying the order of the containers that are to be discharged/loaded onto the ship. This sequence is mainly determined by the current positions of the containers on the ship, their destinations and contents. Containers can be stacked on top of each other on the ship. Thus, the sequence in which containers will be discharged is based on the containers' current positions on the ship. Similarly, the sequence in which containers will be uploaded onto the ship is based on the contents of the containers (i.e., if a container is carrying delicate items, then no container can be stored on top of it on the ship; thus, this container must be stored at the top of a stack). This information, together with the container's destination, is used to determine an uploading sequence. Finally, of course, the sequence always starts with discharged containers, which are followed by the containers loaded onto the ship.

Thus, at any point in time, the quay crane operator has information on the next container he/she is going to work on. If this is a container to be discharged from the ship, then the crane sequence will identify a number of potential storage locations, typically two to four, in the yard for this container. If this is a container to be loaded onto the ship, then the crane sequence also identifies the current location of the container in the yard area.

It is no surprise that managing, controlling and operating such a system is very complex. At the operational level the questions are clear: how should vehicles be dispatched to containers, what is an optimal location for a container discharged from the ship, how should vehicles be routed in this complex network, and what is an effective traffic control mechanism? Similarly, at the strategic level, the issues include optimizing the number of quay cranes, vehicles and yard cranes.

Evidently, these issues are interrelated. Unfortunately, solving a single integrated model that addresses, for instance, all the operational decisions, is well beyond today's computing capability. For that reason, in this research we decompose the problem into several related models: dispatching vehicles to containers, assigning discharged containers to specific locations, and routing vehicles. Our approach is to analyze each model separately in order to gain an insight into the system (see [4]). In this study, we focus on the problem of dispatching vehicles to the containers for a single ship, assuming that a fleet of vehicles are already assigned to this ship. In doing so, we treat other aspects of the system management as given inputs. This includes selecting an appropriate location for a discharged container, vehicle routing, traffic control, etc. Specifically, we focus on the impacts of vehicle deployment on the system throughput. Our objective is to find *easily implementable* vehicle dispatching policies that *minimize the ship makespan*, which is the time the last vehicle returns to the ship area after all containers are discharged from the ship and are taken to their storage locations in the yard, and all new containers are uploaded onto the ship. We refer to this problem as the *vehicle dispatching problem*.

This paper is organized as follows. In the next section, we give a brief review of the related literature. In Section 3, we consider the vehicle dispatching model for a single ship with a single quay crane, and analyze the performance of different vehicle dispatching policies on discharging job sequences, uploading job sequences, and combined job sequences. Based on the insights obtained for these simple models, in Section 4 we analyze a more general model of a single ship with multiple quay cranes, and test the performance of the proposed heuristics using computational analysis. Finally, in Section 5, we discuss future research directions and extensions to the vehicle dispatching problem.

2 Literature review

Problems associated with dispatching and routing vehicles arise frequently in *logistics systems*, see, for instance, Bramel and Simchi-Levi [5]. Thus, these problems have been extensively studied in the operations research/management science literature under different settings including, but not limited to, vehicle fleet management, truck routing, and warehouse management. Unfortunately, most of this research is not directly applicable to a container terminal operation due to its unique characteristics. This, in turn, requires the development of algorithms that take into account the special characteristics and constraints associated with container terminals.

This review is not meant to be exhaustive, but rather indicative of the recent developments that are most related to the problem analyzed here; see Bish [2] and Bish et al. [3] for more extensive reviews of the other related areas, such as material handling systems and resource-constrained scheduling, and Steenken, Voss and Stahlbock [16] and Vis and De Koster [19] for overviews of container terminal operations research.

Most of the literature on container terminals has used queuing theory to analyze terminal operations. These queuing models focus on strategic issues such as determining the equipment capacity, both on the water-side (such as berth capacity), and on the land-side (such as the number of quay cranes, vehicles, and yard cranes); see, for instance, Daganzo [7]. Several researchers focus on the operational level issues, such as scheduling the cranes and determining storage locations for the unloaded containers (see, for instance [2,3,6,10–15]).

Most recently, Kim and Bae [9] develop vehicle dispatching methods in container terminals by utilizing information on locations and times of future delivery tasks. They develop a mixed-integer programming model for assigning optimal delivery tasks to vehicles. Since the mathematical model requires an excessive amount of computational time, they also propose a heuristic algorithm; their numerical study indicates that the proposed heuristic is quite effective. Vis, De Koster and Savelsbergh [18] also consider the transport of containers between the ship and the yard, with the objective of minimizing the number of vehicles used. These two papers assume that each vehicle has a unit-load capacity. Grunow, Gunther and Lehmann [8] further analyze dispatching methods for multi-load vehicles in highly automated container terminals. This stream of research focuses on equipment allocation and dispatching problems, while Vis and Harika [17] and Yang, Choi and Ha [20] evaluate the relative performance of AGVs (Automated Guided Vehicle) and ALVs (Automated Lifting Vehicle) at container terminals.

In this paper, our objective is to develop algorithms that are easy to implement, especially for *large problem sizes*, and whose effectiveness can be characterized analytically. For this purpose, we focus on *simple* vehicle dispatching rules, and develop analytical bounds on the deviation of the heuristic solution from the optimal solution for any problem instance as well as for large problem instances, and complement our analysis with a numerical study.

3 The vehicle dispatching problem: A single crane model

In what follows, we first analyze the vehicle dispatching problem by focusing on a single ship single quay crane model, and obtain insights into the effectiveness of various algorithms for different instances of this problem. In Section 4, we use these insights to analyze a more general problem with multiple quay cranes.

Thus, we first consider a single ship served by a single quay crane with a fixed number, k, of vehicles assigned to it. We assume that all the vehicles are initially at the ship area and return to the ship area after completing the discharging and uploading of the ship. Throughout, we use the terms dispatching policy and algorithm interchangeably, and we refer to each container as a *job*. Throughout the paper, we assume that the yard cranes are always available (similar assumptions are used in other papers on container terminal operations; see, for instance, Kim and Bae [9]) and all operation times are deterministic (however, as will be shown in the sequel, some of our results still hold even when these operation times are random).

As mentioned above, our objective is to find an effective dispatching policy that assigns vehicles to jobs so as to *minimize the makespan*. In the vehicle dispatching problem, makespan is the time the last vehicle returns to the ship area after all containers are discharged and are taken to their locations in the yard, and after all containers are uploaded onto the ship.

Associated with the quay crane is a predetermined *crane job sequence*,

$$J_{-/+} : \{J_1, J_2, \cdots, J_n\},$$

with J_i, $i = 1, 2, \ldots, n$, being either a job to be discharged from the ship (denoted by a "–" job) or a job to be loaded onto the ship (denoted by a "+" job). The job sequence may consist of only "–" jobs, in which case it is denoted by J_-, only "+" jobs, in which case it is denoted by J_+, or a mix of "–" and "+"jobs, in which case it is denoted by $J_{-/+}$.

If the job sequence is a $J_{-/+}$ sequence, then it consists of two parts: the first part includes all the jobs to be discharged from the ship, that is, all the "–" jobs, while the second part includes all the jobs to be loaded onto the ship, that is, all the "+" jobs.

Obviously, this predetermined job sequence imposes precedence constraints among the jobs. That is, a "–" job cannot be discharged until all "–" jobs preceding it in the job sequence are discharged; in other words, the quay crane cannot start the task of discharging a specific "–" job until all its predecessor "–" jobs have been discharged from the ship. Similarly, the quay crane cannot load a "+" job until all "+" jobs preceding it in the job sequence are loaded onto the ship. Finally, a "+" job cannot be loaded until all "–" jobs in the sequence have been discharged.

Each "–" ("+") job requires a crane movement that will *lift* it up from the ship (or the vehicle), and *place* it onto a vehicle (or the ship). Clearly, a vehicle needs to be available by the crane only during the time the crane is *placing* the job onto the vehicle. Thus, the total crane processing time of a job consists of two components: one is the lifting time, the other is the placing time (during which a vehicle is needed by the crane). For notational convenience and simplicity, we do not distinguish between these two components, and assume that each vehicle needs to be available by the crane throughout the discharging/uploading process. However, the subsequent analysis can be easily modified to handle the case where there are two separate components for crane processing, and a vehicle needs to be available only during the placing time. In our analysis, we assume that the time required to discharge/upload a container by the quay crane is deterministic, and is the same for all the jobs. We denote this time by s.

Containers are carried between the ship area and the yard using a fleet of vehicles, each of which can carry one container at a time. Without loss of generality, we assume that each vehicle travels at unit speed, i.e., each vehicle travels one unit of distance per unit time, all vehicle travel times between the ship area and a specific location in the yard are deterministic and are known in advance.

To simplify the analysis, we assume, throughout the paper, that there is always an available yard crane ready to respond to a service request of any vehicle. Thus, the time it takes a yard crane to load or unload a container is assumed to be incorporated into the container travel times between the ship area and the yard area. Therefore, throughout, the term *crane* will always refer to a quay crane.

Associated with each "–" ("+") job is a predetermined drop-off (pick-up) point in the yard, called the *location* of the job. Let d_i be the travel time from the ship to job J_i's location; i.e., the drop off location of J_i if it is a "–" job, or the pick up point of job J_i if it is a "+" job. We refer to d_i as *travel time* or *distance* interchangeably.

We first describe the *greedy algorithm*: The first k (=number of vehicles) jobs are assigned, each to a single vehicle. We then assign the next job to the first available vehicle. Specifically, when assigning a "–" job, the first available vehicle that arrives at the quay crane will be dispatched to this job. Similarly, when assigning a "+" job, the first available vehicle that can arrive at the job's location at the earliest time will be dispatched to this job. That is, if a vehicle is currently busy with another job assignment, then the time it can be available at the next "+" job's location will be the time it completes its current assignment plus the traveling time from the destination of its current assignment to the next job's location; if a vehicle is currently free, then this time will simply be the traveling time from its current location to the next job's location. Based on these times, we then select the vehicle that can arrive at the next job's location at the earliest time.

In the following sections, we present our results for different cases of the vehicle dispatching problem.

3.1 Analysis of various dispatching policies

3.1.1 J_- Job sequences

Consider a J_- job sequence. Note that for such a job sequence, once a vehicle takes a "–" job, it has to drop the job to its location in the yard, and then it has to make an empty trip back to the ship area to take its next job. We apply the greedy algorithm defined above to dispatch vehicles to jobs. We have the following result, whose proof is straightforward and is thus omitted.

Theorem 1 *For any J_- job sequence, the greedy algorithm is optimal, that is, the greedy algorithm minimizes the makespan.*

Remark The greedy algorithm is still optimal even if crane processing times are job-specific, that is, the quay crane time associated with job J_i is s_i, and not a constant s. Similarly, it is optimal even when the vehicle travel time and quay crane processing times for each job are random variables.

To illustrate the greedy algorithm, consider an example with four "–" jobs:

$J_- : \{J_1, J_2, J_3, J_4\}$,

with $d_1 = 1$, $d_2 = 5$, $d_3 = 1$, $d_4 = 5$ and $s = 2$ (all in minutes). Let $k = 2$ and in what follows we use V_1 and V_2 to denote the two vehicles. The greedy algorithm works as follows. First, assign V_1 to J_1 and V_2 to J_2. After $s = 2$ minutes, V_1 leaves the crane with J_1, and the crane starts discharging J_2. After another 2 minutes, V_2 leaves with J_2. Now the first available vehicle for jobs J_3 and J_4 is clearly V_1 by times 4 and 8, respectively. The dispatching solution can be represented as $V_1 : \{J_1, J_3, J_4\}$, and $V_2 : \{J_2\}$. The completion time for this J_- sequence is $20(= 8 + 2 + 10)$ minutes.

3.1.2 J_+ Job sequences

Consider now a J_+ job sequence. For such a job sequence, once a vehicle is assigned to a "+" job, it makes an empty trip to the job's location starting from the ship area, takes the job, and returns back to the ship area with the job. It is easy to see that the greedy algorithm does not necessarily generate an optimal strategy for a J_+ job sequence.

Given a job sequence $J_+ : \{J_1, J_2, \ldots, J_n\}$, consider the following polynomial time algorithm, called the *reversed greedy algorithm*. The reversed greedy algorithm works as follows: Replace each "+" job by a "–" job with the same location, that is, if location p is the pick-up point for a specific "+" job, the associated "–" job has location p as its drop-off point. Now, reverse the order to get the *reversed job sequence* $J_-^R : \{J_n, J_{n-1}, \ldots, J_2, J_1\}$. Apply the greedy algorithm to this reversed list (of "–" jobs), to obtain a set of jobs assigned to each vehicle. For instance, the jobs assigned to vehicle l, $l = 1, 2, \ldots, k$, are given by $V_l : \{J_{l_1}, J_{l_2}, \ldots, J_{l_{f_l-1}}, J_{l_{f_l}}\}$ and they are served by the lth vehicle following that order. The final step of the algorithm is to reverse again the sequence of jobs assigned to each vehicle. That is, vehicle l will serve this set of jobs assigned to it following the order: $\{J_{l_{f_l}}, J_{l_{f_l-1}}, \ldots, J_{l_2}, J_{l_1}\}$.

We have the following result (please see Appendix for the proof):

Theorem 2 *The reversed greedy algorithm is optimal for any J_+ instance.*

Remark Theorem 2 still holds when the crane processing times are job-specific.

Although we have identified the optimal vehicle dispatching rule for uploading job sequences, it is interesting to study how well the simple greedy algorithm, introduced in the previous section, would perform for such job sequences. This is

Table 1. Average percent deviation of the greedy algorithm's makespan from optimality for uploading job sequences

	500 Uploading jobs				
Spread a		2	6	10	16
k	4	1.8 %	4.6 %	5.8 %	8.6 %
	5	2.3 %	5.5 %	9.6 %	10.1 %
	6	2.6 %	6.3 %	9.8 %	9.9 %
	7	2.4 %	6.6 %	10.5 %	11.2 %
	8	2.6 %	6.4 %	11.0 %	12.1 %

because the greedy algorithm is an appealing dispatching policy due to its ease of implementation, flexibility, and robustness (i.e., minor disruptions to the schedule can be easily handled by the greedy algorithm, which schedules jobs one at a time, following the order of the job sequence. This, however, is not true for the reversed greedy algorithm, in which small changes to the schedule would require the entire schedule to be regenerated by the reversed greedy.) For this purpose, we have conducted computational experiments, the results of which are presented in the next section.

3.1.3 Computational analysis of the greedy algorithm for J_+ sequences

In this section, we analyze the effectiveness of the greedy algorithm for uploading job sequences. We consider sequences each consisting of 500 "+" jobs. We set the crane discharging/uploading time to be 3 minutes ($s = 3$). The traveling time of each job (between the ship area and its location in the yard area) is generated from a uniform distribution. To determine the impact of job distribution in the yard on the performance of the greedy algorithm, we use four different sets of range (*spread*) in our uniform distribution: in the first set, traveling times are uniformly distributed between 2 and 4 minutes (with a *spread*, a, of 2 minutes), in the second set, between 2 and 8 minutes ($a = 6$), in the third, between 2 and 12 minutes ($a = 10$), and in the fourth, between 2 and 18 minutes ($a = 16$). Thus, as a increases, job locations become more spread apart from each other in the yard area. We replicate each scenario 500 times and determine the percent deviation of the makespan obtained by the greedy algorithm from the optimal makespan (obtained by the reversed greedy algorithm) over the 500 problems. The results are reported in Table 1.

As can be seen from the table, the greedy algorithm generates schedules with a makespan of at most 12% over the optimal makespan. For a fixed number of vehicles, k, the ratio increases with the spread, a. Thus, the gap between the optimal makespan and the makespan of the greedy algorithm increases as jobs get more spread out in the yard area. Similarly, for a fixed value of a, the performance of the greedy algorithm generally deteriorates as the number of vehicles increases.

In practice, terminal dispatchers shelf most "+" jobs that will be loaded onto a particular ship in adjacent clusters in the yard area. Thus, the "spread" for these jobs is usually small. Consequently, we believe that the greedy algorithm would provide a rather efficient solution to "+" job sequences, and therefore, is a desirable approach due to its simplicity and flexibility.

3.1.4 $J_{-/+}$ job sequences

Consider now a general $J_{-/+}$ job sequence. As mentioned above, if the schedule of a vehicle ends with a "–" job, then the vehicle has to make an empty trip back to the ship area after dropping its last "–" job in the yard area. Similarly, if the schedule of a vehicle begins with a "+" job, then the vehicle has to make an empty trip from the ship area to the yard area to take the "+" job. However, if the schedule of a vehicle is such that the vehicle takes its first "+" job after dropping its last "–" job, then the vehicle saves these two empty trips, and instead, it travels from the location of its last "–" job to the location of its first "+" job. These travel times are sequence dependent, since they depend on the order of the jobs taken by the vehicle.

The optimality of the greedy algorithm for J_- job sequences and the optimality of the reversed greedy algorithm for J_+ job sequences suggest the following algorithm for a $J_{-/+}$ job sequence.

We start with the greedy algorithm applied to the first part of the job sequence, which consists of all the "–" jobs. We then apply the reversed greedy algorithm to the second part of the job sequence which consists of all the "+" jobs. Finally, we combine the two schedules. We refer to this algorithm as the *combined algorithm* (please see Bish et. al [1] for details).

We let Z^C and Z^* respectively denote the makespan obtained by the combined algorithm and the optimal makespan and n denote the number of jobs in the sequence. The next theorem characterizes the effectiveness of the combined algorithm (see Bish et. al [1] for its proof).

Theorem 3 *For every finite instance of a $J_{-/+}$ job sequence, we have*

$$\frac{Z^C}{Z^*} \leq 3.$$

In addition,

$$\lim_{n\to\infty} \frac{Z^C}{Z^*} = 1.$$

The asymptotical performance of the algorithm is especially important, since in practice the number of jobs is in thousands. In addition, in Bish et. al [1] we provide a pseudo-polynomial time algorithm that is optimal for any instance of $J_{-/+}$ job sequences.

In the next section, we use the insights obtained for the single crane model to analyze the vehicle dispatching problem with multiple quay cranes.

4 The vehicle dispatching problem: A multiple crane model

In the previous sections we focused on a single crane model, and showed that the greedy algorithm is optimal (i.e., it minimizes the ship makespan) for a discharging

(J_-) job sequence, and the reversed greedy algorithm is optimal for an uploading (J_+) job sequence. Our next objective is to extend this analysis to the more general case, where multiple quay cranes are assigned to serve a single ship. Associated with each quay crane is a job sequence and the objective is to assign vehicles to containers so as to minimize the time all jobs are done. That is, the objective is to minimize the makespan over all quay cranes. This objective is consistent with a terminal's objective of releasing ships at the earliest possible time.

Thus, the next question is whether the greedy and reversed greedy algorithms continue to be optimal for discharging and uploading job sequences, respectively, when there are multiple cranes. We first focus on situations in which each quay crane has a J_- job sequence. In the multi-crane environment, the greedy algorithm should be interpreted as assigning an available vehicle to the first available ship crane. In this case, however, it is easy to construct examples that demonstrate the greedy algorithm not necessarily to be optimal.

In practice, however, the greedy algorithm is an appealing solution procedure due to its simplicity and flexibility. Therefore, we now use a simulation study to investigate the performance of the greedy algorithm for a multiple crane model with discharging job sequences. Based on this analysis, we, then, refine the greedy algorithm so as to improve its performance for the multiple crane model. Due to the symmetricity between the greedy algorithm and a discharging job sequence, and the reversed greedy algorithm and an uploading job sequence, as observed in the previous section, the performance of the reversed greedy algorithm for a multiple crane model with uploading job sequences will be similar.

In what follows, we first describe the design of our computational experiments and then discuss our findings.

4.1 Design of the computational experiments

Our objective in this section is to evaluate the performance of the greedy algorithm for the multiple crane vehicle dispatching problem with discharging job sequences. For this purpose, we compare the makespan obtained by the greedy algorithm with that of the optimal makespan, obtained by solving a Mixed Integer Program (MIP); the MIP formulation is given in Bish et. al [1]. However, it takes the MIP on the order of a couple of hours on a Sun Sparc 10 workstation to find the optimal solution, even for small sized problems consisting of only 4 vehicles, 2 cranes, and 20 jobs on each crane. Thus, it is not a practical approach for actual dispatching purposes, especially when problems with 500-2500 containers are common in practice.

For this reason, we limit our computational analysis to cases with only 4 vehicles, 2 cranes, each with a job list of $8-12$ jobs, and solve 200 such problems. For each job, we generate a traveling time between the ship area and the job's location based on a uniform distribution in the range of 1 to 17 min. We assume that it takes a crane 2 minutes to lift a container from the ship (or the vehicle), and it takes 1 minute to place (pick) the container on (from) the vehicle (observe that letting the first time component to zero reduces the formulation to the model addressed in the previous sections). Thus, a vehicle needs to be available by a crane only during the last minute of job discharging/uploading.

Table 2 summarizes the percent deviation of the makespan obtained by the greedy algorithm from the optimal makespan over the 200 problems: for each range of deviations, we report the number of problem instances with deviation in that range. Table 2 shows that the greedy algorithm performs reasonably well in most cases (in almost 80% of the instances, the deviation from the optimal solution is less than 10%), with an average deviation of 7% from the optimal solution. The next question is whether the performance of the greedy algorithm could further be enhanced by small refinements that are not computationally expensive to implement. This is discussed in the next section.

Table 2. Percent deviation of the heuristic makespan from optimality

% deviation from optimality	< 1%	1-3 %	3-5 %	5-10 %	> 10%
# of instances in this range (out of 200)	3	26	38	88	45
Average deviation = 7%					

4.2 *A refined greedy algorithm*

Clearly, the main reason for the poor performance of the greedy algorithm is its "myopic" nature. To overcome its "myopic" nature, we propose an enhancement to the greedy algorithm, and include a simple look-ahead rule, described below.

Let $J_{i,j}$, $i = 1, 2,$ and $j = 1, 2, ...,$ denote the jth job in the sequence of crane i. In what follows, we represent each job in terms of its traveling time (between the ship area and its location). Let l_i be the number of jobs in the job list of crane i. Given a fixed $p \leq l_i$, we assign a weight $w_{i,j} = \sum_{k=j}^{\min\{j+p,l_i\}} J_{i,k}$ to each job $J_{i,j}$, for $j = 1, \ldots, l_i$. Thus, the weight of each job represents the minimum time required to complete the remaining jobs on crane i's list, which excludes crane and queuing times. When a vehicle arrives at the ship area, it determines the job(s) that are available at the earliest time for pick-up (which is determined by the earliest available time of the corresponding crane). If there is only one such job, then it selects that job for pick-up (as in the greedy algorithm). If, on the other hand, there are multiple jobs available at the same time, then the vehicle selects the job with the maximum weight. Thus, in the latter case, the vehicle will give higher priority to the job with a longer traveling time, or to the crane job sequence with a longer time for the remaining jobs.

Finally, we further modify the greedy algorithm by the following *enhancement*: When there are a certain number, x, of jobs left in the system, we perform an explicit enumeration to determine the best schedule for these remaining jobs. Clearly, x should be a very small number. Presumably, this last enhancement is not as effective for reasonably long job sequences. To confirm this, we tested a few examples with 20 jobs. It was found that this refinement "enhanced" efficiency by at most 0.3%

in those examples. (The main reason for us to use such an additional enhancement is to remove the "ending" effect which may arise in small-sized problems.)

Next, we tested the performance of the refined greedy algorithm on the same set of 200 problem instances (with $p = 8, x = 4$). The results show that the refined greedy algorithm generated near-optimal solutions for most instances, with an average deviation from optimality of 1.55%, and a standard deviation of 2.61%. The result is summarized in Table 3, shows the distribution of this deviation for the refined greedy algorithm. As can be seen from the table, the refined greedy algorithm performs much better than the greedy algorithm.

Table 3. Percent deviation of the heuristic makespan from optimality

% deviation from optimality	$< 1\%$	1-3 %	3-5 %	5-10 %	$> 10\%$
# of instances in this range (out of 200)	22	95	40	43	0
Average deviation = 1.55%, standard deviation = 2.61%.					

5 Conclusions and future research directions

Our goal in this research is to come up with simple, easily implementable vehicle dispatching policies that generate *good* makespan values for the vehicle dispatching problem.

The greedy algorithm is an appealing solution due to its simplicity and flexibility. Therefore, in this analysis, we considered the greedy algorithm, together with the reversed greedy algorithm, the combined algorithm and the combined greedy algorithm, all of which are based on the greedy algorithm. By considering a single-ship/single-crane model, we were able to prove the optimality of the greedy algorithm for a discharging job sequence, the optimality of the reversed greedy algorithm for an uploading job sequence, the asymptotic optimality of the combined algorithm together with the optimality of the combined greedy algorithm for a combined job sequence. Based on these results, we, then, analyzed a more general problem of a single ship with multiple cranes, and tested the performance of the greedy algorithm for this problem through computational analysis. The results show that, although not optimal, the greedy algorithm performs reasonably well for a multiple crane vehicle dispatching problem with discharging job sequences. We further enhanced the performance of the greedy algorithm by including a look-ahead type of rule, which we refer to as the refined greedy algorithm. Computational analysis reveals that the performance of the refined greedy algorithm is very satisfactory: an average deviation of 1.55% deviation from the optimal solution over all problems tested.

We must note, however, that this research is only a start to analyze the operational issues in container terminals, and there are still many open issues that need to be analyzed.

In practice, other issues need to be incorporated into the analysis and addressed by the algorithms. One important issue is how to determine a storage location for each discharged container. In the model considered here, the storage location of each discharged container is assumed to be given. This problem, where the location of each discharged container is also a decision variable, has been analyzed in [3] and [4]. Another issue would be identifying routes for each vehicle so as to avoid congestion. There is also the issue of coordinating yard crane work load, etc. Yet another important research direction would be to extend this analysis to a multiple ship model. This direction has been studied in several recent papers; see, for instance, Bish [2] and Kim and Bae [9]. Bish [2] considers the vehicle dispatching and container location problem for a multi-ship multi-crane model, develops a heuristic algorithm, which assigns locations to containers based on a transshipment problem and dispatches vehicles to jobs based on a modified version of the greedy algorithm, and analyzes the effectiveness of the heuristic from both worst-case and computational points of view. Her results suggest that a modified version of the greedy algorithm works very well in a multi-ship setting as well. However, analytical results are presented only for a two-ship model and need to be extended to consider any number of ships. On the other hand, Kim and Bae [9] develop a mathematical programming formulation for a multi-ship multi-crane model, suggest a heuristic algorithm, and analyze its performance through a numerical study. We believe that this line of work needs to be extended to analytically characterize the effectiveness of simple heuristics, such as modified versions of the greedy algorithm discussed in this paper, in the context of a multi-ship model.

Although we considered a simplified model in this research, the insights gained in this paper proved to be helpful in analyzing more complex situations at terminal ports.

Appendix: Proof of Theorem 2

Consider any job sequence consisting of jobs $\{J_1, J_2, \cdots, J_n\}$. For dispatching policy π, we refer to the time a vehicle is assigned to J_i, $i = 1, 2, \cdots, n$, as the *start time* of J_i, and denote it as $ST_i(\pi)$. Similarly, we refer to the time J_i, $i = 1, 2, \cdots, n$, is completed under that policy as the *completion time* of J_i, and denote it as $CT_i(\pi)$.

Thus, in a J_- job sequence, the start time of a "–" job is the time the crane starts the task of discharging the job to a vehicle, and the completion time of a "–" job is the time the vehicle returns to the ship area after carrying the discharged job to its location in the yard. In a J_+ job sequence, the start time of a "+" job is the time a vehicle is dispatched to the job's location to bring the job to the ship, and the completion time of a "+" job is the time the quay crane finishes loading the job onto the ship. We will omit the policy parameter and use ST_i and CT_i, when the policy is obvious from the context or when a specific property must hold for all policies.

As stated before, job precedence constraints for a $J_- : \{J_1, J_2, \cdots, J_n\}$ job sequence imply that

$$ST_i \geq ST_{i-1} + s \quad i = 2, \cdots, n,$$

whereas, for a $J_+ : \{J_1, J_2, \cdots, J_n\}$ job sequence we must have

$$CT_i \geq CT_{i-1} + s \quad i = 2, \cdots, n.$$

To prove the Theorem, we need the following lemma.

Lemma 4 *Consider a dispatching policy π_+ applied to a "+" job sequence J_+, with a makespan of $Z(\pi_+)$. There exists a dispatching policy π_- applied to the reversed job sequence J_-^R associated with J_+ that achieves the same makespan, i.e., $Z(\pi_-) = Z(\pi_+)$.*

Proof. Consider a "+" job sequence $J_+ : \{J_1, J_2, \cdots, J_n\}$, and a dispatching policy π_+. We let $V_l : \{J_{l_1}, J_{l_2}, \cdots, J_{l_{f_l}}\}$ denote the job sequence assigned to vehicle $l, l = 1, 2, \cdots, k$, under this dispatching policy. The precedence constraints for this J_+ job sequence imply that job i, $i = 2, \cdots, n$, cannot be completed until all its predecessors in J_+, i.e., jobs $J_1, J_2, \cdots, J_{i-1}$, are completed. Hence, we have

$$CT_i(\pi_+) \geq CT_{i-1}(\pi_+) + s \quad i = 2, \cdots, n. \tag{1}$$

Clearly, the makespan for this dispatching policy is $Z(\pi_+) = CT_n(\pi_+)$.

Now consider the corresponding reversed "–" job sequence, $J_-^R : \{J_n, J_{n-1}, \cdots, J_2, J_1\}$. Our objective is to find a dispatching policy π_- for the reversed job sequence with a makespan of $Z(\pi_-)$ such that $Z(\pi_-) = Z(\pi_+)$.

For this purpose, consider the dispatching policy π_- obtained as follows. Reverse the job sequence $V_l, l = 1, 2, \cdots, k$, defined as above, and denote the resulting sequence as $V_l^R : \{J_{l_{f_l}}, J_{l_{f_l-1}}, \cdots, J_{l_1}\}$. Under this policy, vehicle l starts with job $J_{l_{f_l}}$, continues with job $J_{l_{f_l}-1}$, and so on. Start job J_n, a "–" job now, at time $Z(\pi_+) - CT_n(\pi_+) = 0$, job J_{n-1} at $Z(\pi_+) - CT_{n-1}(\pi_+)$, $\cdots$, and job J_1 at $Z(\pi_+) - CT_1(\pi_+)$.

Now, if we can show that the schedule obtained by dispatching policy π_- satisfies (i) the precedence constraints for the J_-^R job sequence, and (ii) vehicle capacity constraints, then we have a dispatching policy for which

$$Z(\pi_-) = Z(\pi_+) - CT_1(\pi_+) + 2d_1 + s = Z(\pi_+)$$

and we are done.

Consider jobs J_{i-1} and J_i, $i = 2, \cdots, n$. Under dispatching policy π_-, we have:

$$\begin{aligned} ST_{i-1}(\pi_-) - ST_i(\pi_-) &= Z(\pi_+) - CT_{i-1}(\pi_+) - Z(\pi_+) + CT_i(\pi_+) \\ &= CT_i(\pi_+) - CT_{i-1}(\pi_+) \geq s, \quad \text{from Equation (1)} \end{aligned}$$

and hence the precedence constraints for the J_-^R job sequence are satisfied.

Next, we have to show that the schedule obtained by dispatching policy π_- also satisfies the vehicle capacity constraints. Vehicle $l, l = 1, 2, \cdots, k$, can serve jobs

$\{J_{l_{f_l}}, J_{l_{f_l}-1}, \cdots, J_{l_1}\}$ under dispatching policy π_-, since for jobs J_{l_h-1} and J_{l_h}, $h = 2, \cdots, f_l$, we have the following:

$$\begin{aligned} ST_{l_h-1}(\pi_-) - ST_{l_h}(\pi_-) &= Z(\pi_+) - CT_{l_h-1}(\pi_+) - Z(\pi_+) + CT_{l_h}(\pi_+) \\ &= CT_{l_h}(\pi_+) - CT_{l_h-1}(\pi_+) \geq 2d_{l_h} + s, \end{aligned}$$

indicating that vehicle capacity constraints are satisfied. Hence, dispatching policy π_- generates a feasible schedule for the J_-^R job sequence with a makespan of $Z(\pi_+)$. This completes the proof. □

Lemma 4 leads to the following corollary:

Corollary 5 *Consider a dispatching policy π_- applied to a job sequence J_-, with a makespan of $Z(\pi_-)$. There exists a dispatching policy π_+ for the reversed "+" job sequence associated with J_- such that it achieves exactly the same makespan, i.e., $Z(\pi_+) = Z(\pi_-)$.*

Now we are ready to prove Theorem 2.

Proof of Theorem 2. Let π_+^* be the optimal dispatching policy for a J_+ job sequence, with a makespan of $Z(\pi_+^*)$. By Lemma 4, we can find a dispatching policy π_- for the corresponding reversed "–" job sequence J_-^R such that $Z(\pi_-) = Z(\pi_+^*)$.

Now consider the optimal dispatching policy π_-^* for this J_-^R job sequence with a makespan of $Z(\pi_-^*)$. By Corollary 5, we can find a dispatching policy π_+ for the corresponding reversed "+" job sequence J_+, which is the original J_+ job sequence, such that $Z(\pi_-^*) = Z(\pi_+)$.

By the optimality of $Z(\pi_-^*)$ for the J_-^R job sequence, we have:

$$Z(\pi_+^*) = Z(\pi_-) \geq Z(\pi_-^*) = Z(\pi_+) \tag{2}$$

On the other hand, the optimality of $Z(\pi_+^*)$ for the J_+ job sequence implies that

$$Z(\pi_+^*) \leq Z(\pi_+),$$

and hence, Equation (2) holds as equality. That is,

$$Z(\pi_+^*) = Z(\pi_-^*) \tag{3}$$

Finally, Theorem 1 tells us that the greedy algorithm is an optimal dispatching policy for the J_-^R job sequence. Thus, given a J_+ job sequence, we can obtain the reversed job sequence J_-^R associated with J_+ and find the optimal schedule for this J_-^R job sequence by applying the greedy algorithm. Given the optimal schedule for the J_-^R job sequence, we can construct a schedule for the original J_+ job sequence as done in the proof of Lemma 4. That is, we can do that by reversing the sequence of jobs assigned to each vehicle. Furthermore, this must be (one of) the optimal schedule(s) for the J_+ job sequence, since $Z(\pi_-^*) = Z(\pi_+^*)$ by Equation(3). Observing that this is the reversed greedy algorithm completes the proof. □

References

1. Bish EK, Chen FY, Leong YT, Nelson BL, Ng JW, Simchi-Levi D (2000) Dispatching vehicles in a mega container terminal. Unabridged Technical Report, Virginia Polytechnic Institute and State University, Dept of Industrial and Systems Engg
2. Bish EK (2003) A multiple-crane-constrained scheduling problem in a container terminal. European Journal of Operational Research 144: 83–107
3. Bish EK, Leong T, Li C, Ng JWC, Simchi-Levi D (2001) Analysis of a new scheduling and location problem. Naval Research Logistics 48: 363–385
4. Bish EK (1999) Theoretical analysis and practical algorithms for problems in a mega container terminal. Ph.D. Dissertation, Northwestern University
5. Bramel J, Simchi-Levi D (1997) The logic of logistics. Theory, algorithms, and applications for logistics management. Springer, New York, NY
6. Castilho BD, Daganzo CF (1993) Handling strategies for import containers at marine terminals. Transportation Research 27B(2): 151–166
7. Daganzo CF (1990) The productivity of multipurpose seaport terminals. Transportation Science 24: 205–216
8. Grunow M, Günther HO, Lehmann M (2004) Dispatching multi-load AGVs in highly automated seaport container terminals. OR Spectrum 26: 211–235
9. Kim KH, Bae JW (2004) A look-ahead dispatching method for automated guided vehicles in automated port container terminals. Transportation Science 38: 224–234
10. Kim KH, Kang JS, Ryu K-R (2004) A beam search algorithm for the load sequencing of outbound containers in port container terminals. OR Spectrum 26: 93–116
11. Kim KH, Park YM, Ryu K-R (2000) Deriving decision rules to locate export containers in container yards. European Journal of Operational Research 124: 89–101
12. Kim KH, Kim HB (1999) Segregating space allocation models for container inventories in port container terminals. International Journal of Production Economics 59: 415–423
13. Kim KH, Kim KY (1999) An optimal routing algorithm for a transfer crane in port container terminals. Transportation Science 33(1): 173–176
14. Kim KH (1997) Evaluation of the number of rehandles in container yards. Computers and Industrial Engineering 32(4): 701–711
15. Kim KY, Kim KH (1999) A routing algorithm for a single straddle carrier to load export containers onto a containership. International Journal of Production Economics 59: 425–433
16. Steenken D, Voss S, Stahlbock R (2004) Container terminal operation and operations research – A classification and literature review. OR Spectrum 26: 3–49
17. Vis IFA, Harika I (2004) Comparison of vehicle types at an automated container terminal. OR Spectrum 26: 117–143
18. Vis IFA, De Koster R, Savelsbergh MWP (2004) Minimum vehicle fleet size under time window constraints at a container terminal. Transportation Science (forthcoming)
19. Vis IFA, De Koster R (2003) Transshipment of containers at a container terminal: an overview. European Journal of Operational Research 147: 1–16
20. Yang CH, Choi YS, Ha TY (2004) Simulation-based performance evaluation of transport vehicles at automated container terminals. OR Spectrum 26: 149–170

Dirk Briskorn · Andreas Drexl · Sönke Hartmann

Inventory-based dispatching of automated guided vehicles on container terminals

Abstract This paper deals with automated guided vehicles (AGVs) which transport containers between the quay and the stack on automated container terminals. The focus is on the assignment of transportation jobs to AGVs within a terminal control system operating in real time. First, we describe a rather common problem formulation based on due times for the jobs and solve this problem both with a greedy priority rule based heuristic and with an exact algorithm. Subsequently, we present an alternative formulation of the assignment problem, which does not include due times. This formulation is based on a rough analogy to inventory management and is solved using an exact algorithm. The idea behind this alternative formulation is to avoid estimates of driving times, completion times, due times, and tardiness because such estimates are often highly unreliable in practice and do not allow for accurate planning. By means of simulation, we then analyze the different approaches. We show that the inventory-based model leads to better productivity on the terminal than the due-time-based formulation.

Keywords Container logistics · Container terminal · Automated guided vehicles · Dispatching · Assignment problem · Simulation

---

D. Briskorn (✉) · A. Drexl
Institut für Betriebswirtschaftslehre, Lehrstuhl für Produktion und Logistik,
Christian-Albrechts-Universität zu Kiel, 24098 Kiel, Germany
E-mail: briskorn@bwl.uni-kiel.de, andreas.drexl@bwl.uni-kiel.de

S. Hartmann
HPC Hamburg Port Consulting GmbH, HHLA Container Terminal Burchardkai,
21129 Hamburg, Germany
E-mail: s.hartmann@hpc-hamburg.de

S. Hartmann
HHLA Container Terminal Altenwerder GmbH, Bei St. Annen 1, 20457 Hamburg, Germany

1 Introduction

In various regions of the world, double-digit growth rates in container handling have been common during the last years and, hence, a substantial number of container vessels is built each year. In addition, new vessels are often larger than older ones—currently, modern vessels can carry more than 9,000 standard containers (20-foot equivalent unit, TEU), and even larger ships are already planned. Thus, the capacity of the worldwide container vessel fleet increases year by year. This development puts pressure on container terminal operators to enlarge terminal capacities to avoid congestion in ports. As a consequence, more container terminals are built, and existing ones are expanded. For reasons of efficiency and stacking density, new and extended terminal facilities increasingly make use of automated equipment. This leads to the necessity of complex terminal control systems which allow for an optimized utilization of the automated resources.

Due to its practical relevance, container terminal logistics has been a prominent field of research. A comprehensive-literature survey has recently been given by Steenken et al. (2004). Further overviews have been provided by Meersmans and Dekker (2001), Vis and de Koster (2003), as well as Vis (2006). Important optimization problems include berth planning (see Guan and Cheung 2004; Imai et al. 1997, 2001; Lim 1998; Park and Kim 2003), quay crane planning (see Daganzo 1989; Peterkofsky and Daganzo 1990), and straddle carrier scheduling (see Böse et al. 2000; Kim and Kim 1999b; Steenken et al. 1993). Moreover, approaches for locating containers in the yard have been developed (see de Castilho and Daganzo 1993; Kim and Kim 1999a; Kim et al. 2000; Taleb-Ibrahimi et al. 1993; Zhang et al. 2001).

Several papers have studied specific optimization problems arising in container terminals with automated equipment. Automated guided vehicles (AGVs) have been studied by Bae and Kim (2000). Bish et al. (2005) propose a greedy dispatching method for AGVs. Grunow et al. (2004) consider double load AGVs, that is, AGVs that can carry two 20-ft containers at a time. A general model for scheduling equipment such as AGVs or automated stacking cranes (or non-automated resources such as straddle carriers and reefer mechanics) has been proposed by Hartmann (2004). Meersmans and Wagelmans (2001) discuss an integrated scheduling approach for automated stacking cranes and AGVs. A simulation study to compare AGVs and automated shuttle carriers has been given by Vis and Harika (2004). Kim et al. (2001) employ simulation to provide a test bed for the control system of an automated container terminal. There are numerous papers in which resource allocation/dispatching rules have been applied in different manufacturing settings. For the sake of brevity, we do not go into details in this paper and refer the reader to, e.g., Hwang and Kim (1998), de Koster et al. (2004) and Vis (2006).

In this paper, we focus on highly automated terminals which employ AGVs. This study has been carried out in cooperation with the HHLA Container Terminal Altenwerder in Hamburg, Germany [for details on this terminal, see Baker (1999)]. We consider a container terminal configuration similar to the Altenwerder terminal that employs quay cranes, AGVs, and automated stacking cranes. Quay cranes are used to discharge containers from and load containers onto vessels. AGVs are means for horizontal transport of containers between the stacking area and the quay, and they are unable to load or unload themselves. The yard is organized in a number of stacks, and each stack (or yard block) is served by one or more stacking cranes. The terminal layout considered throughout this paper is displayed in Fig. 1. In this paper, we only deal with the waterside, that is, containers arriving by a vessel which have to be brought to the stacking area and containers being picked up by a vessel which have to be brought from the stack to the quay (the landside with its outside truck and rail operations is not considered, hence, it is not shown in Fig. 1).

The goal of the paper is to present a method for assigning AGVs to transportation jobs that is applicable to real-world container terminals. Therefore, the main requirements for the method are high waterside productivity, very short

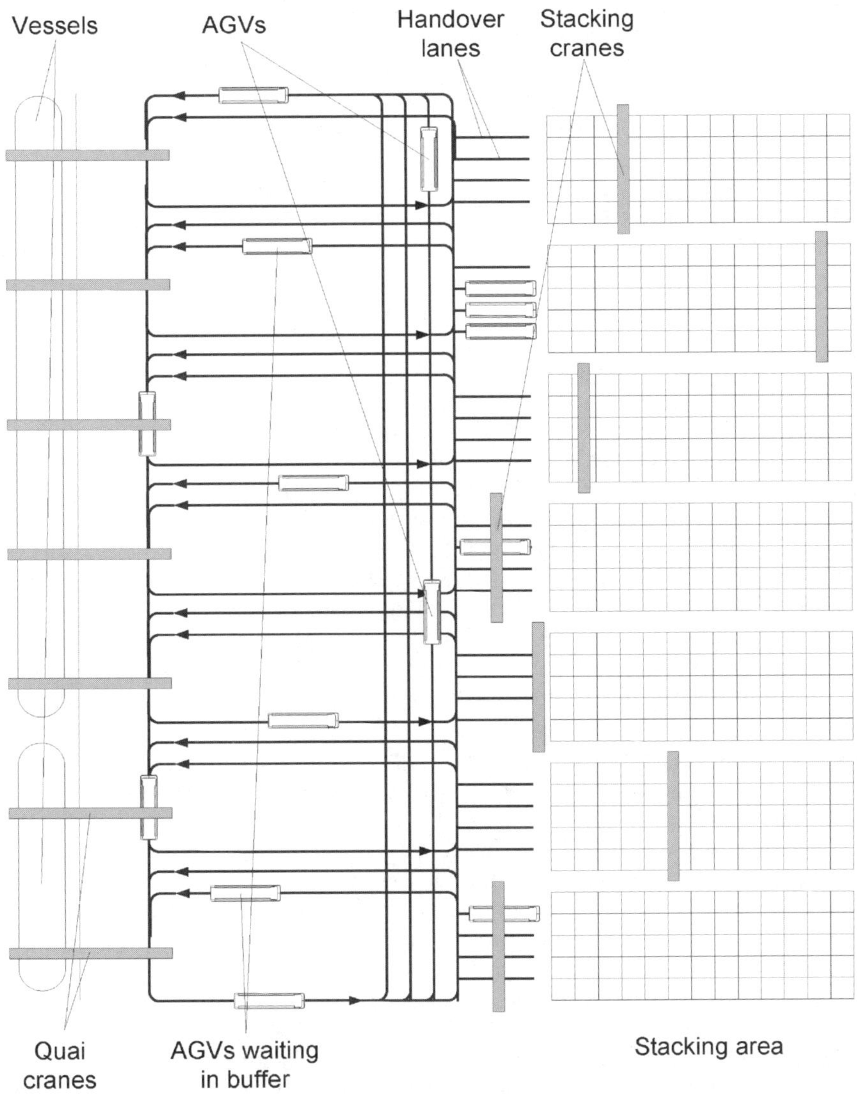

Fig. 1 Layout of the container terminal

computation times, and robustness. High productivity means that the number of container transported per hour should be as high as possible. Short computation times are necessary to allow for real-time application within a terminal control system. Robustness means that the method should perform well in a rather unpredictable environment (which is typical in practice due to quay crane delays, inaccurate estimates for AGV travel times, manual interference, etc.).

The outline of the paper is as follows. We first describe a rather conventional approach to the AGV assignment problem, which is based on due times and an earliness–tardiness objective. This formulation will be solved both by a greedy heuristic [such simple methods are often used in practice and are also discussed in the scientific literature; see Bish et al. (2005)] and an optimal algorithm. Subsequently, we propose a new approach to the AGV assignment, which introduces the idea of inventory related to quay cranes. The motivation for this is to provide a problem formulation that avoids to employ time estimates as the latter are typically inaccurate on real world terminals. Our goal is to define a method that is more robust than a time-based one and, thus, leads to higher productivity. The approaches are then compared in a simulation study. We first point out how much the terminal productivity can be improved by using an optimal algorithm instead of a simple heuristic in the conventional time-based formulation. Then, we indicate the improvement that can be obtained from using the inventory-based formulation instead of the time-based one.

2 General problem description

We consider the problem of assigning jobs to AGVs. Each job corresponds to the transportation of a container from a pick-up location to a delivery location. An AGV can be assigned one job (and, thus, a single container) at a time. After completing a job, an AGV can start another job. A job consists of an empty drive from its last position to the pick-up location, a hand-over time at the pick-up location, a drive to the delivery location, and a hand-over time at the delivery location. Two types of processes are distinguished, namely, discharging and loading a vessel. For a job related to a discharging operation, the pick-up location is a quay crane and the delivery location is a stack. Analogously, for a job related to a loading operation, the pick-up location is a stack and the delivery location is a quay crane. For each job, the locations are fixed (specific quay crane or specific stack). Estimates of driving times between any two locations on the layout, as well as estimates of the hand-over times are assumed to be given (if needed by the actual solution approach).

Depending on the vessel's stowage plan and operational strategies, some container i may have to arrive at the quay crane before some container j when loading a vessel. That is, there may be precedence relations between some (but usually not all) of the jobs related to the same loading quay crane. There are no precedence relations between discharging jobs.

The problem essentially consists of a number of AGVs and a number of jobs. We consider n AGVs, namely, those which are currently available and those which will soon complete their current job (note that this means we have a look-ahead in the assignment process). For these AGVs, an estimated waiting time for availability is given. Without discussing the details in this paper, it should be mentioned that determining the n AGVs to be considered is not based on a horizon but on conditions related to certain events in the progress of the current job (only the occurrence of these events allows for a relatively good estimation of the availability time). Due to the problem-inherent rolling planning horizon, only the n most urgent jobs are considered when computing for an assignment of jobs to AGVs.

The main goal when assigning jobs to AGVs is to maximize the waterside productivity, that is, the number of containers handled per hour by the quay cranes. This goal cannot be used directly as an objective function for the AGV assignment problem. In fact, different objective functions can be defined to achieve the productivity goal. Two such approaches will be discussed in the following sections. In general, one may achieve high productivity by employing goals such as minimization of the quay crane waiting times for AGV (when AGVs arrive too late), minimization of the AGV waiting times at quay cranes (when AGVs arrive too early), minimization of the empty travel times, and an even distribution of AGVs among the quay cranes. (Note that the loaded travel times cannot be influenced by assignment decisions because the pick-up and delivery locations of each job are fixed.)

The AGV assignment problem is embedded into an overall terminal control system. Whenever a certain event occurs, a new AGV assignment is calculated. The main event is the completion of a job. Thus, frequent re-planning is done. If the assignment procedure assigns a job to an AGV that is currently available, this assignment is fixed and the AGV starts this job. Otherwise, if the assignment procedure assigns a job to an AGV that is not yet available, the assignment is not fixed. In the latter case, the job and the AGV will be considered again when the assignment procedure is started after the next event. This way, the decision to actually execute a job is made as late as possible. This allows for decisions based on actual data, which is important as data are frequently changing in practice due to delays, etc. In fact, frequent changes in the data and the inaccuracy of time estimates (which are typical in practice) lead to a short planning horizon and to an assignment problem in which an AGV obtains only one job (instead of a scheduling problem with a sequence of jobs).

In Sections 3 and 4, we present two different formulations of the problem setting described above. Both approaches have essentially the same structure as they are both assignment problems with n jobs and n AGVs (i.e., each AGV must be assigned exactly one job and vice versa) and with an objective to minimize the total assignment costs. They differ only in the way of selecting the n jobs to be assigned and in the definition of the costs c_{ja}, which evaluate the assignment of an AGV a to a job j.

3 Due-time-based approach

3.1 Problem formulation

In this section, we provide a formulation of the AGV assignment problem that makes use of due times for the jobs. This approach is similar to the formulation of Hartmann (2004) and will be summarized briefly.

Each quay crane is associated with a sequence of either loading or discharging jobs. Considering the time the quay crane needs for loading or discharging one container, we can define a due time d_j for each job j. The due time reflects the time at which an AGV should arrive at a quay crane either empty (discharging operation) or with a container (loading operation). Note that a job always has a later due time than all of its predecessors.

As AGVs are unable to load and unload themselves, they should arrive at the quay cranes just in time. Early arrival implies that the quay crane is not yet ready and that the AGV has to wait, which is a waste of AGV capacity. Late arrival means that the quay crane has to wait for the AGV, which decreases its productivity. This leads to a traditional earliness–tardiness objective function. Moreover, one may wish to obtain short empty travel times (to save fuel costs and to save AGV capacity for future jobs). Thus, our objective function minimizes the weighted sum of earliness, tardiness and empty travel time.

For a more formal definition, let J be the set of the jobs to be assigned, and let α_E, α_T, and α_e be the weights for earliness, tardiness, and empty travel time, respectively. Moreover, let f_j^q be the estimated arrival time of job j at the quay crane resulting from the assignment, and let e_{ja} denote the empty travel time of job j when assigned to AGV a. Now the costs c_{ja} of assigning AGV a to job j are defined as

$$c_{ja} = \begin{cases} \alpha_E \cdot (d_j - f_j^q) + \alpha_e \cdot e_{ja} & \text{if } f_j^q < d_j \\ \alpha_T \cdot (f_j^q - d_j) + \alpha_e \cdot e_{ja} & \text{otherwise.} \end{cases} \tag{1}$$

Note that the due time d_j does not refer to the completion of the job but to the arrival time f_j^q at the quay crane. In case of a discharging job, the latter corresponds to the end of the drive to the pick-up location. Let us consider a discharging job j with assigned AGV a and waiting time for availability w_a of AGV a (we have $w_a = 0$ if AGV s is currently available). Then, we obtain $f_j^q = w_a + e_{ja}$ for discharging jobs. In case of a loading job, however, the due time refers to the end of the drive to the delivery location. Let h_{SC} be the estimated hand-over time at the stacking crane, and let t_{ja} be the estimated transportation time from the pick-up to the delivery location. Then, we have $f_j^q = w_a + e_{ja} + h_{SC} + t_{ja}$ for loading jobs.

We consider n jobs and n AGVs for the assignment problem. As outlined in Section 2, the n AGVs are those that are currently available and those which will soon complete their current job. The n jobs are given as the n most urgent jobs that are not yet in process, that is, the n jobs with the earliest due times among those jobs that have not yet been started.

3.2 Solution methods

To solve the due time based assignment problem, we employ two procedures. Both start by computing the set of jobs J and the set A of AGVs to be assigned, as described in the previous subsection.

The first approach is the Hungarian method of (Kuhn 1955) which is implemented as described in (Munkres 1957). This algorithm leads to an optimal assignment with respect to the due-time-based assignment costs given in Eq. 1.

The second approach is a simple greedy heuristic that will be used to provide benchmark results for the comparison. We employ a priority rule-based procedure similar to that of Hartmann (2004). The procedure repeatedly applies the following steps until each job has been assigned to an AGV, that is, until $J = \emptyset$ and $A = \emptyset$.

1. Select the job j to be assigned next as the most urgent job, that is, the job with the smallest due time $d_j = \min\{d_i \mid i \in J\}$.
2. Select the AGV a that leads to the smallest increase in the objective function, that is, the lowest possible costs $c_{ja} = \min\{c_{jb} \mid b \in A\}$ for job j.
3. Assign AGV a to job j.
4. Remove AGV a from A and job j from J, respectively.

3.3 Implications for stacking-crane decisions

The AGV assignment problem decides which empty AGV carries out which job, but it should not decide which container the AGV will actually receive. Consider two empty AGVs a and b with waiting times for availability $w_a < w_b$. Moreover, consider two jobs i and j with the same stack as pick-up location and with due times $d_i < d_j$. Let us assume that the AGV assigment decision was to assign job i to AGV a and job j to AGV b. It may happen that AGV b arrives at the stack before a (a may have been delayed due to congestion on the layout). Now the stacking crane should put container i on AGV b because container i is more urgent (note that one could say that AGVs a and b switch their jobs).

The stacking-crane decisions (i.e., which container is to be moved next) is based on various goals and requirements such as high waterside and landside productivity, short empty travel times, AGVs, short waiting times for external trucks, etc. Considering the interface to the AGVs, we assume that the stacking cranes make use of rules analogous to those employed for the AGVs when deciding which AGV should receive which container. This means that the stacking cranes prefer containers with earlier due times (in addition to their further goals). This does not have an impact on the AGV assignment problem itself, but is important when testing the AGV assignment approach in a simulation study as will be done in Section 5.

4 Inventory-based approach

4.1 Basic idea

At each quay crane, there is a waiting buffer for AGVs, that is, an area in which arriving AGVs have to wait until the quay crane is ready to serve them. This buffer can be seen as a storage. In this analogy, the quay cranes are customers which have to be supplied with goods. These goods correspond to AGVs. A loading quay crane requires AGVs with containers to be loaded, while a discharging quay crane requires empty AGVs on which a discharged container can be put. Like in inventory management, the task is to make sure that no customer has to wait for goods, i.e., the inventory level should not be zero. On the other hand, the inventory level should not be too high. In our case, the latter is especially important because not only containers but also AGVs are tied up in stock. Hence, if queues become too long, there is a negative effect on the system's behavior because less transportation capacity is available.

Considering the AGV buffer as an inventory, we say that the inventory level of a quay crane is the number of AGVs in the buffer. Furthermore, the inventory level plus those AGVs on their way to the quay crane's buffer can be seen as the quay crane's net inventory level. To keep the analogy, we define a special net inventory level for our problem. We consider Q quay cranes $q = 1, \ldots, Q$. For each quay crane q, the inventory level for assignment decisions ila_q is defined as the number of AGVs that are busy with a job of a quay crane q and have not yet reached q. Furthermore, we denote the set of AGVs belonging to ila_q as ILA_q, that is, we have $ila_q = |ILA_q|$. Note that for a loading quay crane q, ILA_q consists of the AGVs that are either waiting in the buffer at q, transporting a container towards q, waiting for a container for q at a stack, or driving to a stack where a container for q is to be picked up. For a discharging quay crane p, ILA_p contains those empty AGVs that are either waiting in the buffer at p or driving towards p (observe that AGVs transporting a container picked up at p do not belong to ILA_p).

Considering the analogy described above, the basic idea for assigning AGVs to jobs can be stated as follows: Whenever an AGV a should get a new job, assign a to the first unassigned job of the quay crane q whose buffer is most probably empty when a would arrive at q. According to the analogy to inventory management, we choose the quay crane q with the smallest ila_q. In other words, the next job of that quay crane q for which ila_q is minimal is the most urgent job. One may also say that quay crane q is the most urgent quay crane to receive an AGV. A methodology to assign jobs to AGVs, which is based on this basic idea, will be presented in Section 4.2.

There is another motivation for this idea: If we want to reduce waiting times of AGVs at quay cranes, we have to shorten the waiting queues. By sending the AGV to quay crane with lowest ila_q, we select the shortest expected waiting queue for the AGV to queue into.

However, the inventory levels ila_q, as described above, are not yet suitable for directly comparing the current needs of the quay cranes for further AGVs with each other. Obviously, the time an AGV needs to arrive at the quay crane is much longer for loading quay cranes than for discharging ones. In the former case, it includes driving to the stacking crane, waiting for service, and driving to the quay crane, while in the latter case there is just a direct drive to the quay crane. Naturally, to reach the same supply level for all quay cranes (or, in other words, the same productivity), the inventory level of loading quay cranes must be higher than that of discharging ones. Therefore, we introduce a parameter ϕ called phase factor by which the inventory level of loading quay cranes must be higher. We consider adapted inventory levels for loading quay cranes q by defining $ila'_q = ila_q/\phi$. The inventory levels of discharging quay cranes are not modified, that is, we set $ila'_p = ila_p$ for each discharging quay crane p. The urgency with which a quay crane requires an AGV is now measured by inventory levels ila'_q for all quay cranes q.

So far, we have defined a quay crane q with inventory level ila'_q to be more urgent than that of a quay crane p if we have $ila'_q < ila'_p$. Finally, we consider quay cranes having the same inventory level, that is, $ila'_q = ila'_p$. To resolve such a tie, we define the quay crane for which the last AGV was started a longer time ago to be more urgent.

Note that ila'_q can further be modified to reflect operational issues in practice. One might wish to prioritize some quay crane q, e.g., if q has the longest remaining job list and must be accelerated to finish the vessel on time. This can be achieved by reducing ila'_q. This makes the jobs of quay crane q appear more urgent and, thus, leads to more AGVs for quay crane q. This should provide a higher productivity of q (although, of course, the productivity of the remaining quay cranes may decrease). This example shows the straightforward applicability of the inventory idea with respect to practical needs.

4.2 Assignment procedure

First, we determine all AGVs, say n, which are currently free or will be free within a short time as described in Section 2. Next, we find n jobs to be assigned to those available AGVs. At this point, we employ our basic idea as described in Section 4.1: The most urgent job is a job which belongs to the quay crane q, which has the lowest inventory level ila'_q. Among all those, we select a job all predecessors of which are already assigned to an AGV, are in transport, or are finished. By paying attention to the precedence relations when assigning AGVs to jobs, we reduce the risk of AGV waiting times at a quay crane that are caused by delayed predecessor containers. We then note the job just chosen as assigned, temporarily increase the corresponding ila'_q by one and, once again, determine the most urgent job based on the new inventory levels. This process loops until we have n jobs.

To assign the n jobs to the n AGVs, we create a standard linear assignment problem. The costs c_{ja} of assigning job j to AGV a consist of three components:

- An AGV a may have a current job that must be completed before it can start the next empty travel. The estimated waiting time for availability w_a obviously influences the duration until the next job j can be started as well as the duration until the AGV can arrive at the related quay crane. Note that w_a is zero if AGV a does not have a current job.
- According to the pick-up location of job j and the current position of AGV a, there is an expected empty travel time e_{ja} if j is assigned to a. This empty travel time affects the arrival of the AGV at the quay crane.
- We define $1 \leq o_j \leq n$ as the ordinal number of job j according to the order in which the jobs were chosen for assignment. That is, job j with $o_j = 1$ is the most urgent job with respect to the inventory levels ila'_q, job i with $o_i = 2$ is the second most urgent job and so on. (Note that o_j corresponds to the due time d_j in the due time based approach as both reflect the urgency of a job to receive an AGV.)

Now we define the cost as follows:

$$c_{ja} = (\lambda \cdot (n - o_j) + 1) \cdot (w_a + e_{ja})$$

The first part of the formula covers the job's urgency (λ is a weight to adjust the impact the job's urgency has on the costs). The urgency value of the least urgent job (ordinal number $o_j = n$) is 1. The next (more urgent) jobs have coefficients $1 + \lambda, 1 + 2 \cdot \lambda, 1 + 3 \cdot \lambda$ and so on. The second part of the formula reflects the estimated time that will pass until the container related to job j would be picked up if AGV a is assigned to job j. Multiplying both parts means that a very urgent job j (i.e., with a low o_j) and an AGV a that would need a long time to pick up the container related to job j have high assignment costs c_{ja}.

Having determined the costs c_{ja}, we solve the resulting assignment problem by the Hungarian method of (Kuhn 1955), designed as an executable in (Munkres 1957). This algorithm leads to an optimal assignment in terms of our objective to minimize the total assignment cost.

4.3 Implications for stacking-crane decisions

As already discussed in Section 3.3, stacking cranes are involved in the decision of which container to load on an AGV. Therefore, we describe a rule for loading containers which is, analogously to the assignment rule, based on net inventory levels.

We distinguish the loading decisions to be made when an AGV receives a container from a stacking crane, and those to be made when the AGV receives a container from a quay crane. In the latter case, the AGV simply receives an arbitrary container from the quay crane it is waiting at. In the former case, this decision is much more difficult: The stacking crane may have containers required by different quay cranes, thus, it has to decide which to pick first. To support the selection, we introduce a further inventory level. The inventory level for transport decisions ilt_q of a loading quay crane q is defined as the number of AGVs driving straight towards q after picking up a container for q at the stacking area. Additionally, we define the corresponding set of AGVs as ILT_q.

Then, we select the quay crane in a way similar to the assignment decision: We assume (see Section 3.3) a stacking crane to consider the quay crane q with the lowest ilt_q among all loading quay cranes having containers at the specific stacking crane as the most urgent quay crane. Again, we want to respect the precedence relations, namely only pick up containers whose predecessors are already picked up. However, it is possible that none of the containers to be loaded fulfills this precedence condition because we consider a subset of the containers. For example, it might occur that each container has at least one predecessor not picked up yet which stands at another stacking crane. Then, to prevent congestion as much as possible, we propose to start with strong requirement formulations and lower them step by step, if no container fulfills them. As soon as we find some containers, we select the one belonging to the most urgent quay crane.

Sending an AGV to a quay crane q with low ilt_q is motivated by reducing waiting times of quay cranes and AGVs. This idea directly corresponds to the one for selecting containers for the assignment process described in Section 4.2.

4.4 Enforcing dual cycles

An AGV's drive to the pick-up location is often necessary but worth avoiding if possible. It ties up the AGV capacity and, moreover, leads to more traffic in the terminal so the risk of congestion increases. Therefore, we provide a feature to be plugged into the decision process described so far.

A constellation of an AGV transporting a container to its destination and receiving a new job with a pick-up location equal to the previous job's delivery location is called a dual cycle. Dual cycles are possible only at stacks where quay cranes are either loading or discharging, which means they do not discharge a container immediately after loading another one in the same ship bay.

The assignment process described above arranges dual cycles only if there is a container with sufficient urgency at a stack where an available AGV is located. To suppress more empty drives, we take into account containers stored at a stack which would be ignored when creating the assignment problem in Section 4.2 because of a lack of urgency. Hence, we state an assignment rule as follows: If an AGV is available at a stack, it is assigned to the most urgent job located at this specific stack and whose predecessors already have been assigned or completed. As a result, we might assign a container which would not be considered by the basic method of Section 4.2 but offers a profitable dual cycle. This assignment process is executed right before the basic assignment process in Section 4.2. The jobs and AGVs assigned by this procedure are deleted from the corresponding sets. For the remaining AGVs, the assignment problem is created, solved, and evaluated as stated in Section 4.2.

Note that the AGV process in case of a dual cycle differs from the standard process only in that the empty travel to the pick-up location is actually a dummy drive-obviously, it takes no time because the last delivery location of the AGV corresponds to its next pick-up location. Afterwards, we decide which container to load on the AGV and select the most urgent one as described in Section 4.3. Therefore, we always arrange a dual cycle for the most urgent container of the specific stack (to be accurate, the AGV assignment procedure can only decide to leave the empty AGV at that stack, but we assume that the stacking crane scheduling selects the most urgent container with respect to the second inventory level ilt_q). Unfortunately, although this rule reduces empty travel times, it can also lead to undesirable effects which can be resolved as follows:

- As outlined in Section 4.1, we aim at inventory levels as similar as possible. By partially ignoring the urgency of jobs, we risk to disturb this balance. Therefore, we introduce two parameters $0 \leq \sigma, \tau \leq 1$ to prevent the balance from getting too much disturbed. Furthermore, we calculate the current minimum and maximum inventory levels among all quay cranes, that is, $ila_{all}^{min} = \min\{ila'_q \mid q = 1, \ldots, Q\}$ and $ila_{all}^{max} = \max\{ila'_q \mid q = 1, \ldots, Q\}$. Analogously, the current minimum and maximum inventory levels $ilt_{loading}^{min}$ and $ilt_{loading}^{max}$ among the loading quay cranes are calculated. We employ them to formulate two conditions for a dual cycle concerning a specific candidate job j and its quay crane's q_j inventory levels ila'_{q_j} and ilt_{q_j}:

$$ila'_{q_j} \quad \leq \quad (1-\tau) \cdot ila_{all}^{min} + \tau \cdot ila_{all}^{max} \tag{2}$$

$$ilt_{q_j} \quad \leq \quad (1-\sigma) \cdot ilt_{loading}^{min} + \sigma \cdot ilt_{loading}^{max} \tag{3}$$

Following these conditions, we only choose a container for a dual cycle if it belongs to one of the more urgent quay cranes.

- Dual cycles only support loading quay cranes by more efficient use of AGVs (the AGV driving time for loading quay cranes is shortened on the average). Moreover, the dual cycle approach assigns AGVs to loading quay cranes that otherwise might have been assigned to discharging ones. Again, this disturbs the balance between loading and discharging quay cranes. Hence, we have to adapt the phase factor ϕ described in Section 4.1 to readjust that balance.

5 Simulation study

To compare and evaluate the two assignment approaches given in Sections 3 and 4, we developed a simulation model. In the following, we give some details of the simulation model, summarize the parameters employed, and, finally, discuss the results.

5.1 Model

According to our problem setting, we identify three substantial material flow components of the considered container terminal configuration (for a sketch of the terminal layout in the simulation model, we refer again to Fig. 1).

- Quay cranes load containers onto a vessel or discharge them from it. We can look at their life cycle as an endless loop of either waiting for AGVs or handling containers. When a quay crane holds a container to set down on an AGV or waits for a container to load on the vessel, it has to wait until an AGV arrives at the quay crane. After a quay crane's interaction with an AGV, it either transports the container onto the vessel (if loading) or picks the next container from it (if discharging). To characterize the quay crane's behaviour, we employ three distributions: hand-over time for AGVs to be loaded with discharged containers, hand-over time to get containers from AGVs to load them on a vessel, and the time the quay crane requires before it is ready for the next hand-over. The former two contain the processes of picking the container and lifting it up to a height that allows the AGV to leave (if loading) and putting the container down and releasing it (if discharging), respectively. The latter includes the container's travel to or from the vessel.

- Stacking cranes manage the stacking area and, therefore, receive containers from AGVs after they were discharged from vessels. Additionally, stacking cranes provide containers for AGVs to be loaded onto vessels. Both processes are modeled by distributions for the transfer times, that is, the times the AGVs have to wait at the stacks. As the behaviour of the stacking cranes is not modelled explicitly, these distributions implicitly contain all other activities such as shuffling containers and serving the landside.
- AGVs transport containers from quay cranes to the stacking area and vice versa. Their only activity to be modeled is driving. Therefore, a distribution for the driving time from each possible starting position to each possible destination position is registered in the model. These distributions cover interferences of AGVs on the layout, especially congestion. Moreover, for hand-over at the quay cranes and stacking cranes, an estimated availability time for an AGV is generated. Both the time at which the estimate is generated in advance and the error of the estimate (i.e., deviation from actual availability time) are controlled by distributions.

The simulation model has been implemented in Desmo-J, a discrete event-based simulation framework in Java (see Page et al. (2000)). A more detailed presentation of the simulation model can be found in Briskorn and Hartmann (2006).

5.2 Experimental design

To evaluate our approach, we compare four different methods to assign jobs to AGVs. First, we implemented the greedy heuristic described in Section 3, which we will refer to as "dueTimePrio." Our own approach, which was described in Section 4, was realized both with ("invDualCycle") and without forcing dual cycles ("inv"). Because we want to get results concerning the different methods to select containers for assignment, namely, the due-time-based rule and the inventory-based idea, we have to eliminate effects caused by different assignment methods. We achieve this by using the Hungarian method for assigning containers selected by the due-time idea in a fourth method, "dueTimeHung."

We apply these approaches to scenarios that differ by the structure of the containers' precedence relations. Varying this structure gives a hint about the capability of an approach because the structure defines the degrees of freedom which are left for it. Obviously, precedence relations between containers i and j can solely exist if i and j belong to the same quay crane. We considered five structures of precedence relations:

- The lowest requirement level is given in a scenario without precedence relations. The approaches can randomly choose containers to load or discharge when available.
- The strongest requirement level is given by "linear" precedence relations between the containers of each quay crane. Then, at each point of time there is just a single container for each quay crane, which can be loaded or discharged.

- In addition, we have three settings with partial precedence relations. They are different with respect to the number of precedence relations per job, leading to scenarios with "many," "medium," and "few" precedence relations per job.

In each scenario there are 20 stacking cranes and 40 AGVs. Ten quay cranes, of which five are loading and five are discharging, are randomly distributed on the 20 possible positions. We created 60 jobs per hour and quay crane. Note that this roughly corresponds to the maximum technical productivity of a quay crane. This way, the actual throughput results from the AGV dispatching strategy under consideration. The distributions for the material flow behavior mentioned in Section 5.1 were taken from statistics of the Container Terminal Altenwerder. The original statistics were modified for reasons of confidentiality, but the resulting distributions still allow for a realistic simulation.

For the simulation runs, we identify four goals resulting from the discussion in Section 2. We use them to compare the approaches:

- Increasing the container terminal's waterside productivity, i.e., the number of containers loaded onto and discharged from vessels per hour, is the main goal of our approach.
- Waiting times of quay cranes increase the time of the vessels in port. Hence, we want to reduce them.
- Waiting times of AGVs tie up capacity without having any positive effect on the system's productivity, so we want to reduce them.
- Empty travel times should be shortened because, like waiting times, they tie up capacity without supporting the main goal. Besides, they increase traffic on the AGV layout and, therefore, the probability of congestion.

We carried out two series of simulation runs. In preliminary experiments, we tested a broad variety of values for each parameter while fixing others. After evaluating these runs, we fixed all parameters to their best settings for further experiments. Tables 1 and 2 give the fixed values of essential parameters. Note that phase factor ϕ has to be adapted according to Section 4.4 when dual cycles are forced.

For each approach, we performed 100 simulation runs with a simulation time of 11 h per run, which were preceded by 2 h to get the system in balance and followed by 2 h to make sure containers were not running out in the period to be evaluated. Solely, the period of 11 h is evaluated by means of statistics.

Table 1 Parameters for due time approach

Parameter	Symbol	Value
Earliness weight	α_E	1
Tardiness weight	α_T	7.5
Empty driving weight	α_e	1

Table 2 Parameters for inventory approach

Parameter	Symbol	Value
Phase factor	ϕ	1.6
Cost step	λ	3
Dual cycle	τ	1
Dual cycle	σ	0.5

5.3 Comparison of the approaches

In the following, we present the results of the simulation runs, taking into account our four approaches and five different scenarios.

Table 3 gives an overview of the productivity resulting from the different approaches. The productivity is measured as the average number of containers loaded or discharged per hour and quay crane. Although we used neither the original approach employed at the Container Terminal Altenwerder nor the original statistics, we cannot give absolute productivity figures in this paper to avoid misinterpretations. Therefore, the results are given as relative figures. We selected "dueTimePrio," the simplest method in our study, as a base and set its productivity index to 1.0 for each of the five scenarios. The productivity resulting from the other methods are given relative to those of "dueTimePrio" (e.g., 1.015 of "dueTimeHung" for the "medium" scenario indicates a productivity improvement of 1.5% over "dueTimePrio").

One can observe that productivity using "dueTimeHung" is slightly higher in each scenario than when "dueTimePrio" is applied. Remember that these approaches only differ in the algorithm, not in how the most urgent jobs are determined or how job assignments are evaluated. The results show that the Hungarian method is better suited than the greedy heuristic, although the productivity is increased only by 1.0–1.8%. Furthermore, "inv" reaches a higher productivity than "dueTimeHung". These two approaches make use of the same algorithm (i.e., the Hungarian method) but employ different problem formulations. Therefore, we can say that the inventory-based concept is more promising than the due-date approach. In particular, we can see that the improvement due to the inventory concept is higher than the improvement that can be obtained from using an optimal algorithm in the due-time-based model. When comparing "inv" and "invDualCycle", we observe that using the option to enforce dual cycles in the inventory-based approach seems to be extremely promising. Also, note that the

Table 3 Quay crane productivity

Precedence relations	dueTimePrio	dueTimeHung	Inv	invDualCycle
Linear	1	1.010	1.050	1.049
Many	1	1.014	1.046	1.059
Medium	1	1.015	1.045	1.183
Few	1	1.014	1.075	1.229
Without	1	1.018	1.047	1.190

superiority of the dual cycle approach further increases, if there are less precedence relations.

Table 4 gives an impression of the influence the approaches have on the total empty travel time of AGVs. Again, the Hungarian method in "dueTimeHung" is superior to the simple priority rule in "dueTimePrio". The inventory-based approach leads to smaller empty travel times than the due-time-based approach. Obviously, enforcing dual cycles strongly reduces empty driving times. The effect of dual cycles on the empty travel times increases with decreasing number of precedence relations. This is because less precedence relations make it more likely to fulfill the conditions for arranging dual cycles on a higher requirement level (see Section 4.3), which will reduce congestion in front of the quay crane.

Table 5 shows the waiting times of the AGVs in the buffer at the quay crane. Recall that AGVs have to wait in this buffer, if more AGVs than the quay crane can handle have been assigned to this quay crane, or if AGVs have to wait for delayed predecessors. We can see that the inventory-based approach reduces waiting times of AGVs significantly. If the dual cycle extension is considered, the waiting times of the AGVs are higher than otherwise. The latter results from the drawback discussed in Section 4.4: By enforcing dual cycles, we partially ignore the urgency of containers. Therefore, it becomes more likely that we send AGVs to quay cranes with higher ila_q. Hence, AGV queues get longer and waiting times in the buffer increase.

The waiting times of the quay cranes are given in Table 6. Again, the inventory-based idea leads to better results than the due-time approach. Enforcing dual cycles reduces the quay crane waiting times even further.

5.4 Impact of the look-ahead

As described in Section 2, the assignment procedures consider the AGVs that are currently free and those that will soon be available. This implies that we have a certain look-ahead which gives us more degrees of freedom for finding good assignments. On the other hand, considering AGVs that are not available yet means that we have to take estimated availability times into account, and the quality of such estimates is not so good in practice. Thus, to validate the look-ahead approach, we compare it with a version without look-ahead that considers only AGVs that are currently available.

The results can be found in Table 7. We compare the inventory-based approach with and without look-ahead and report relative productivity (with the version without look-ahead being the base). The version with look-ahead leads to 10%

Table 4 Empty travel times of AGVs

Precedence relations	dueTimePrio	dueTimeHung	Inv	invDualCycle
Linear	1	0.955	0.906	0.876
Many	1	0.951	0.906	0.814
Medium	1	0.943	0.924	0.580
Few	1	0.952	0.914	0.559
Without	1	0.950	0.919	0.529

Table 5 AGV waiting times in buffer at quay crane

Precedence relations	dueTimePrio	dueTimeHung	inv	invDualCycle
Linear	1	1.032	0.860	0.944
Many	1	1.027	0.881	1.036
Medium	1	1.075	0.666	0.696
Few	1	1.039	0.911	0.964
Without	1	1.007	0.601	1.052

higher productivity. This confirms that having more degrees of freedom for optimization is more important than the often low quality of the availability time estimates.

Finally, it should be mentioned that the inventory approach with look-ahead considers 4.6 AGVs on average when executing the assignment procedure. In the version without look-ahead, however, usually only one AGV is considered in cases of high workload (which are simulated in this case and which are most important in practice). This is because the assignment procedure is executed whenever an AGV has completed its last job (cf Section 2). In case of high workload, all other AGVs are busy, hence, that AGV is the only one free when the assignment procedure is executed.

5.5 Impact of precedence relations

Finally, we have a brief look at the impact of the precedence relations on the productivity, empty travel times, waiting times of AGVs in the quay crane buffer, and quay crane (QC) waiting times for AGVs. The results are displayed in Table 8. We consider only the greedy priority-rule-based heuristic for the due date approach (dueTimePrio), which has been the benchmark in our study. As in the previous tables, we give relative results. Here, we have selected the linear precedence relations as a basis for the comparison.

We observe a significant influence of the precedence relations' density on the results. In particular, having less precedence relations leads to higher productivity. If we have no precedence relations at all, the productivity (with the same heuristic) is 11.8% higher compared to the case of linear precedence relations. This is because less precedence relations make it less likely that an AGV has to wait for a delayed predecessor in the buffer at a loading quay crane. This is confirmed by Table 8,

Table 6 Quay crane waiting times for AGVs

Precedence relations	dueTimePrio	dueTimeHung	Inv	invDualCycle
Linear	1	1.032	0.860	0.944
Many	1	0.990	0.974	0.955
Medium	1	0.982	0.957	0.800
Few	1	0.990	0.962	0.837
Without	1	0.987	0.977	0.838

Table 7 Impact of look-ahead on productivity (method: inv)

Precedence relations	Only free AGVs	Free and soon available AGVs
Linear	1	1.090
Many	1	1.097
Medium	1	1.099
Few	1	1.095
Without	1	1.107

which shows that the AGV waiting times in the quay crane buffer decrease drastically when we have less precedence relations.

5.6 Computation times

We close this section with a brief look at the times required to compute one assignment. In both the due time and the inventory approach, the average computation time for one execution of the Hungarian method has been below 0.001 s. The maximum computation time for one execution has been 0.016 s. The experiments were carried out on an Athlon XP 2200+ computer with 512 MB RAM. These computation times show that the approaches presented in this paper are well suited for application in a terminal control system, which requires decisions in real time.

6 Conclusions and outlook

In this paper, we proposed an approach to schedule container transports between quay cranes and the stacking area. We captured the problem of assigning transportation jobs to AGVs by introducing a concept related to inventory management. The essential idea is to assign an AGV to a job that belongs to a quay crane to which a relatively small number of AGVs is currently assigned. This problem formulation was compared to a more traditional formulation that is based on due times for the jobs and an earliness–tardiness objective. Both formulations differ only in how the jobs to be considered are determined and in the way the assignment costs of jobs to AGVs are calculated, but not in the underlying mathematical structure.

Table 8 Impact of precedence relations (method: dueTimePrio)

Precedence relations	Productivity	Empty travel	AGV waiting	QC waiting
Linear	1	1	1	1
Many	1.035	1.000	0.744	0.982
Medium	1.065	0.981	0.370	0.973
Few	1.071	0.993	0.256	0.964
Without	1.118	0.989	0.242	0.943

In a simulation study, we found that the problem formulation has an impact on the resulting terminal productivity. Even when both problem formulations are solved with the same algorithm (the well-known Hungarian method), the inventory-based concept outperformed the due-time-based approach with respect to waterside productivity (although only by a few percent). At first glance, the due-time approach seems to allow for more precise scheduling because it accurately plans events and durations on the terminal. However, our results indicate that the bad time estimates, which are common in practice (and which were considered in our simulation model in a realistic way), lead to suboptimal decisions in the due-time approach and, thus, to lower productivity. The inventory-based approach avoids the use of estimated times to a large extent. Hence, it appears to be more robust and, thus, better suited for application in practice. Besides, it leads to a simpler terminal control system because frequent updates of times are not necessary.

Additionally, we introduced a feature to enforce dual cycles of AGVs at stacks (that is, a stacking crane unloads a container from the AGV and puts another on the AGV). This allows reduction of the empty travel times of the AGVs and, as shown by our results, leads to higher waterside productivity.

Furthermore, we analyzed the impact of the precedence relations both on the productivity and on the performance of the different approaches. Less precedence relations between containers to be loaded onto vessels lead to higher productivity. This is due to more degrees of freedom for the AGVs, that is, in case of fewer precedence relations, AGVs can directly proceed to the quay crane without having to wait for a delayed predecessor to pass. Moreover, the additional productivity gain of the dual cycle extension increased with a decreasing number of precedence relations.

Considering the good results of the inventory-based concept for AGV dispatching, an objective of further research should be the application of this approach to other types of equipment for container handling. In particular, inventory-based optimization would be promising for stacking cranes and straddle carriers. In both cases, the inventory idea would have to be adapted to reflect the specific requirements of those types of equipment.

References

Bae JW, Kim KH (2000) A pooled dispatching strategy for automated guided vehicles in port container terminals. Int J Manag Sci 6:47–70

Baker C (1999) Altenwerder—the details. Port Dev Int (07/08):24–25

Bish EK, Chen FY, Leong YT, Nelson BL, Ng JWC, Simchi-Levy D (2005) Dispatching vehicles in a mega container terminal. OR Spectrum 27:491–506

Böse J, Reiners T, Steenken D, Voß S (2000) Vehicle dispatching at seaport container terminals using evolutionary algorithms. In: Sprague RH (ed) Proceedings of the 33rd Annual Hawaii International Conference on System Sciences. IEEE, Piscataway, pp 377–388

Briskorn D, Hartmann S (2006) Simulating dispatching strategies for automated container. GOR Proceedings 2005 (in press)

Daganzo CF (1989) The crane scheduling problem. Transp Res B 23:159–175

de Castilho B, Daganzo CF (1993) Handling strategies for import containers at marine terminals. Transp Res B 27:151–166
de Koster RBM, Le-Anh T, van der Meer JR (2004) Testing and classifying vehicle dispatching rules in three real-world settings. J Oper Manag 22:369–386
Grunow M, Günther H-O, Lehmann M (2004) Dispatching multi-load AGVs in highly automated seaport container terminals. OR Spectrum 26:211–235
Guan Y, Cheung RK (2004) The berth allocation problem: models and solution methods. OR Spectrum 26:75–92
Hartmann S (2004) A general framework for scheduling equipment and manpower at container terminals. OR Spectrum 26:51–74
Hwang H, Kim SH (1998) Development of dispatching rules for automated guided vehicle systems. J Manuf Syst 17:137–143
Imai A, Nagaiwa K, Tat CW (1997) Efficient planning of berth allocation for container terminals in Asia. J Adv Transp 31:75–94
Imai A, Nishimura E, Papadimitriou S (2001) The dynamic berth allocation problem for a container port. Transp Res B 35:401–417
Kim KH, Kim HB (1999a) Segregating space allocation models for container inventories in port container terminals. Int J Prod Econ 59:415–423
Kim KY, Kim KH (1999b) A routing algorithm for a single straddle carrier to load export containers onto a container ship. Int J Prod Econ 59:425–433
Kim KH, Park YM, Ryu K-R (2000) Deriving decision rules to locate export containers in container yards. Eur J Oper Res 124:89–101
Kim KH, Won SH, Lim JK, Takahashi T (2001) A simulation-based test-bed for a control software in automated container terminals. Proceedings of the international conference on computers and industrial engineering, Montreal, Canada, pp 239–243
Kuhn HW (1955) The Hungarian method for the assignment problem. Nav Res Logist Q 2:83–97
Lim A (1998) The berth planning problem. Oper Res Lett 22:105–110
Meersmans PJM, Dekker R (2001) Operations research supports container handling. Technical report EI 2001-22. Econometric Institute, Erasmus University, Rotterdam
Meersmans PJM, Wagelmans APM (2001) Effective algorithms for integrated scheduling of handling equipment at automated container terminals. Technical report EI 2001-19. Econometric Institute, Erasmus University, Rotterdam
Munkres JR (1957) Algorithms for the assignment and transportation problems. J Soc Ind Appl Math 5:32–38
Page B, Lechler T, Claassen S (eds) (2000) Objektorientierte simulation in Java mit dem Framework DESMO-J. BoD GmbH, Norderstedt
Park Y-M, Kim KH (2003) A scheduling method for berth and quai cranes. OR Spectrum 25: 1–23
Peterkofsky RI, Daganzo CF (1990) A branch and bound solution method for the crane scheduling problem. Transp Res B 24:159–172
Steenken D, Henning A, Freigang S, Voß S (1993) Routing of straddle carriers at a container terminal with the special aspect of internal moves. OR Spectrum 15:167–172
Steenken D, Voß S, Stahlbock R (2004) Container terminal operations and operations research—a classification and literature review. OR Spectrum 26:3–49
Taleb-Ibrahimi M, de Castilho B, Daganzo CF (1993) Storage space vs. handling work in container terminals. Transp Res B 27:13–32
Vis IFA (2006) Survey of research in the design and control of automated guided vehicle systems. Eur J Oper Res 170:677–709
Vis IFA, de Koster R (2003) Transshipment of containers at a container terminal: an overview. Eur J Oper Res 147:1–16
Vis IFA, Harika I (2004) A comparison of vehicle types for automated container terminals. OR Spectrum 26:117–143
Zhang C, Liu J, Wan Y-W, Murty KG, Linn RJ (2001) Storage space allocation in container terminals. Technical report, Hong Kong University of Science and Technology

Matthias Lehmann · Martin Grunow · Hans-Otto Günther

Deadlock handling for real-time control of AGVs at automated container terminals

Abstract In automated container terminals, situations occur where quay cranes, stacking cranes, and automated guided vehicles (AGVs), directly or indirectly request each other to start a specific process. Hence, all of the affected resources are blocked, possibly leading to the complete deadlock of individual cranes or AGVs. Particularly, AGVs are liable to deadlocks because they always need a secondary resource, either a quay crane or a stacking crane, to perform the pick-up and drop-off operations. Because usually no buffering of containers takes place at the interfaces between AGVs and cranes, the consequences of deadlocks are rather severe. Two different methods for the detection of deadlocks are presented. One is based on a matrix representation of the terminal system. The other directly traces the requests for the individual resources. To resolve deadlock situations arising in an automated container terminal, three different procedures are proposed. These procedures aim to modify the sequence of handling operations or to assign them to alternative resources so that conflicts between concurrent processes are resolved. The suitability of the concept is demonstrated in an extensive simulation study.

Keywords AGV dispatching · Container terminals · Deadlock detection and resolution

---

M. Lehmann
Department of Production Management, Technical University Berlin, Wilmersdorfer Str. 148, D-10585 Berlin, Germany E-mail: Matthias.Lehmann@tu-berlin.de

M. Grunow
Department of Manufacturing Engineering and Management, Technical University of Denmark, Building 423, 2800 Kgs., Lyngby, Denmark E-mail: grunow@ipl.dtu.dk

H.-O. Günther (✉)
Department of Production Management, Technical University Berlin, Wilmersdorfer Str. 148, D-10585 Berlin, Germany
E-mail: Hans-Otto.guenther@tu-berlin.de

1 Introduction

According to Kim et al. (1997), a deadlock can be defined as "a situation where one or more concurrent processes in a system are blocked forever because the requests for resources by the processes can never be satisfied". Deadlocks result from decentralized planning, which is the only realistic mode to govern large-scale logistics systems with highly dynamic interactions and incomplete knowledge about future events. To meet online requirements for logistics control of container terminals, the decomposition of the entire logistics control system into various modules for the different types or groups of resources is inevitable.

Whereas deadlocks in manufacturing systems have been investigated by many researchers (see, e.g., Egbelu and Tanchoco 1984; Lim et al. 2003; de Koster et al. 2004; Le-Anh and de Koster 2005), deadlocks in highly automated seaport container terminals have hardly gained attention until now. This paper focuses on deadlocks occurring in automated seaport container terminals, where quay cranes unload containers from vessels and place them on automated guided vehicles (AGVs). Containers are then transported to the storage area where they are collected by stacking cranes. For export containers, these operations are performed in the reverse order (for a detailed description of the handling and transportation processes in container terminals, cf. Steenken et al. 2004). Unlike the situation in manufacturing systems, there is no buffer between the cranes and the AGVs, which makes these container terminals susceptible to deadlocks. In this paper, we alternatively consider terminal configurations with single-load and dual-load AGVs. The latter vehicle type may transport up to two standard 20-ft containers at a time.

Deadlocks usually occur in the execution phase of the logistics processes; for instance, if a request is launched by the control system of a specific resource, e.g., an AGV, for another resource, e.g., a quay crane, which is scheduled to unload a container from the AGV. However, because of the dynamic interactions within the terminal, the random variations in the processing times, and the concurrency of a large number of different processes, the quay crane may also request the same AGV for loading another container onto that AGV. A typical deadlock situation as shown in Fig. 1 occurs, when neither quay crane nor AGV can proceed with their actual operation, especially because no buffers are available to decouple the mutually dependent operations (note that in the example, both containers are 40-ft containers, such that the AGV can carry only one of them at a time). Hence, in

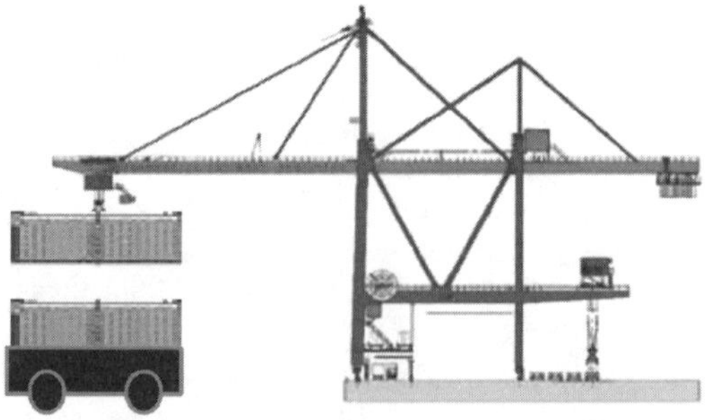

Fig. 1 A typical deadlock situation in a container terminal

automated seaport container terminals, the interactions between quay cranes, stacking cranes, and AGVs has to be monitored carefully. These monitoring systems include, as a major constituent, adequate deadlock handling procedures.

One major cause for a deadlock to arise is discrepancies between the schedules of quay cranes, stacking cranes, and AGVs. In Fig. 2, the planning steps to get from the stowage plan of a vessel to the AGV schedules are briefly summarized. Note that the crane operation sequences for quay and stacking crane are determined independently before the AGV schedule is determined and cannot be changed in the course of the AGV dispatching. The AGV is merely considered a service device with the aim to meet the target times resulting from quay and stacking crane schedules. The entire procedure consists of the following steps:

1. Determination of the quay crane sequences: In the first step, for each quay crane the sequence of the individual operations is derived from the stowage plan of the vessel. In the example, quay crane Q2 has to perform first the pick-up of container 3.
2. Determination of the quay crane schedules: The quay crane sequences are converted into quay crane schedules in the next step, considering average handling times (in the example, handling times of 1 time unit for quay cranes and 3 time units for stacking cranes are assumed). Some quay cranes may not yet be ready for service, e.g., because they need to be repositioned first (in the example, Q2 becomes ready for service at t_0=2).
3. Determination of the stacking crane sequences: Once the quay cranes have been scheduled, the earliest drop-off times of the containers at the stacking cranes (or the related pick-up times in case of an export container) can be calculated. Because the storage location for each container has been determined in advance, these times can be calculated by adding (subtracting) the average travel times (given by a time matrix) to the times of the quay crane schedules. The resulting times are used to determine the sequence of the stacking crane operations.
4. Determination of the stacking crane schedules: Considering the handling times at the stacking cranes, the corresponding schedules are derived. To obtain a feasible schedule, some operations may have to be right-shifted to avoid overlapping. The shift operations have been adapted to the characteristics of the problem at hand in the following manner: If after step 3 a pick-up operation and a drop-off operation overlap in a way that the pick-up would start while a drop-off is performed, then the pick-up is scheduled *before* the drop-off (even though it would have been started later). Thus, at the stacking cranes, pick-up operations are prioritized over drop-off operations. The reason for this is that drop-offs are the last operations for each container, while pick-ups are the first operation and a delay would also hold up the drop-off at the quay crane.
5. AGV dispatching: The schedules of quay cranes and stacking cranes are the inputs to AGV dispatching. The AGV system is considered as the service resource with the objective of performing all necessary transportation operations while minimizing the late arrival of AGVs, especially at the quay cranes. In the dispatching step, orders are assigned to AGVs and combined to an AGV tour. In the example, orders 1 and 4 are assigned to AGV V1. Note that, because of the individual sequence, operations may be delayed again compared to the crane schedules. For example, the drop-off of container 4 is scheduled at

time 4:35 for stacking crane S2. However, according to the schedule of AGV V1, container 4 has not even been picked up by that time. These discrepancies arise because of the hierarchical determination of the schedules of cranes and AGVs. The schedules of the cranes are not adapted to the results of the AGV dispatching.

This planning procedure roughly models the requirements of the container terminal practice where a large number of additional side constraints need to be considered. However, these aspects are not relevant for our study, which focuses on deadlocks caused by AGV dispatching.

As stated before, the discrepancies between the crane and AGV schedules may finally lead to a deadlock situation. In our simulation study (cf. Grunow et al. 2006), we detected the occurrence of deadlocks, which in some cases caused the complete standstill of one or more quay cranes—the worst possible scenario for a container terminal. Therefore, in a simulation study as well as in a real logistics control system, it is important to take deadlocks into consideration, because they do arise and impair the performance of the AGV system significantly.

To illustrate the development of a deadlock, consider the situation for the schedules derived in Fig. 2. The corresponding section from a terminal configuration with quay cranes Q1 and Q2, stacking cranes S1 and S2, and vehicles V1 and V2 is outlined in Fig. 3.

The distances (travel times) between quay and stacking cranes are indicated in Fig. 3a by the time matrix. Operations are encoded by their type ("p" for pick-up and "d" for drop-off) and the corresponding transportation order, the scheduled

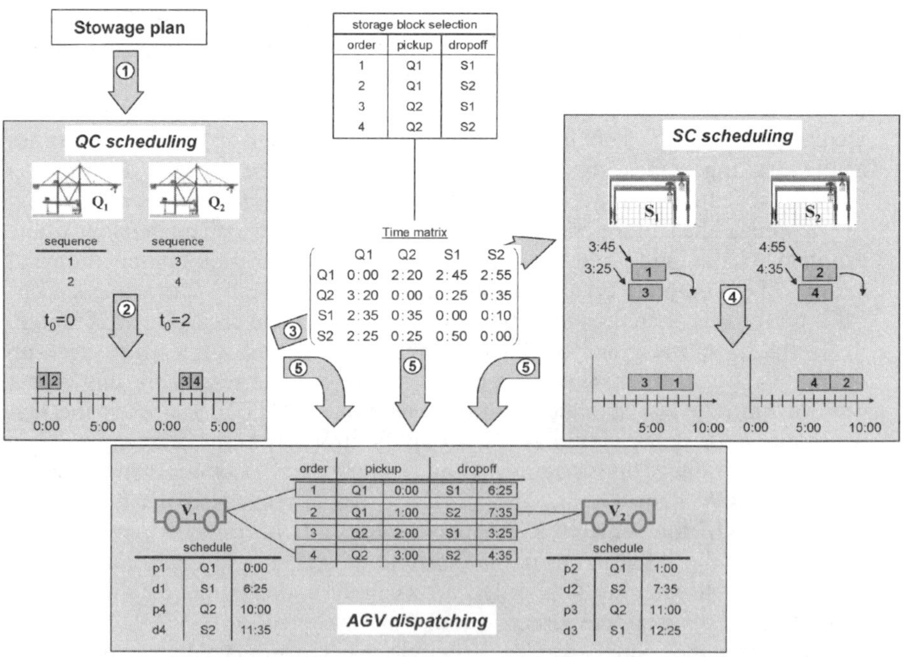

Fig. 2 Main planning steps to get from the stowage plan to an AGV schedule

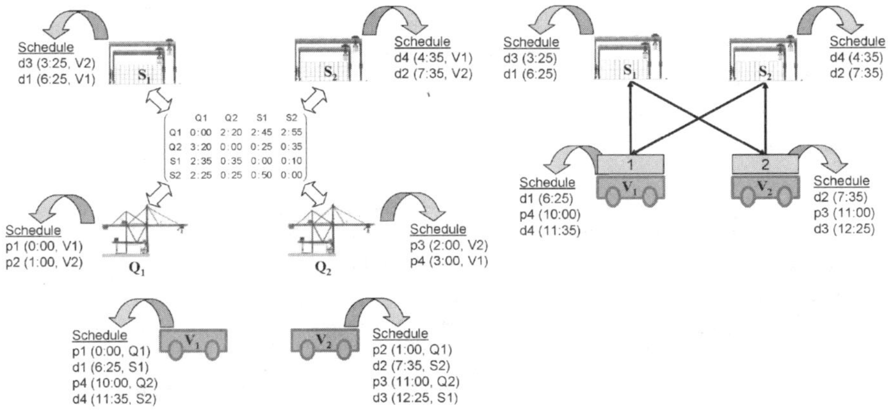

a Situation at time t = 0:00 b Situation at time t = 2:00 (deadlock)

Fig. 3 Development of a deadlock situation

starting time, and the secondary resource, i.e., the vehicle that transports the container or the crane that loads or unloads the container. For example, the first operation "d3 (3:25, V2)" in the schedule of stacking crane S1 indicates that S1 plans to start serving vehicle V2 for performing the drop-off operation d3 of container (transportation order) 3 at time t=3:25. For the handling times, 1 time unit is assumed for quay cranes and 3 time units is assumed for stacking cranes. Consider the situation at time t=0:00 depicted in Fig. 3a.

The current schedule of vehicle V1 at this point in time is:

1. t=0:00: start the pick-up of container 1 at Q1 (the pick-up is finished at t=1:00) and travel to S1 (arrival at t=3:45)
2. t=6:25: drop off container 1 and travel to Q2 (arrival at t=10:00)
3. t=10:00: pick up container 4 and travel to S2 (arrival at t=11:35)
4. t=11:35: drop off container 4

The schedules of vehicle 2, quay cranes Q1 and Q2, and stacking cranes S1 and S2 can be interpreted accordingly.

Figure 3b shows the situation at time t=2:00 after the first pick-up operations p1 and p2 have been completed by vehicles V1 and V2, respectively. The next operation of vehicle V1 is to drop off container 1. To carry out this operation, stacking crane S1 is required. However, stacking crane S1 waits for vehicle V2, which waits for stacking crane S2, which in turn waits for vehicle V1. Hence, a deadlock is given.

This paper shows how to detect deadlocks and how to resolve them in a container terminal application. This paper is closely linked to the paper on AGV dispatching strategies by Grunow et al. (2006). The following section provides an overview of different types of deadlocks and corresponding handling techniques. Next, in Section 3, two different deadlock detection methods are introduced. In Section 4, three different procedures are proposed to resolve deadlock situations arising between handling units in an automated container terminal. These procedures aim to modify the sequence of handling operations or to assign them to alternate resources so that no conflicts between concurrent processes occur.

2 Classification of deadlocks and deadlock handling techniques

The deadlock problem is—as stated by Coffman et al. (1971)—a logical problem that may arise in different contexts. Hence, before developing a deadlock-handling strategy for a specific problem, it is useful to take a closer look at the general characteristics of deadlocks. According to Coffman et al. (1971) and Liu and Hung (2001), for a deadlock to occur, four conditions must hold simultaneously:

1. "Mutual exclusion" condition: no resource can be shared by more than one task.
2. "Wait for" condition: tasks hold resources already allocated to them while waiting for additional resources.
3. "No preemption" condition: resources are not accessible until they are released by the task using them.
4. "Circular wait" condition: a circular chain exists, such that each task holds one or more resources that are being requested by the next task in the chain.

Each deadlock handling approach aims at breaking at least one of these conditions or ensuring that all of them are never fulfilled at the same time. Because the first condition holds for most systems, research is focussed on the three latter ones. The usual way to avoid the second condition is to require that a task must request all resources needed for its completion at once. A prominent example for this strategy is the use of the *Banker's algorithm* as proposed, for example, by Kim et al. (1997). Condition 3 can be overcome if a task is able to release all currently allocated resources as soon as its requests for further resources are denied. However, this possibility is only given in very few systems. Finally, a circular wait as stated in the fourth condition can be avoided by imposing a linear ordering of resources. A task is then only allowed to request resources following the currently allocated ones in the ordering.

In general, deadlocks can be addressed by three approaches: prevention, avoidance, and detection and resolution (cf. Venkatesh and Smith 2003; Liu and Hung 2001).

1. *Deadlock prevention* is an offline approach, aiming at the complete avoidance of any situation that may lead to a deadlock. This is usually accomplished by establishing a set of generic rules ensuring that the four necessary conditions for deadlocks cannot be simultaneously satisfied.
2. *Deadlock avoidance* dynamically allocates the system resources by using a suitable online control policy, seeking to avoid deadlock situations resulting from the next event in the operation of the handling system.
3. The third approach of *deadlock detection and resolution* does not attempt to prevent deadlocks in advance, which is usually impossible in a highly dynamic application environment. Instead, deadlocks are detected as soon as possible and then resolved.

In AGV systems employed in automated container terminals, two different types of deadlock may occur. They can come about within the AGV system or through the interaction between the AGV system and other equipment units (e.g., cranes) of the material handling system. The first type of deadlock concerns the routing of AGVs. Deadlocks may arise between two or more vehicles blocking each other in the guide path because they enter a segment of the guide path from opposite directions. Apparently, all of the four conditions stated above are fulfilled: two AGVs usually may not share a guide path segment (condition 1), an AGV blocks a segment while waiting for the next to be free (condition 2), a segment is only released when the AGV has completely left it (condition 3) and a circle of AGVs waiting for each other can occur (condition 4). It is easy to see that for the problem at hand the first three conditions are satisfied permanently. So, only the fourth condition can be avoided. A simple deadlock prevention strategy is to design the guide path as a single loop, thus making the formation of a circular wait impossible. In most seaport container terminals, a loop-like guide path is employed, which makes a mutual blocking of AGVs nearly impossible. However, in the case of a more complex guide path, other deadlock handling techniques must be employed.

In the academic literature, deadlock avoidance approaches for vehicle routing problems have been proposed. Moorthy et al. (2003), as well as Yeh and Yeh (1998), used a graph-based heuristic to predict a deadlock. If a vehicle would create a circular wait by entering a guide path segment, it has to wait until the situation has changed or a rerouting must be performed. Wu and Zeng (2002) propose a Petri-net-based approach to predict (and avoid) possible deadlocks. In Fanti (2002), digraphs are used to control the path assignment to the vehicles and their moves in the system. Finally, in Reveliotis (2000), an adaptation of the *Banker's algorithm* for robust AGV conflict resolution is suggested. Whereas the former approaches clearly aim at deadlock avoidance, the last one can as well be considered a quite flexible deadlock prevention algorithm. Deadlocks within the AGV system have to be avoided by adequate procedures embedded into the routing and traffic control software supplied by the provider of the AGV system. Routing and traffic control, however, are outside the scope of this paper. Hence, we do not consider these types of deadlocks in the sequel.

The second important category of deadlocks is caused through the interaction between the AGV system and other handling equipment. Such deadlocks may arise in situations where resources are waiting for requests from each other. This type of deadlock has been observed especially in the operation of flexible manufacturing systems (e.g., Kim and Kim 1997; Venkatesh and Smith 2003). Equipment may also be blocked in this manner in a container terminal. Particularly, AGVs are liable to this kind of deadlock because they always require a secondary resource, either a quay crane or a stacking crane, to perform the pick-up and drop-off operations. Because usually no buffering of containers takes place at the interface between the AGVs and the cranes, the consequences of deadlocks are rather severe. Again, all four deadlock conditions are satisfied: a crane or an AGV cannot perform more than one operation ("mutual exclusion"), a resource is blocked while it is waiting for another resource to complete the task ("wait for") as in Fig. 1, a resource can only be released by its current task ("no preemption") and circular wait relations may occur as in Fig. 3b. To our knowledge, however, no paper exists that focuses on deadlock problems in container terminals caused by the interaction of the AGV system with the quay and stacking cranes. This may be due to the fact that the problem is less relevant when simple online dispatching rules are applied and the AGVs are operated in single-load mode, as is currently practiced in existing automated container terminals. However, Grunow et al. (2004, 2006) show that significant performance gains can be obtained when more elaborate offline heuristics are employed which do utilize the multi-load capabilities of the AGVs. The numerical results of this paper will show that deadlocks do become an important issue under these circumstances.

Principally, all three proposed methods may be used to handle this second type of deadlock in container terminals. However, to fully utilize the approach of *deadlock prevention*, complete information about all future events in the system is required and a static application environment has to be assumed. For the container terminal application considered here, none of these conditions hold. Equally, as *deadlock avoidance* is basically applicable for systems with incomplete information, a static environment still has to be assumed. In the case of stochastically varying handling times, as these handling times are typical of seaport container terminals, the applicability of this approach is rather limited. According to Kim et al. (1997) *deadlock detection and resolution* is most effective regarding utilization rates of the resources in the system. On the other hand, this approach requires procedures of significantly higher complexity to be embedded into the logistics control software of the terminal and thus higher effort associated with the handling of deadlocks.

Kim et al. (1997) consider deadlock prevention (referred to as "prevent deadlock without look-ahead") as the most conservative approach, followed by deadlock avoidance (referred to as "prevent deadlock with look-ahead"), and, finally, deadlock detection and resolution (or recovery) is deemed the most flexible approach. For the above reasons, we decided to develop a deadlock detection and resolution approach for AGV systems in highly automated container terminals.

3 Deadlock detection

The main challenge with deadlock detection is to find a realistic and compact representation of the system's actual status. In manufacturing systems, the predominant representation scheme is the resource allocation graph, which can be adapted to the container terminal environment (see Section 3.1). Regardless of which representation is used, it must be updated whenever the system status significantly changes. Because frequent updates are needed, they must be realized without considerable computational effort. For the detection of deadlocks, two different techniques are presented: One technique is based on a matrix representation of the terminal system (see Section 3.2). The other technique is a graph-oriented procedure, which directly traces the requests for the individual resources (see Section 3.3).

3.1 Adaption of the resource allocation graph to container terminals

A popular representation scheme of the system state is the so-called resource allocation graph (cf. Peterson and Silberschatz 1989; Elmasri and Navathe 1989;

Kim and Kim 1997). This graph consists of vertices for each part and each resource of the system. Parts are represented by circles, whereas resources are represented by rectangles. There are two types of edges in this graph: *request edges* and *assignment edges*. A request edge $p \rightarrow r$ (directed from part p to resource r) is introduced, if part p is currently waiting for resource r. In contrast, an assignment edge $r \rightarrow p$ (directed from resource r to part p) is added, if resource r has been allocated to part p. In some applications, parts are identified with tasks. In a container terminal, cranes and AGVs are resources, while handling operations (pick-up or drop-off) can be considered as parts.

As mentioned above, the dynamics of the system make it extremely difficult to completely avoid deadlocks. We therefore decided to employ the resource allocation graph mainly to examine the status of the container terminal during the simulation phase and identify deadlocks as soon as they arise, as well as to trigger the deadlock resolution procedures introduced in Section 4. Yet, the sheer number of operations to be performed would lead to a rather large resource allocation graph. We therefore tried to modify the resource-allocation-graph concept to match the special requirements of the problem at hand. An important property of our problem is that each operation requires exactly two resources—one AGV and one crane. In flexible manufacturing systems, where the resource-allocation-graph concept originates from, a task could also require more than two resources.

Secondly, remember that we only seek for a deadlock detection and resolution approach and do not aim at the prevention or avoidance of deadlocks. It is therefore sufficient to consider only the next operation in the schedule of each resource. For the problem at hand, we say that an operation is assigned to a resource if and only if it is the next operation of this resource. Now, if an operation only requires two resources, there can be the three main situations for each operation represented in Fig. 4 (apart from trivial cases, in which one or both of the resources are not yet assigned to the task).

In the first case of Fig. 4a, the operation is assigned to both resources, while in the latter two cases either the AGV or the crane is requested but not yet assigned (in the sense that the operation is not the first operation in the respective resource's

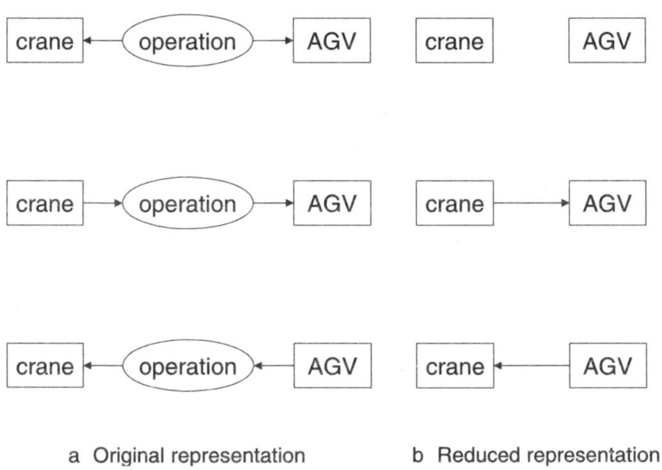

Fig. 4 Representation of waiting relationships

current schedule). It is now easy to see that operations need not be considered at all if the aim is to detect a current deadlock. The only important information is if a resource is waiting for another resource or not. Thus, the latter two situations in Fig. 4a can be described as "*the crane is waiting for the AGV*" or "*the AGV is waiting for the crane*," as is depicted in Fig. 4b. If a resource is waiting for another resource, this implies that the next operation in the schedule of this resource is not identical to the next operation in the schedule of the other resource. In the first case of Fig. 4a, the operation is the next one in the schedules of both the crane and the AGV. In other words, none of the resources is waiting for another resource (cf. Fig. 4b).

For the problem at hand, it is irrelevant which operation actually causes the waiting relation. The representations shown in Fig. 4a,b are equivalent. In addition, the representation of Fig. 4b considerably reduces the matrix size, especially if numerous transportation orders have to be monitored, as is the case in a container terminal.

3.2 Matrix-based deadlock detection

3.2.1 Matrix representation of the resource allocation

Belik (1990) introduced a matrix representation of the resource allocation graph and suggested matrix operations to update the matrix after the insertion or deletion of an edge as well as a cycle test. Kim and Kim (1997) applied this concept to avoid deadlocks in flexible manufacturing systems. A similar approach can be applied for deadlock handling in a container terminal, additionally making use of the reduced representation of waiting relations proposed in Section 3.1.

For the matrix representation of the waiting relations resulting from the current resource allocation, some information has to be stated explicitly, which is already implicitly available in the graph. As an example, consider the situation shown in Fig. 5a. Quay crane Q1 is waiting for vehicle V1, which in turn waits for stacking crane S1. As a result, Q1 is also waiting for S1. Although this relationship is easy to deduce from the graph, it has to be coded explicitly in the corresponding matrix (see Fig. 5b). Matrix D consists of $\{0,1\}$-elements, where $D(i,j)=1$, if and only if resource i is waiting (directly or indirectly) for resource j. As $D(\text{Q1,V1})=1$ and $D(\text{V1,S1})=1$ represent the immediate waiting relation, $D(\text{Q1,S1})$ must also be set to 1 to reflect the indirect waiting relation between Q1 and S1. The possible transitivity

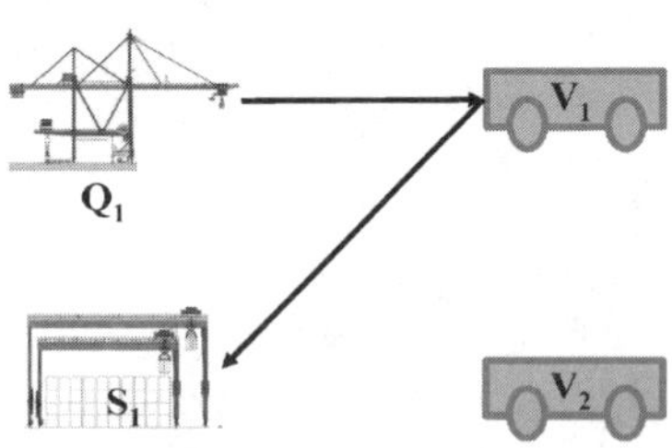

a Graph representation

$$D = \begin{array}{c} \\ Q1 \\ S1 \\ V1 \\ V2 \end{array} \begin{array}{c} \begin{array}{cccc} Q1 & S1 & V1 & V2 \end{array} \\ \begin{pmatrix} 0 & \textcircled{1} & 1 & 0 \\ 0 & 0 & 0 & 0 \\ 0 & 1 & 0 & 0 \\ 0 & 0 & 0 & 0 \end{pmatrix} \end{array}$$

b Matrix representation

Fig. 5 Resource allocation graph and matrix representation of waiting relations

of the “waiting for” relationship needs to be considered for each edge that is going to be inserted or deleted.

3.2.2 Matrix operations

There are two basic events that induce changes in the matrix representation of the resource allocation: the release of a new schedule and the completion of a task, either pick-up or drop-off. When a new schedule is released, most of the waiting relations between the resources are no longer up-to-date. Hence, it seems adequate to create a new matrix from all of the waiting relations in the updated schedule. In contrast, the completion of a task calls for rather small changes in the matrix. Assume, for instance, that task A has just been completed by use of AGV V1 and crane S1. Let the successors of task A in the respective schedules of resources V1 and S1 be task B and task C, respectively. Clearly, there can be at most four possible conditions (see Fig. 6 for a graphical representation).

1. If task B is not the next one in the schedule of the corresponding crane, V1 has to wait for this crane (cf. Fig. 6a).
2. If the corresponding crane of task B had been waiting for V1 so far, it no longer does so, because V1 is now ready to perform task B (cf. Fig. 6b).
3. If task C is not the next one in the schedule of the corresponding vehicle, S1 has to wait for this vehicle (cf. Fig. 6a).
4. If the corresponding vehicle of task C had been waiting for S1 so far, it no longer does so, because S1 is now ready to perform task C (cf. Fig. 6b).

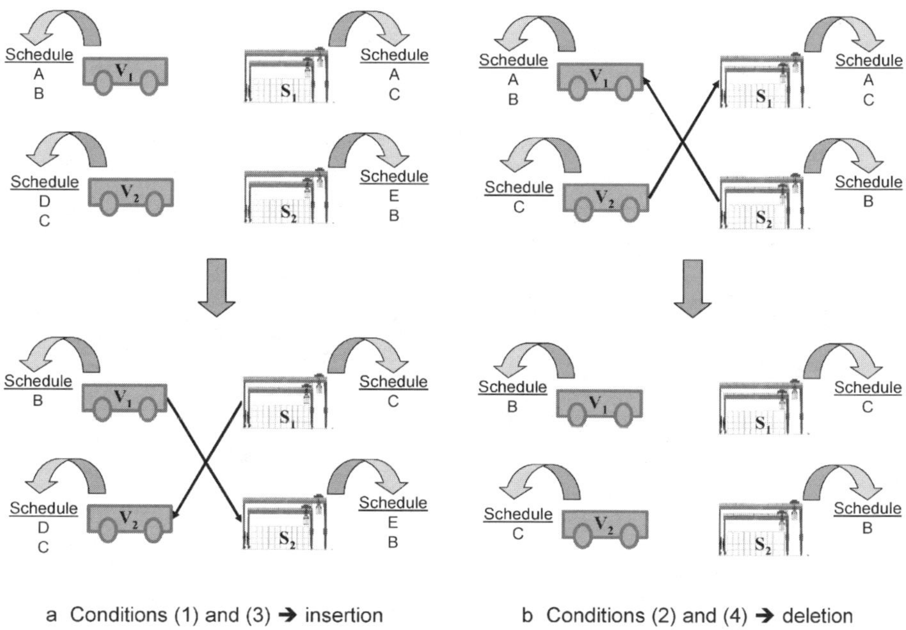

Fig. 6 Update of waiting relations after completion of a task

Cases 1 and 3 require the insertion of a new waiting relation, while in cases 2 and 4 an existing waiting relation has to be removed. The insertion and deletion of edges in the resource allocation graph are realized by matrix operations. In this paper, the matrix operations are only described briefly. For a detailed explanation, the reader is referred to Kim and Kim (1997).

Figure 7 gives an example of the insertion and deletion of an edge. In the first step, matrix $\Psi_{u,v} = (\Pi_u + I_u) \otimes (\Pi_v^T + I_v^T)$ is calculated, which identifies the changes to matrix D caused by the insertion or deletion of some edge (u,v) in the resource allocation graph. Π_u is the u-th column vector of D and Π_v^T is the v-th row vector of D. I_u and I_v^T denote the unit column and row vectors of u and v, respectively.

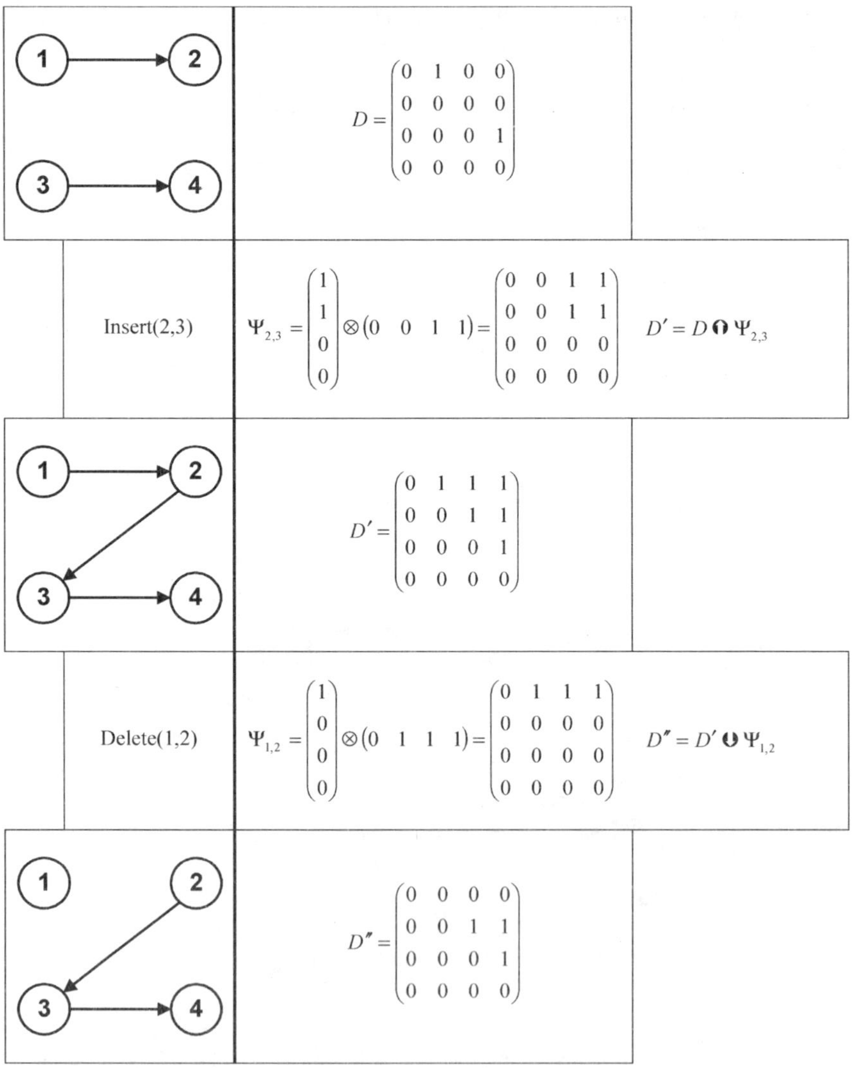

Fig. 7 Example insert and delete operations

In the second step, matrix D is updated with matrix $\Psi_{u,v}$. If edge (u,v) has to be inserted, the updated matrix D' is calculated by $D'=D \Uparrow \Psi_{u,v}$, where $\Uparrow$ indicates that the element-wise maximum of both matrices is determined. In the case of insertion, matrix D' is calculated by use of the respective minimum function $D'=D \Downarrow \Psi_{u,v}$.

As can be seen, the transitivity is maintained by applying these operations, e.g., node 4 becomes a successor of node 1, after the insertion of edge (2,3). Note that the operations necessary to update the matrix only work properly if the graph is acyclic (cf. Belik 1990). It is therefore important to perform a cycle test after each insert operation. This cycle test merely consists of checking the diagonal elements of D. If one of them is positive, i.e., the corresponding resource is waiting for itself, a deadlock has been detected. Creating a new matrix D (e.g., after the release of a new schedule) only requires a succession of insert operations, each of which is followed by a cycle test.

Note that with the deadlock detection method presented in this section, the current state of the entire system is reflected in every moment. Some information is essential, for instance the answer to the question of whether there is a deadlock cycle in the system. In addition, there is information that is not necessary, but can be used to improve minor decisions of the approach. For instance, it can be derived from the matrix how many other resources are currently waiting for a particular crane or AGV. This information could be used to select the crane to which a specific deadlock resolution procedure is applied (see the development of deadlock resolution techniques in the next section). It actually happens that more than one resource of the same type (quay crane or stacking crane) is involved in a cycle. In such a case, priority in the deadlock resolution procedure could be given to the crane that is currently requested by the largest number of other resources.

3.3 Graph-oriented deadlock detection

As mentioned in the previous section, the major drawback of the matrix-based deadlock detection procedure is the necessity to reflect the entire system state by a single matrix and to apply the update operations to the entire matrix. Hence, in the case of a large number of equipment units, this deadlock detection procedure is computationally rather demanding. The computational complexity of the matrix-based deadlock detection procedure can be assessed as follows: Let *nbCranes* be the overall number of quay and stacking cranes used in the terminal and *nbAgvs* be the number of AGVs employed. Then, the dimension of the waiting relationship matrix is *nbCranes+nbAgvs*, hence, each matrix update is of complexity $O\left((nbCranes + nbAgvs)^2\right)$. Each time a new schedule is released, one waiting relation has to be inserted for each equipment unit. The computational effort is therefore $O\left((nbCranes + nbAgvs)^3\right)$ for each scheduling request. On the other hand, the completion of each order requires the insertion of a waiting edge of complexity $O\left((nbCranes + nbAgvs)^2\right)$. However, because the number of scheduling requests is at least as high as the number of orders handled per day (because each new order triggers a scheduling request), the latter effort can be neglected. In a real terminal application, *nbCranes*=50 and *nbAgvs*=50 are common dimensions, and the number of scheduling requests easily exceeds 10,000 per day, while in a manufacturing environment, these numbers are an order of magnitude smaller. Furthermore, real-time requirements are much more restrictive in container terminals, in which runtimes of 1 s are desired. Therefore, the matrix-based deadlock detection method appears to be attractive for applications in small-sized container terminal configurations.

In any case, the complete waiting relationship matrix provides useful information about the actual system state. However, for deadlock detection, only part of the matrix is actually needed. To detect a new deadlock caused by the insertion of an edge into the waiting relationship graph, the deadlock detection procedure can be simplified as follows:

Assume that edge R1→R2 has to be inserted into the waiting relationship graph. First, the successor of R2 in the graph, say R3, is identified. The procedure further identifies the immediate successor of the current resource R3, say R4, and continues until the iteration limit of *maxNbNodes* is reached. In each step, it is checked whether the successor resource is identical with R1, the origin of the edge to be inserted into the graph. Should this be the case, then a cycle in the graph has been found and a deadlock has been detected. Otherwise, the graph is proven to be acyclic, i.e., no deadlock currently exists. Note that at most $maxNbNodes = 2 \cdot \min\{nbCranes, nbAgvs\}$ nodes must be checked because in the deadlock graph, crane nodes and AGV nodes alternate. Thus, after *maxNbNodes* nodes, either all AGVs or all cranes have been checked. Figure 8 outlines the complete procedure in pseudocode.

In the example of Fig. 3b, *maxNbNodes* is equal to 4, indicating that each cycle can be, at most, of length 4. If the algorithm would be employed for that example, e.g., starting with AGV V1, stacking crane S1 would be identified as successor because a waiting relation exists between V1 an S1. As the actual node (S1) does not equal the start node (V1), and *maxNbNodes* is not exceeded, a new iteration is started. In the end, the algorithm terminates for *actNode*=*startNode*=V1, reporting the cycle V1→S1→V2→S2→V1.

```
Algorithm: Deadlock Detection

1) actNode := startNode,
   nbNodes := 1,
   maxNbNodes := 2·min{nbCranes, nbAgvs}.

2) IF (successor exists) THEN
      actNode := succ(actNode),
      nbNodes := nbNodes + 1.
   ELSE
      return "no cycle" and stop.
   END.

3) IF (actNode = startNode) THEN
      return "no cycle" and stop.
   ELSE
      IF (nbNodes > maxNbNodes) THEN
         return "no cycle" and stop.
      ELSE
         go to 2.
      END.
   END.
```

Fig. 8 Algorithm for graph-oriented deadlock detection

The correctness of the proposed deadlock detection procedure is based on certain properties of the problem at hand:

1. Each resource has at most one successor in the waiting relationship graph. Recall that resource *v* is called successor of resource *u* if the next operation of *u* is not the next operation of *v*, but *v* is required to perform this operation. Because each resource has at most one next operation, it can also have no more than one successor.
2. If an acyclic graph becomes cyclic after the insertion of edge (*u*,*v*), then it has at most one cycle and (*u*,*v*) is part of this cycle.

Property 2 can be proved indirectly. Suppose (for the purpose of contradiction) that two cycles would be created by inserting (*u*,*v*). Then, each of them must include (*u*,*v*). Otherwise, the cycle would have existed before the insertion of this edge. As a result, the situation displayed in Fig. 9 arises. Starting with *u* and proceeding from each node to its successor node, a node *w* with at least two successors must be reached before a cycle is closed. Otherwise, (*u*,*v*) cannot belong to the second cycle because there is no connection between (*u*,*v*) and that cycle. But *w*, having more than one successor, is a contradiction to property 1. Therefore, property 2 holds.

It is important to keep in mind that the proposed deadlock detection procedure—like the matrix-based deadlock detection—assumes an *acyclic* graph. Hence, the detection of a cycle (i.e., a deadlock) cannot be postponed, but must take place immediately as soon as the deadlock occurs. Nonetheless, after the release of a new schedule, there can be more than one cycle in the real system. To understand this, one has to remember that there is the real system on one hand and its internal representation on the other hand. Once a new schedule is released, the internal representation of the new situation is build up step by step, inserting one waiting relation after the other, maintaining the property of being acyclic. Each time a deadlock is detected, both the real system and the internal representation are changed accordingly. In the end, the real system is deadlock-free and the internal representation correctly reflects the actual situation. In Section 4.5, after having introduced the deadlock resolution procedures, an example is given to illustrate the congruence of real system and internal representation.

The major advantage of the graph-oriented procedure is that it does not need elaborative matrix operations. The procedure simply passes through the chain of waiting relations each time an edge is inserted. Note that the deletion of an edge does not need to be considered because it cannot create a cycle. Instead of examining the complete matrix, only the waiting relations of the two affected

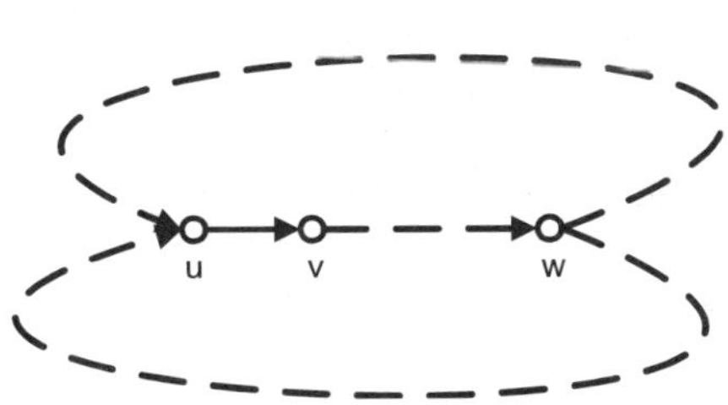

Fig. 9 Section from the waiting relationship graph after the insertion of edge (*u*,*v*)

resources need to be updated. When a new schedule is released, one waiting relation has to be inserted for each resource. The complexity of the deadlock detection algorithm is thus reduced to $O(\min\{nbCranes, nbAgv\} \cdot (nbCranes + nbAgvs))$ for each scheduling request.

Partly, the higher performance of the graph-based approach is achieved by sacrificing some nonessential information, as is the number of resources currently waiting for each AGV or crane. This information is used in the matrix-based approach to select the most constrained crane in the deadlock cycle to which the deadlock resolution procedure (explained in the next section) should be applied. It is, however, easy to determine the number of AGVs physically waiting in front of each crane. Thus, in the graph-oriented approach, the most constrained crane is chosen as the one with the most (physically) waiting AGVs.

4 Deadlock resolution

In a deadlock situation there exists a cycle of waiting relations between cranes and vehicles in the resource allocation graph. The links between two corresponding resources consist of operations for handling the transportation orders in the crane's and the vehicle's schedules. Hence, to resolve the deadlock, the schedule of at least one of the affected resources has to be modified. Of course, the effort required for changing the schedule depends on the type of resource. For instance, resequencing the transportation orders in the schedule of a quay crane would require extremely high effort. Quay cranes typically represent the bottleneck resources of the terminal because they directly influence the turnover-time of the vessels. Moreover, it is often quite difficult to reshuffle containers inside the vessel, which might be necessary if the quay crane had to deviate from its predetermined loading or unloading schedule. Hence, it is a strong precondition that the schedules of the quay cranes remain unchanged.

In contrast, repositioning containers is much easier accomplished at the storage blocks by use of the stacking cranes. Such operations are less time consuming and are frequently carried out during the normal operation of the terminal. In many cases, it is even possible to pick up another container in the crane's schedule without reshuffling, if the container is stored on top of a pile in the block. Changing a stacking crane's schedule to resolve a deadlock is especially appropriate because it has no side effects apart, of course, from a longer handling time.

Changing a vehicle's schedule, however, is not that easy. Because a vehicle is involved in both pick-up and drop-off operations, there is less flexibility in changing its schedule. Depending on the current status of the vehicle, certain operations may not be feasible. For instance, a pick-up operation can only be performed if the vehicle is not fully loaded, while dropping off a container requires that this container is already loaded on the AGV. Another obvious constraint is that a drop-off operation can never be performed before the corresponding pick-up operation. All these limitations restrict the possibility of changing the vehicle's schedule. Additional constraints arise from the fact that an AGV is unable to load a container without the aid of a crane. Therefore, the interdependency between the vehicle's and the crane's schedules must be considered.

In our deadlock resolution approach, we first try to resolve a deadlock situation by changing the stacking crane's schedule. If this is not feasible, changing the vehicle's schedule is the only remaining option. We propose three different deadlock resolution procedures, namely:

1. *modifyScSequence*
2. *advanceOrder*
3. *reassignOrder*

The first procedure requires the least intervention into the operations of the terminal system, while the last one has the strongest impact on the schedules of the different resources.

4.1 Procedure *modifyScSequence*

This procedure is the easiest to implement, because it requires the least changes in the schedules of the resources. It is called for a vehicle and a stacking crane in the deadlock cycle, when the vehicle is waiting for the crane. An exemplary situation is shown in Fig. 10. The edges in the graph indicate the scheduled interaction between stacking crane S1, quay crane Q1, and vehicles V1 and V2. The schedule of each crane or AGV is shown beside the respective resource. As an example, for stacking crane S1, the pick-up of container 1 is scheduled after the pick-up of container 2. Remember that for a vehicle the consecutive pick-up of two (20-ft) containers (as p4 and p2 for V2) is feasible if dual-load carriers are employed, as in the example.

Suppose that in the situation depicted in Fig. 10a, the waiting edge (Q1,V1) has to be inserted into the resource allocation graph (e.g., as a result of the completion of the task that had been scheduled before p3). Obviously, this insertion would result in a circular relationship, indicating a deadlock between the four resources. In this case, the procedure *modifyScSequence* can be called up for S1. By means of this procedure, the operation p1, which V1 is waiting for, is advanced from the second to the first position in the schedule of S1. As a result, V1 no longer has to wait for S1 because its next operation can now be processed. In addition, S1 for the moment ceases to wait for V2 because its current next operation (p1) can be processed as well. Only after resolving the impending deadlock, edge (Q1,V1) can be inserted without causing a cycle (see Fig. 10b).

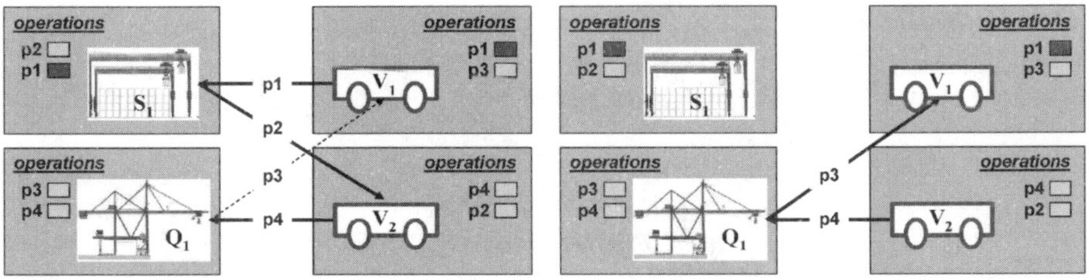

a Situation before *modifyScSequence(S1)* b Situation after *modifyScSequence(S1)*

Fig. 10 Procedure *modifyScSequence*

Note that by applying the *modifyScSequence* procedure, no additional waiting relation is created. Instead, two former waiting relations have been removed. Hence, no further deadlock can be caused by the use of this procedure.

4.2 Procedure *advanceOrder*

Procedure *advanceOrder* is applied in a situation in which a quay crane is immediately waiting for an AGV and both resources are involved in a deadlock cycle. As an example, consider the situation shown in Fig. 11. Assume that a new edge (Q2,V1) has to be inserted into the resource allocation graph depicted in Fig. 11a. Clearly, a cycle would be created. To prevent deadlocks, the schedule of vehicle V2 has to be changed, i.e., its four operations are resequenced. Suppose that the next operation of Q1—being either a pick-up or a drop-off operation from the viewpoint of the corresponding vehicle V2—could be advanced to the first position in the schedule of V2. In the case of a pick-up operation, this procedure is feasible, if V2 currently provides sufficient unused loading capacity. On the other hand, in the case of a drop-off operation, the respective container must already have been loaded onto V2. In the example of Fig. 11a, the next operation p2 of Q1 meets the first condition because V2 is empty in the current situation. Thus, the deadlock is resolved by advancing both the pick-up and the drop-off operations, p2 and d2, of the corresponding container to the first and second positions in the schedule of V2 (for a pick-up, we always move both the pick-up and the corresponding drop-off to the top of the schedule to avoid infeasible schedules, e.g., the vehicle's capacity is exceeded). As a result, the waiting relations between Q1 and V2, as well as those between V2 and Q2, can be deleted and the edge (Q2,V1) can now be inserted without creating a cyclic graph (see Fig. 11b).

Note again that no additional waiting relation is created. However, as a consequence of the change in the schedule of V2, this vehicle may need to deviate from its current trip. Should no online information about the vehicle's current position and route in the guide path be available, the vehicle can only be redirected to the new destination Q1 after the arrival at its former destination Q2. While, in the dispatching phase, rerouting vehicles from their current mission is considered infeasible, rerouting might be unavoidable to resolve a deadlock situation.

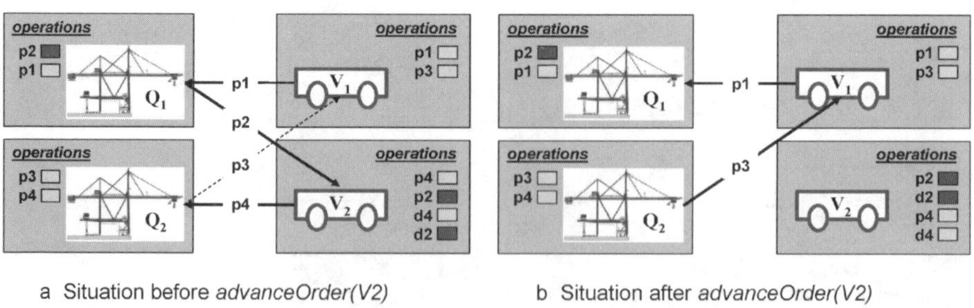

Fig. 11 Procedure *advanceOrder*

4.3 Procedure *reassignOrder*

Although most deadlocks can be resolved by the use of the two procedures presented in the previous subsections, situations may occur in which none of these procedures can be applied. For instance, in the situation depicted in Fig. 11a, it might happen that the next operation of the quay crane could not be carried out immediately, even if it is located at the first position in the schedule of the corresponding AGV, because of two possible reasons. The first reason is that the next operation is a pick-up, but the vehicle's loading platform is blocked by other containers. The second reason is that the next operation is a drop-off, but the container has so far not been loaded onto the vehicle. In both situations, advancing the corresponding operation in the vehicle's schedule will be of no help because additional operations need to be performed first.

In such a case, a *standby AGV* has to be provided. Standby AGVs are not utilized during regular operating conditions. Instead, they are activated only to resolve deadlock situations. The adequate number of such vehicles is best determined by means of simulation studies. In most terminal configurations, a single standby vehicle is sufficient to resolve all occurring deadlocks. However, in systems vulnerable to frequent deadlocks, a greater number of standby vehicles may be a better choice. In addition, vehicles temporarily unemployed may be used instead of standby AGVs in particular, as some buffer capacity exists in almost every AGV system to be found in manufacturing systems and terminal configurations.

The interventions required for the execution of the *reassignOrder* procedure are usually more severe compared to the other deadlock resolution procedures presented in the previous subsections. As an example, consider the situation shown in Fig. 12a.

Assume again that edge (Q2,V1) has to be inserted. In this case, procedure *advanceOrder* cannot be applied, because Q1 is waiting for vehicle V2 to perform operation p2. The loading platform of vehicle V2, however, is blocked by the 40-ft container 1, which needs to be dropped off first. For the same reason, operation p4 cannot by advanced in the schedule of V1. To resolve the deadlock, which arises after the insertion of the edge (Q2,V1) into the chain V1–Q1–V2–Q2, the pick-up

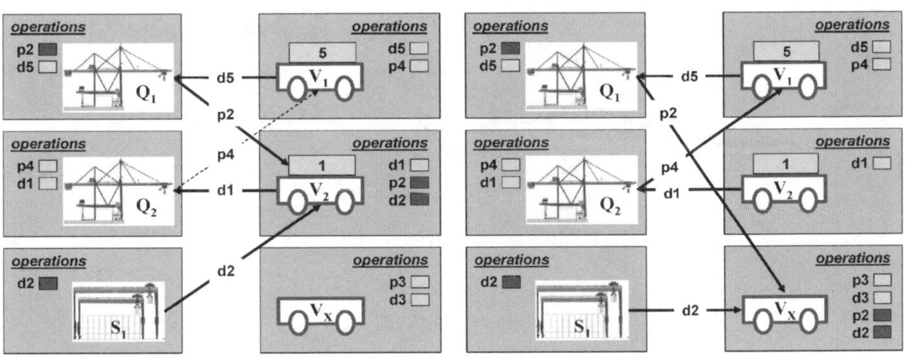

a Situation before *reAssignOrder(V2)* b Situation after *reAssignOrder(V2)*

Fig. 12 Procedure *reassignOrder*

and drop-off operations p2 and d2 of the container located at the first position in the schedule of Q1 are assigned to a standby AGV. In Fig. 12, this vehicle is denoted by Vx. To avoid being trapped in a deadlock themselves, standby AGVs are only operated in single-load carrier mode, i.e., they carry out one transportation order at a time without interlocking pick-up and drop-off operations of different orders. If a new transportation order is assigned to a standby AGV, it is always appended at the end of its schedule. As a result of this procedure, cranes Q1 and S1 are no longer waiting for V2 and the insertion of edge (Q2,V1) does not create any cycle in the resource allocation graph. The result is shown in Fig. 12b. As can be seen, additional edges (Q1,Vx) and (S1,Vx), indicating the waiting relationship of Q1 and S1 for Vx, have to be introduced. It is therefore important to assure that these new edges do not create further deadlock cycles. As an example, Fig. 13 illustrates the consequences of using the procedure *reassignOrder*.

Let edge 9 [in Fig. 12b labeled (Q2,V1)] be the edge that was to be inserted initially. The insertion of edge 9 would close a cycle. In the course of the *reassignOrder* procedure, edges 1 and 2 have been deleted. Because the next order of Q1 has been reassigned from V2 to Vx, neither Q1 nor S1 is waiting for V2 any more. However, both cranes may now be waiting for Vx and thus require the insertion of edges 5 and 8. This, in turn, could lead to new cycles covering the standby AGV Vx itself. Now let us have a closer look at the individual cycles. If Vx belongs to a cycle, it must be waiting for some crane. This crane cannot be a quay crane, because all quay crane operations assigned to a standby vehicle are at the first position of the quay crane's schedule (see the definition of the *reassignOrder* procedure, which is only used for quay cranes). Hence, Vx must be waiting for a stacking crane. This means that at least one stacking crane is included in the cycle. Thus, the *modifyScSequence* procedure can be applied to the

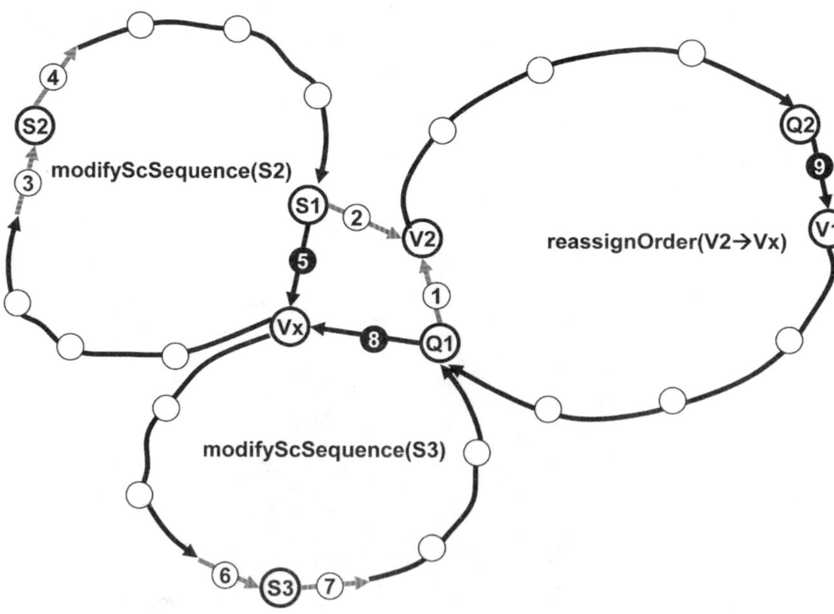

Fig. 13 Correctness of the procedure *reassignOrder*

most constrained of the stacking cranes in the cycle. In Fig. 13, these cranes are supposed to be S2 and S3 for the respective cycles. The application of the procedure *modifyScSequence* leads to the deletion of a pair of edges, 3 and 4 or 6 and 7. As a result, edges 5 and 8 can be inserted without creating a new cycle. No further waiting relations are created. Hence, the initial deadlock is finally resolved in, at most, two steps.

4.4 The deadlock resolution algorithm

The three basic deadlock resolution procedures represent the core modules of the comprehensive deadlock resolution algorithm outlined in Fig. 14. The deadlock resolution must be called if new waiting relations arise. There are only two events that may cause new waiting relations. Firstly, after an operation has been finished, the next operations of the affected AGV and the affected crane move to the top of their schedules. If these operations are not on top of the schedules of the corresponding resource (a crane for the AGV and an AGV for the crane), new waiting relations are created. On the other hand, each new dispatching may create several new waiting relations by thoroughly changing the existing schedules. In both cases, first the matrix-based or the graph-oriented deadlock-detection procedure is employed. Once a deadlock has been detected, the corresponding cycle of resources waiting for each other is identified. Next, the type of deadlock is analyzed and it is decided which deadlock resolution method has to be applied. If the cycle contains a standby AGV, this deadlock has to be resolved first. As shown in Section 4.3, there always exists a stacking crane in the cycle, for which the procedure *modifyScSequence* can be applied. If the cycle does not contain a standby AGV, the cycle is checked for stacking cranes. If there is more than one stacking crane in the cycle, the one showing the largest total number of waiting relations is chosen and the procedure *modifyScSequence* is applied to this stacking crane. If neither standby AGVs nor stacking cranes can be found in the cycle, the procedure *advanceOrder* or the procedure *reassignOrder* is applied to the quay cranes in the cycle, depending on the feasibility of the next order.

Algorithm: Deadlock Resolution

Triggering event: Deadlock occured

1) Detect deadlock cycle.
2) ***IF*** (cycle contains standby AGV) ***THEN***
   Call procedure *modifyScSequence* and stop.
   END.
3) ***IF*** (cycle contains stacking crane) ***THEN***
   Call procedure *modifyScSequence* and stop.
   END.
4) Select quay crane in cycle
   IF (next order in quay crane schedule can be advanced) ***THEN***
   Call procedure *advanceOrder* and stop.
   ELSE
   Call procedure *reassignOrder* and stop.
   END.

Fig. 14 Deadlock resolution algorithm

It should be noted that the deadlock resolution algorithm presented here is complete in the sense that any type of deadlock, in the boundaries specified in this paper, is guaranteed to be resolved. To prove this, it must be shown that for any possible deadlock one of the three solution procedures can be applied and that the application of any solution procedure does not lead to another deadlock.

The first property immediately follows from the design of the deadlock resolution algorithm (cf. Fig. 14), because for any possible structure of the deadlock cycle exactly one of the procedures is called up. To show the second property, remember that the application of the procedures *modifyScSequence* and *advanceOrder* cannot lead to new cycles (see Sections 4.1 and 4.2). From Section 4.3, it follows that procedure *reassignOrder* can in fact create new deadlocks, but each of them could be resolved by an additional application of the procedure *modifyScSequence*.

4.5 Congruence between resource allocation graph and system schedule

As mentioned in Section 3.3, after a new schedule is released the resource allocation graph is constructed anew step by step until it completely resembles the system's schedule. Once a cycle is detected, that cycle is resolved and the construction of the resource allocation graph continues until the entire system is exactly mirrored. In Fig. 15, the construction process is explained by an example.

In Fig. 15a the situation of the real system is illustrated after a new schedule has been released. It is easy to see from the schedules that the four equipment units are waiting for each other. A circular deadlock exists. However, the resource allocation

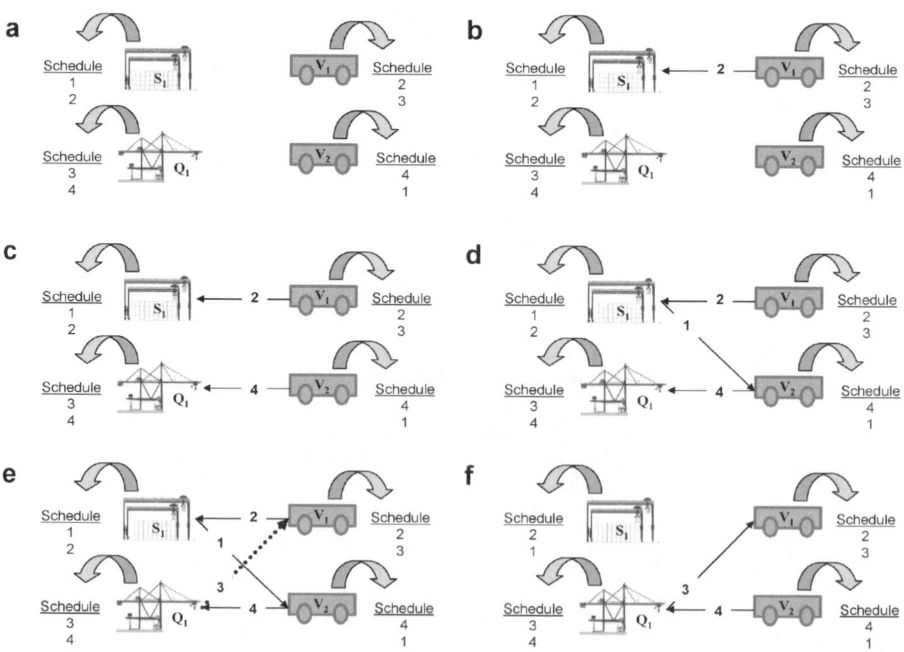

Fig. 15 Internal representation and real system

graph will be maintained cycle-free throughout the whole construction process. Figure 15a–d show the first three steps of the procedure, in which arcs are inserted to represent the waiting relations caused by the insertion of orders 2, 4, and 1. The system schedule does not change meanwhile. In Fig. 15e, the attempt to insert an edge for the waiting relation of order 3 would lead to a cycle, which is detected in the resource allocation graph. The procedure *modifyScSequence* is applied to both the system schedule and the graph. As a result, the schedule of S1 is changed, which leads to the resolution of the deadlock in the real system. The resource allocation graph is modified by deleting edges 1 and 2, as well as by inserting edge 3. All deadlocks are now resolved, and the internal representation is equivalent to the situation in the real system, as shown in Fig. 15f.

5 Numerical results

To prove the effectiveness of the deadlock handling approach presented in this paper, we conducted numerical tests using an event-driven simulation model. The main experimental parameters were the size of the container terminal, the degree of stochasticity for the handling times of the cranes, and the approach used for dispatching AGVs. We defined small, medium, and large terminal configurations, the latter comprising 15 quay cranes, 45 stacking cranes, 120 AGVs, and 3,000 transportation orders. Handling times of cranes were analyzed by Vis and Harika (2004) and Yang et al. (2004). Our simulation studies are based on the distribution observed empirically by Vis and Harika (2004). Apart from the deterministic case, we tested three stochastic cases, using the original distribution of Vis and Harika (2004) as well as similar distributions of the same structure and with the same mean value, but with half and double of the original variance value, respectively. For dispatching the AGVs, we used an online heuristic adapted from manufacturing environments. A first-come-first-served rule is used for vehicle-initiated dispatching, and a nearest-vehicle rule for transportation-order-initiated dispatching. These rules were extended to accommodate for the multi-load capability of the AGVs. As an alternative, an offline (so-called pattern-based) heuristic was developed, which generates a schedule for the entire fleet of AGVs for a limited look-ahead horizon. In an iterative approach, the AGVs are assigned to the transportation orders one by one according to ascending target start times. For each order–vehicle combination, feasible sequences of pick-up and drop-off operations are evaluated and the best overall assignment is chosen. Note that the predictive offline approach and the online approach are both designed as real-time dispatching procedures that are executed each time a new dispatching request emerges. These heuristics may treat the AGVs as single-load or dual-load carriers, depending on whether or not the simultaneous transport of two 20-ft containers is permitted. For details regarding the experimental design, please refer to the paper by Grunow et al. (2006), which also provides a detailed description of the dispatching heuristics used in our simulation studies.

As a first observation, we noticed that the number of deadlocks was not significantly influenced by the degree of stochasticity. Apparently, the main reasons for deadlocks are the dynamics of the entire system and the lack of coordination between quay crane and stacking crane schedules, but not the stochasticity in handling times. As long as varying handling times do not change the stacking crane schedules no additional deadlock can be created. In Fig. 16 the occurrence of the different deadlock types—requiring the respective procedures *modifyScSequence*, *advanceOrder*, and *reassignOrder*—is shown. For different terminal sizes and different AGV modes, an offline heuristic (pattern) and a simple online rule (basic) are compared. Each bar in the chart comprises the aggregated results for each of the four stochasticity levels, i.e., 52 different data sets.

From the presented results, the following conclusions can be drawn:

1. The number of deadlocks (per 1,000 transportation orders) decreases with increasing terminal size. Obviously, in larger terminals, more cranes and AGVs are operated, which inhibits the development of cycles of mutually blocked equipment.
2. The total number of deadlocks is small but significant. For realistic terminal configurations, in which about 10,000 containers are moved each day, no more than three or four deadlocks are to be expected. Still, deadlocks have to be accounted for as long as there is even a slight probability for them to happen, because a single deadlock can (and usually will) lead to the complete standstill of the terminal because of the innumerable interactions between the equipment units.
3. However, the effort made to deal with deadlocks should be adequate. Regarding their limited number, it is unnecessary to implement deadlock prevention or avoidance methods. These methods require a more conservative use of the equipment, which leads to lower utilization rates (cf. Section 2 of this paper). The proposed deadlock resolution methods flexibly respond to the (rare) occurrence of a deadlock and are hence more efficient in terms of capacity

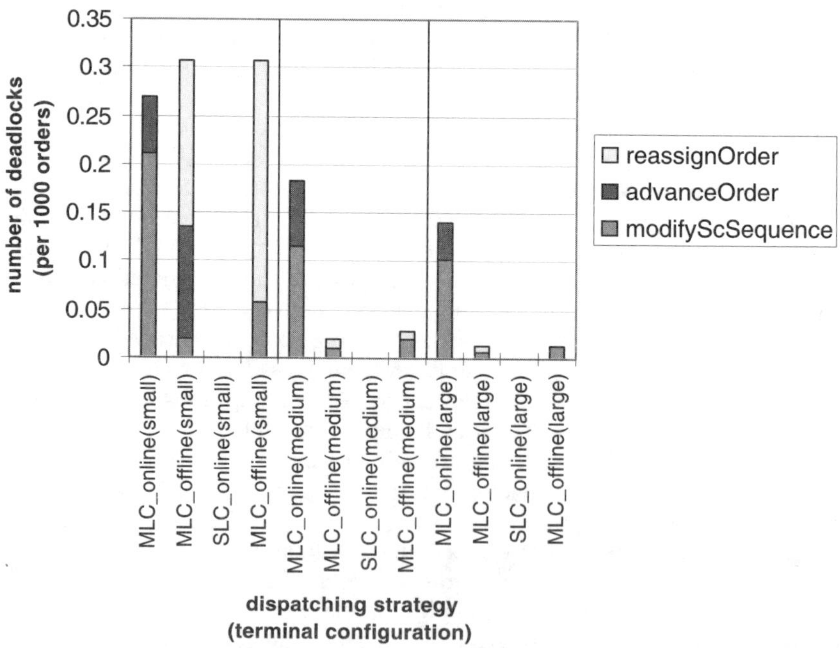

Fig. 16 Distribution of deadlocks

utilization. Furthermore, it is sufficient to employ a single standby AGV (which we did in our tests even for the large terminal configuration).

4. When using the online dispatching rule with the single-load-carrier mode, no deadlocks occurred at all. Under this rule, each AGV is assigned, at most, one order. The potential for deadlocks is extremely small. This also explains why this topic has not yet arisen in the terminal practice where similar rules are used to dispatch AGVs as single-load carriers. Yet, Grunow et al. (2006) showed that applying the offline pattern-based heuristic and utilizing the multi-load capability of the AGVs significantly improves the system performance. However, our simulation results reveal that the problem of deadlock handling will become an important issue when this advanced dispatching heuristic is used.
5. The online dispatching rule combined with dual-load-carrier mode (referred to as MLC, for multi-load carrier, in Fig. 16) results in the most deadlocks for the larger scenarios (medium and large). For this rule, all assignments of transportation orders to AGVs are fixed. Hence, the AGV schedules cannot be changed to respond to changing target times. However, for the online rule with dual-load carriers, no deadlocks that need to be resolved with the "*reassignOrder*" procedure occurred, because under this rule, at most, two orders can be assigned to one AGV.
6. For the pattern-based offline heuristic, deadlocks are only significant for the small scenario. In large terminals' configurations, there is nearly always a possibility to avoid deadlocks through adequate schedules.
7. Finally, if the pattern-based heuristic was employed for single-load-carriers, deadlocks of the type "*advanceOrder*" did not occur. If the unnested schedules produced by this rule lead to a deadlock at all, they usually lead to a deadlock of the type "*modifyScSequence*," or to the most severe type "*reassignOrder*" in its classical form as shown in Fig. 12.

6 Summary and conclusions

In our initial simulation experiments, we observed that the performance of AGV dispatching strategies suffered from the occurrence of deadlock situations. Therefore, a comprehensive scheme to handle deadlocks occurring in the operation of the AGV system had to be developed. The basic module in the deadlock-handling scheme refers to the detection of deadlocks. Two different approaches have been developed. While the matrix-based deadlock detection procedure is more elaborative, it has the advantage of providing a complete representation of the current state of the terminal system, i.e., waiting relationships between all equipment units in the system can be revealed. On the other hand, the graph-oriented procedure is computationally more efficient, especially if the terminal configuration comprises a large number of resources and AGV dispatching has to be performed in real-time.

To resolve deadlocks, an algorithm consisting of three different resolution procedures has been developed. First, the type of deadlock is identified. Then, based on the type of resources involved in the deadlock, individual procedures are applied to reassign transportation orders or to modify the sequence of operations in the schedule of a resource. It has been shown that the algorithm is able to resolve all deadlock situations that may occur in the context of AGV dispatching in automated container terminals. AGVs are especially liable to deadlocks, because they always need a secondary resource to perform the pick-up and drop-off operations. Special emphasis was given to dual-load AGVs, which require more complex dispatching strategies and thus are more easily involved in deadlock situations.

For a long time, the occurrence of deadlocks in automated manufacturing and logistics systems has been recognized only for the routing of AGVs in the guide path, while blocking effects between vehicles and handling units have mostly been overlooked in the academic literature. Hence, we consider the proposed deadlock-handling scheme as a first step towards integrated scheduling and dispatching approaches for equipment units in highly automated container terminals.

The results of an extensive simulation study underline the necessity to develop deadlock-handling strategies. If the multi-load capability of the AGVs or the more advanced offline heuristic shall be used, which significantly enhance the system performance, then deadlocks will become an important practical problem. A deadlock may lead to the standstill of an entire container terminal. The proposed deadlock resolution methods are able to flexibly handle deadlocks occurring in different terminal configurations when different AGV dispatching methods are used. Furthermore, the limited number of deadlocks indicates that our approach of resolving rather than entirely preventing or avoiding deadlocks is the most appropriate alternative, as more conservative approaches would result in lower equipment utilization. Despite their complexity, it is therefore economically viable to include the proposed deadlock detection and resolution methods in the logistics control software of an automated seaport container terminal.

References

Belik F (1990) An efficient deadlock avoidance technique. IEEE Trans Comput 7(39):882–888

Coffman EG, Elphick MJ, Shoshani A (1971) System deadlocks. ACM Comput Surv 3(2):67–78

de Koster RMBM, Le-Anh T, van der Meer JR (2004) Testing and classifying vehicle dispatching rules in three real-world settings. J Oper Manag 22:369–386

Egbelu PJ, Tanchoco JMA (1984) Characterization of automatic guided vehicle dispatching rules. Int J Prod Res 3(22):359–374

Elmasri R, Navathe NB (1989) Fundamentals of database systems. Benjamin Cummings, Redwood City

Fanti MP (2002) Event-based controller to avoid deadlock and collisions in zone-control AGVS. Int J Prod Res 40(6):1453–1478

Grunow M, Günther H-O, Lehmann M (2004) Dispatching multi-load AGVs in highly automated seaport container terminals. OR Spectrum 2(26):211–235

Grunow M, Günther H-O, Lehmann M (2006) Strategies for dispatching AGVs at automated seaport container terminals. OR Spectrum 4(28). DOI 10.1007/s00291-006-0054-3

Kim CO, Kim SS (1997) An efficient real-time deadlock-free control algorithm for automated manufacturing systems. Int J Prod Res 6(35):1545–1560

Kim CW, Tanchoco JMA, Koo PH (1997) Deadlock prevention in manufacturing systems with agv systems: Banker's algorithm approach. J Manuf Sci Eng 119:849–854

Le-Anh T, de Koster MBM (2005) On-line dispatching rules for vehicle based internal transport systems. Int J Prod Res 8(43):1711–1728

Lim JK, Kim KH, Yoshimoto K, Lee JH, Takahashi T (2003) A dispatching method for automated guided vehicles by using a bidding concept. OR Spectrum 1(25):25–44

Liu F-H, Hung P-C (2001) Real-time deadlock-free control strategy for single multi-load automated guided vehicle and a job shop manufacturing system. Int J Prod Res 39 (7):1323–1342

Moorthy RL, Hock-Guan W, Wing-Cheong N, Chung-Piaw T (2003) Cyclic deadlock prediction and avoidance for zone-controlled AGV system. Int J Prod Econ 83:309–324
Peterson JL, Silberschatz A (1989) Operating system concepts. Addison-Wesley, Reading
Reveliotis SA (2000) Conflict resolution in AGV systems. IIE Trans 32:647–659
Steenken D, Voß S, Stahlbock R (2004) Container terminal operation and operations research — a classification and literature review. OR Spectrum 1(26):1–49
Venkatesh S, Smith JS (2003) A graph-theoretic, linear-time scheme to detect and resolve deadlocks in flexible manufacturing cells. J Manage Syst 3(22):220–238
Vis IFA, Harika I (2004) Comparison of vehicle types at an automated container terminal. OR Spectrum 1(26):117–143
Wu N, Zeng W (2002) Deadlock avoidance in an automated guided vehicle system using a coloured Petri net model. Int J Prod Res 40(1):223–238
Yang CH, Choi YS, Ha TY (2004) Simulation-based performance evaluation of transport vehicles at automated container terminals. OR Spectrum 2(26):149–170
Yeh M-S, Yeh W-C (1998) Deadlock prediction and avoidance for zone-control AGVS. Int J Prod Res 36(10):2879–2889

Kap Hwan Kim · Su Min Jeon · Kwang Ryel Ryu

Deadlock prevention for automated guided vehicles in automated container terminals

Abstract Automated guided vehicles (AGVs) are an important component for automating container terminals. When utilizing AGVs to transport containers from one position to another in a container terminal, deadlocks are a serious problem that must be solved before real operations can take place. This study assumes that the traveling area for AGVs is divided into a large number of grid-blocks, and, as a method of traffic control, grid-blocks are reserved in advance when AGVs are running. The first purpose of the reservation is to make room between AGVs and to prevent deadlocks. The objective of this study is to develop an efficient deadlock prediction and prevention algorithm for AGV systems in automated container terminals. Because the size of an AGV is much larger than the size of a grid-block on a guide path, this study assumes that an AGV may occupy more than one grid-block at a time. This study proposes a method for reserving grid-blocks in advance to prevent deadlocks. A graphical representation method is suggested for a reservation schedule and a priority table is suggested to maintain priority consistency among grid-blocks. It is shown that the priority consistency guarantees deadlock-free reservation schedules for AGVs to cross the same area at the same time. The proposed method was tested in a simulation study.

Keywords AGV · Deadlock · Graph

K. H. Kim (✉) · S. M. Jeon
Department of Industrial Engineering, Pusan National University, Jangjeon-dong, Kumjeong-ku, Busan 609-735, South Korea
E-mail: kapkim@pusan.ac.kr, 1006sumin@pusan.ac.kr

K. R. Ryu
Department of Computer Engineering, Pusan National University, Jangjeon-dong, Kumjeong-ku, Busan 609-735, South Korea
E-mail: krryu@pusan.ac.kr

1 Introduction

Recently, automating container terminals became an important issue in worldwide hub ports. The storage in the yard is automated at the Thames Port, at the Parsir Panjang Terminal in Singapore, and at the Kawasaki Container Terminals. Both container transport in the apron and storage in the yard are automated at Rotterdam and Hamburg.

A difficult part of automation is the transportation function during ship operations in which automated guided vehicles (AGVs) are present in all the realized systems. The difficulties arising from using AGVs in container terminals are due to the large number and size of the vehicles, which require highly complicated traffic control methods.

One of the most important issues of an AGV traffic control system is preventing a large number of AGVs from becoming deadlocked. Previous researches (Evers and Koppers Stijn 1996; Lee and Lin 1995; Rajeeva et al. 2003; Reveliotis 2000; Yeh and Yeh 1998) used zone control policies to avoid collisions and deadlocks. In these researches, guide paths were partitioned into zones; each of which is large enough to contain the entire body of an AGV, and each zone is exclusively assigned to only one AGV to avoid collisions. To overcome the inefficient use of space of the zone control method, this paper proposes a method of partitioning a traveling area into grid-blocks, each of which is smaller than the physical size of an AGV. As a result, conventional deadlock prediction algorithms, which utilize the zone control method, cannot be used in the AGV systems of this study. In addition, outbound containers have to be delivered to a quay crane (QC) in the same order as specified in the load sequence list for the QC, which makes the control of AGVs more difficult.

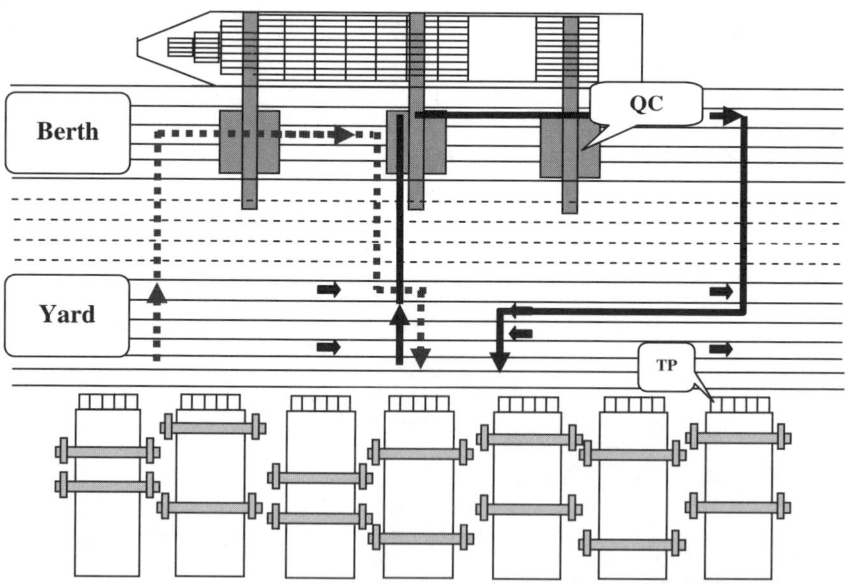

Fig. 1 An illustration of a container terminal

In describing automated container terminals (ACTs), the Europe Container Terminals (ECT) terminal in Rotterdam and the Container Terminal Altenwerder (CTA) terminal in Hamburg will be illustrated as reference models because, in these two terminals, the movement of containers in the apron and the stacking of containers in the yard are automated. The ECT terminal, which is the first ACT in the world, utilizes an automatic stacking crane for each block, whereas the CTA terminal utilizes two rail-mounted gantry cranes (RMGCs) for each block, as shown in Fig. 1.

AGVs transport inbound containers from QCs to RMGCs, and outbound containers from RMGCs to QCs. For each QC, five to eight AGVs usually transport containers between the apron and the marshalling yard. Because of a large number of AGVs, simple closed-loop guide-path networks were used in the initial stage of the ECT. To speed up the deliveries by AGVs, more complicated guide-path networks have been adopted, with the support of more complex control software.

ACTs usually utilize free-ranging AGVs that do not follow permanent physical guide paths like electric wires and thus can travel on guide paths temporarily specified in the memory of a supervising control computer, which we call a "virtual guide path." Figure 1 illustrates the guide paths of an automated guided vehicle system in a container terminal.

The guide path network in Fig. 1 consists of travel lanes under QCs, travel lanes in front of blocks, transfer lanes in front of blocks, and cross lanes in the middle of the apron. The areas in front of blocks, where cross lanes, travel lanes, and transfer lanes merge, have high possibilities for not only congestion but also deadlock.

In the zone control method, which is the most popular in practice, zones, each of which is large enough to accommodate the entire body of an AGV, are prespecified. Only a single AGV is allowed to enter a zone, which prevents collisions between multiple AGVs. Also, AGVs usually travel on unidirectional guide paths in practice. Because the guide paths are unidirectional, the sequences of zones on which AGVs move are consistent with each other. From the consistency in the sequence of zones, the possibility of deadlocks is removed. For multiple AGVs to travel on bidirectional guide paths, detailed movements of each AGV must be scheduled so that collisions and deadlocks between AGVs can be avoided.

However, to improve the utilization of space on the lanes, this study partitions the travel area for AGVs into grid-blocks, each of which is smaller than the size of a vehicle. Therefore, an AGV may occupy more than one grid-block at a time, as shown in Fig. 2. Thus, the problems of detecting and preventing deadlocks in this study are different from those of AGV systems where the zone control method is used. Thus, algorithms developed for the zone control method cannot be utilized in this situation.

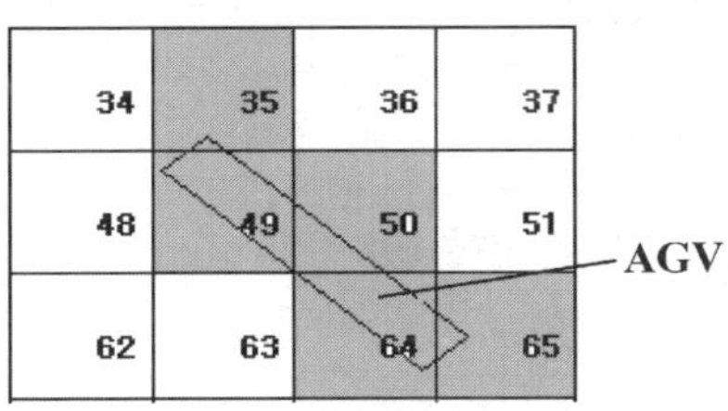

Fig. 2 An illustration of multiple grid-blocks occupied by an AGV

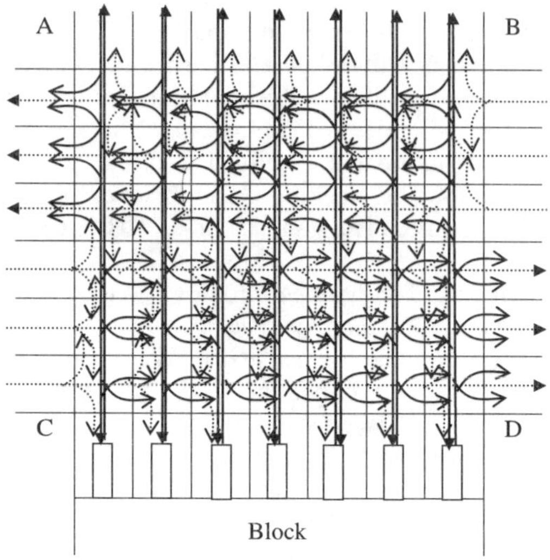

Fig. 3 Travel routes in front of a block

The traffic in front of transfer positions is very congested and has many flows with different moving directions. For example, Fig. 3 illustrates possible travel routes of AGVs in front of transfer points at the end of a block. The size of each grid-block on travel lanes in front of a yard block is 5×7 m.

Because there exist many possible routes on this region, and because a vehicle occupies more than one grid-block during travel, the easiest way to prevent deadlocks of vehicles is to consider the entire set of grid-blocks in the region ABCD as a single zone and to allow a single vehicle to enter the zone. However, in this case, long waiting periods of AGVs for entering the zone may be expected. Thus, to allow more than one vehicle to travel in the zone, methods to prevent collisions and deadlocks are necessary, which is the issue of this study.

The Petri-net has been most widely used for anticipating and avoiding deadlocks. Various methods utilizing the Petri-net have been suggested by Viswanadham et al. (1990), Banaszak and Krogh (1990), Lee and Lin (1995), Wu (1999), and Wu and Zhou (2000). Cho et al. (1995) and Yeh and Yeh (1998) suggested methods for anticipating and avoiding deadlocks by using graph models. Reveliotis (2000) proposed a routing algorithm without deadlocks on bidirectional guide path networks assuming a zone control method. In the study, a modified version of Banker's algorithm was used to check whether a route is deadlock-free. Kim and Tanchoco (1991, 1993) proposed an algorithm for finding conflict-free shortest time routes for AGVs. They used the concept of a time window graph in which a node represents a free time window and an arc represents the reachability between two free time windows. Because the algorithm finds a route through the free time windows in the graph and because the route is conflict-free, deadlocks between AGVs are avoided. Möhring et al. (2004) proposed a conflict-free routing algorithm for AGVs in ACTs, which is based on the studies by Kim and Tanchoco (1991, 1993).

Fanti et al. (1997) defined deadlocks and derived necessary and sufficient conditions for a deadlock occurrence in case that a job occupies a single resource at a specific moment in time. Fanti (2002) applied this general concept to the zone control scheme for the traffic control of AGVs.

Compared to the number of studies on deadlocks of AGVs in manufacturing applications, there have been fewer studies of transportation centers, such as container terminals. Rajeeva et al. (2003) addressed the deadlock problem of AGVs in ACTs. Rajeeva et al. assumed a unidirectional guide-path network and a yard layout in which the yard cranes travel parallel to the berths. Rajeeva et al. proposed algorithms to anticipate and prevent deadlocks and conducted a simulation study to verify the performance of the suggested algorithms. Evers and Koppers Stijn (1996) suggested a traffic control method for AGVs by using the concept of a "semaphore," which represents the maximum capacity of a zone to accommodate multiple AGVs at the same time. Duinkerken et al. (1999) proposed a framework for the traffic control of AGVs called "TRACES" and provided the results of a simulation study to demonstrate the validity of the framework. Vis and Harika (2004) and Yang et al. (2004) compared performances of different types of automated transport vehicles used in port container terminals. Grunow et al. (2004) suggested a dispatching algorithm for AGVs with capacities of two loads in container terminals.

This study is different from previous studies in that a vehicle is allowed to occupy and reserve more than one grid-block at a time, whereas previous studies assumed that a vehicle can occupy, at most, two adjacent zones. The next section illustrates deadlocks and proposes a graphical representation of the reservation schedule, which a vehicle has to construct while moving along its route. Section 3 proposes a method for detecting deadlocks by using the graphical representation of reservations. A method to prevent deadlocks is also presented. Section 4 introduces a simulation study conducted to evaluate the performance of the proposed method. Section 5 improves the method of Section 3 by decomposing the traveling area into multiple modules. Finally, concluding remarks are given in Section 6.

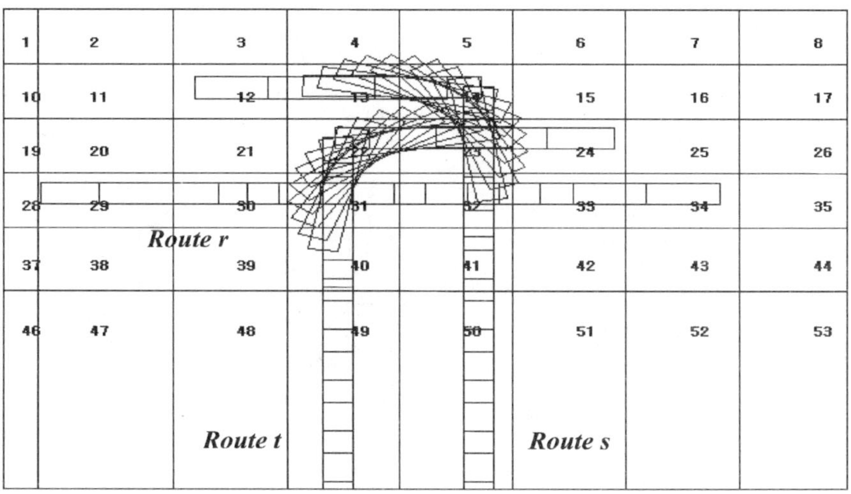

Fig. 4 An illustration of three different routes

2 A graphical representation of reservations

When a route is assigned to a vehicle, the vehicle occupies a series of grid-block sets in a sequence. To prevent collisions and deadlocks, a vehicle must reserve one or more grid-blocks before occupying them in addition to the grid-blocks it occupies. Occupied grid-blocks are those physically covered by an AGV (refer to Fig. 4). Reserved grid-blocks are those that will be exclusively occupied by an AGV, and no other AGV can be allowed to reserve or occupy these grid-blocks. If a vehicle cannot reserve its required grid-blocks, the vehicle is not allowed to move to the next position. In that case, the vehicle must wait until all the required grid-blocks become available for reservation.

Suppose that, in front of the transfer points of a block, three vehicles are expected to pass the area as shown in Fig. 4. Then, the set of occupied grid-blocks at each stage can be summarized as in Table 1. Each stage corresponds to a different set of occupied grid-blocks (Table 1). For an AGV to travel a route, because a grid-block can be occupied by only one AGV, the AGV must reserve new grid-blocks whenever required. A reservation is necessary in the three following cases: new grid-blocks are required for an AGV to move on these blocks (case 1); additional grid-blocks are required between the position of an AGV and the position where the AGV is supposed to stop, because time is needed for an AGV to come to a stop (case 2); additional blocks must be reserved to prevent deadlocks (case 3), which is the main issue of this study. Also, grid-blocks must be released for reservation by an AGV as soon as they become unoccupied. Table 2 illustrates how grid-blocks are reserved and released according to the reservation requirements for case 1. Occupied grid-blocks must also include the blocks required for case 2, as well as those for case 1, although Table 1 does not include them. Inclusion of grid-blocks for case 2 into occupied blocks does not change the discussion in the following.

This paper suggests a directed graph model in which a node denotes a stage of a vehicle during travel. Figure 5 illustrates a node representing a stage that a vehicle passes through while in transit. The route of a vehicle can be represented by a series of stages. The attributes of a stage consist of the sets of reserved grid-blocks, grid-

Table 1 Occupied grid-blocks in stages of three routes

Stages (i)	Route s	Route t	Route r
1	50	49	33, 34, 35
2	41, 50	40, 49	33, 34
3	32, 41, 50	31, 40, 49	32, 33, 34
4	23, 32, 41	22, 31, 40, 49	31, 32, 33
5	14, 23, 32, 41	22, 31, 40	31, 32
6	14, 23, 32	13, 22, 31, 40	30, 31, 32
7	13, 14, 23, 32	13, 22, 31	29, 30, 31
8	5, 13, 14, 23, 32	13, 14, 22, 23, 31	29, 30
9	4, 5, 13, 14, 23, 32	14, 22, 23, 31	
10	4, 5, 13, 14, 23	22, 23, 31	
11	4, 13, 14, 23	22, 23	
12	13, 14, 23	22, 23, 24	
13	13, 14		
14	12, 13, 14		

Table 2 Reservation required for an AGV to travel on route *s*

Stage	Occupied grid-blocks	Grid-blocks required for reservation	Grid-blocks to be released
1	50	41	
2	41,50	32	
3	32, 41, 50	23	50
4	23, 32, 41	14	
5	14, 23, 32, 41		41
6	14, 23, 32	13	
7	13, 14, 23, 32	5	
8	5, 13, 14, 23, 32	4	
9	4, 5, 13, 14, 23, 32		32
10	4, 5, 13, 14, 23		5
11	4, 13, 14, 23		4
12	13, 14, 23		23
13	13, 14	12	
14	12, 13, 14		

blocks to be reserved, and grid-blocks to be released. The grid-blocks listed as "to be released" are those that will be released from the reservation once the AGV enters the next stage. The grid-blocks listed as "reserved" are those that have been successfully reserved. Because a vehicle can only occupy grid-blocks it has successfully reserved, grid-blocks that are occupied by a vehicle have already been reserved by the vehicle. The grid-blocks listed as "required for reservation" are those for which reservations are required for the vehicle to proceed to the next stage.

A directed arc is connected from one node to another when the former has a grid-block as "required for reservation" and the latter has the same grid-block as "reserved." However, if a grid-block exists that is listed as "reserved" in both the nodes, then no arc can be drawn between them. Because no grid-block listed "reserved" is shared between two nodes connected by an arc, two vehicles can be simultaneously in the states represented by the two nodes. Also, because the vehicle in the state of the latter node has already reserved a grid-block which is required for the vehicle in the state of the former node, the vehicle of the former node cannot proceed to the next stage unless the vehicle of the latter node moves to another state in which the grid-block is released.

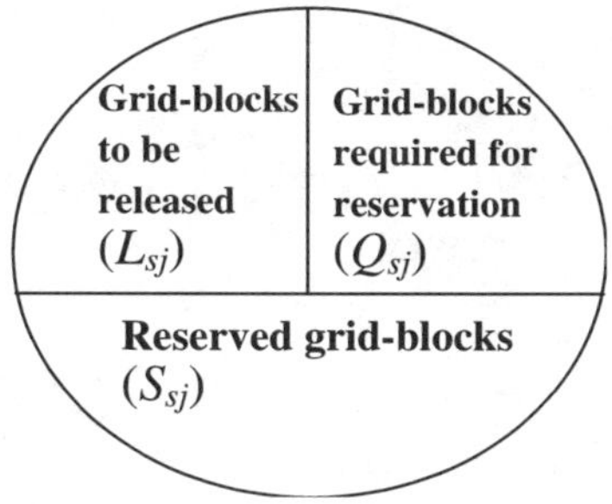

Fig. 5 Node in the reservation graph

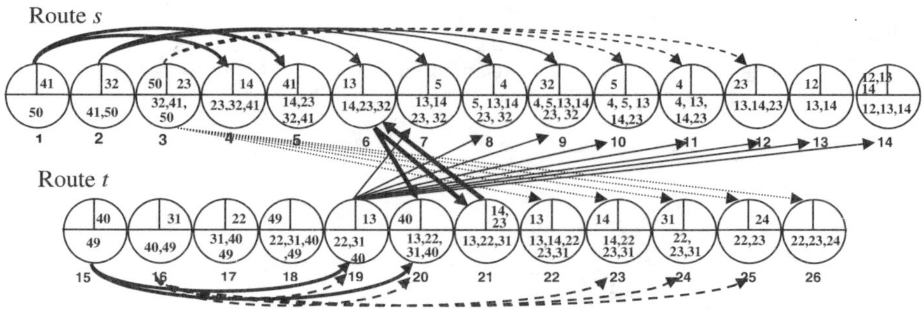

Fig. 6 A reservation graph for routes *t* and *s* in Table 1

Routes *s* and *t* in Table 1 are scheduled to enter to the area in Fig. 4. Also, assume that each vehicle in a stage only reserves the grid-blocks that are necessary for the next stage; that is, only the grid-blocks needed for case 1 are reserved. Then, the reservation schedule of the two vehicles can be drawn as shown in Fig. 6.

Figure 7a illustrates a deadlock situation. AGV 1 attempts to reserve grid-block B for a left turn, whereas AGV 2 attempts to reserve grid-block A for a right turn. Because grid-block A is already reserved by AGV 1 and grid-block B is in the set of reserved grid-blocks of AGV 2, an arc must be drawn from the node for the current stage of AGV 1 to that of AGV 2, and another arc must be drawn in the opposite direction as shown in Fig. 7b. Notice that the graph has a cycle, which implies that the two AGVs are in a deadlock state. Thus, to check whether multiple reservation schedules may cause a deadlock, it is necessary to check whether cycles exist in the reservation graph.

Once a reservation graph is constructed, it is possible to detect possible deadlocks by identifying cycles in the reservation graph. To find cycles in the graph, an algorithm to find components that are strongly connected may be used (Aho et al. 1974). The two following properties are obvious, considering how reservation graphs are constructed.

Property 1 If there exists no cycle in a reservation graph, then the corresponding schedules are deadlock-free.

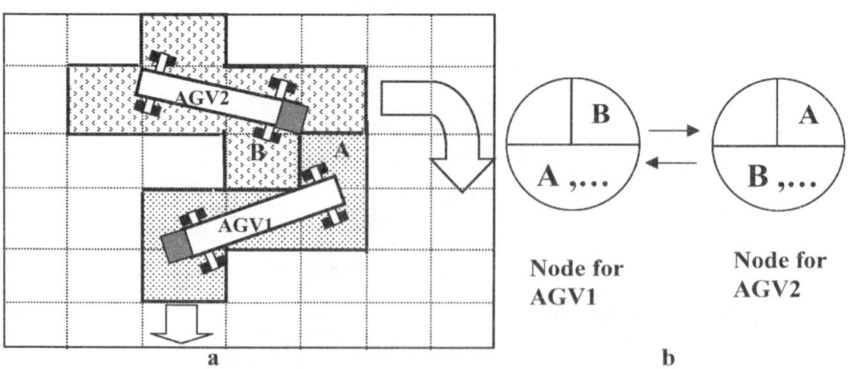

Fig. 7 An illustration of a deadlock

Property 2 If there is no cycle in which the number of nodes is not greater than the number of schedules in a reservation graph, then the schedules are deadlock-free.

The number of schedules in a reservation graph corresponds to the number of AGVs. A deadlock occurs when there is an AGV in every state represented by nodes on the cycle in the reservation graph. If the number of AGVs is smaller than the number of nodes on the cycle, then the deadlock that is represented by the corresponding cycle cannot occur.

3 A method for preventing deadlocks

This section proposes a method to prevent deadlocks between AGVs. The following notations will be used:

- m — The number of routes under consideration
- n_t — The number of stages on route t, which is known when a route is selected for a vehicle. Table 1 shows that routes s, t, and r have 14, 12, and 8 stages, respectively
- O_{tj} — The set of occupied grid-blocks at stage j of route t. $O_{t(n_t+1)} = \varnothing$
- S_{sj} — The set of reserved grid-blocks at stage j in the schedule for route s. $S_{s0}=\varnothing$
- Q_{sj} — The set of grid-blocks for which a reservation is requested at stage j in the schedule for route s. When a grid-block is requested for reservation at stage j, it will be included in $S_{s(j+1)}$, $Q_{s0}=\varnothing$
- L_{sj} — The set of grid-blocks for which releases are requested at stage j in the schedule for route s. When a grid-block is requested for release at stage j, it will not be in $S_{s(j+1)}$, $L_{s0}=\varnothing$
- $a_{(si)(tj)}$ — The directed arc from node (si), which represents stage i in the schedule for route s, to node (tj) on a reservation graph. A directed arc is connected from node (si) to node (tj) if there exists a grid-block, g, such that $g \in Q_{si}$, and $g \in S_{tj}$, and $S_{si} \cap S_{tj}=\varnothing$

The statement $S_{si} \cap S_{tj}=\varnothing$ implies that two vehicles can be at stage i of route s and state j of route t at the same time, respectively. The statement that there exists a grid-block, g, such that $g \in Q_{si}$ and $g \in S_{tj}$ implies that a vehicle at stage i of route s cannot proceed to the next stage as long as another vehicle remains at stage j of route t. Grid-block g is said to precede grid-block q if $g \in S_{si}$ and $q \in Q_{si}$ at node (si) and is represented as $g \rightarrow q$. The fact that $g \rightarrow q$ implies that grid-block g was reserved earlier than grid-block q.

Let P be a set of precedence relationships ($P=\{(a \rightarrow b)\}$). Note that ($a \rightarrow b$) and ($b \rightarrow a$) cannot be in the same set P, simultaneously. The precedence relationship is transitive, that is, the fact that $g_1 \rightarrow g_2$ and $g_2 \rightarrow g_3$ implies that $g_1 \rightarrow g_3$. That is, if $(a \rightarrow b) \in P$ and $(b \rightarrow c) \in P$, then $(a \rightarrow c) \in P$. The following property is useful for constructing deadlock-free routes for AGVs:

Property 3 If there exists a set of precedence relationships in which all the precedence relationships among grid-blocks in multiple schedules are included, then the multiple schedules are deadlock-free.

Proof In a reservation graph, by definition of an arc, a directed arc is the connection between two nodes when one node lists a grid-block as "required for reservation" and that grid-block already been reserved by the other node. The grid-block is called a "connecting grid-block." We define a grid-block, g, as preceding grid-block q if $g \in S_{si}$ and $q \in Q_{si}$ at a node (si). Thus, if there exists a cycle of length n in a reservation graph, then we can find a series of connecting grid-blocks, ($g_{(1)}$, $g_{(2)}, \ldots, g_{(n)}$), each of which corresponds to a node in the order of the directed cycle. Because $g_{(i)}$ is included in $S_{s(i+1)}$ and $g_{(i+1)}$ is included in $Q_{s(i+1)}$, $g_{(i)}$ must precede $g_{(i+1)}$. However, because there is also a directed arc from node (n) to node (1) in the cycle, $g_{(n)}$ must precede $g_{(1)}$ again, which is a contradiction. Thus, if there is a cycle in a reservation graph, then no set of precedence relationships (P) can exist which will satisfy the precedence relationships among grid-blocks in the reservation graph. Thus, the conclusion holds, QED.

3.1 A deadlock-free reservation scheduling (*DFRS*) method

Whenever an AGV begins moving, a reservation schedule for its route is constructed. At the moment, reservation schedules for other AGVs may already be being implemented. The scheduling method in this section is used to add a reservation schedule for a new AGV to the existing reservation graph so that deadlocks are prevented. This approach is practical considering that the algorithm in this paper must be implemented in real time and the revision of the schedules for previously arrived vehicles may be time-consuming.

The method in this section maintains a set of precedence relationships (P), creates a new node in a way that no precedence relationship among grid-blocks in P is violated, and adds new precedence relationships into P whenever a new node in the reservation graph is created. New nodes in a reservation schedule are created in chronological order.

For example, to construct node (t_j), we have to determine S_{tj}, L_{tj}, and Q_{tj}. However, S_{tj} and L_{tj} depend on already known data as follows: $S_{tj} = S_{t(j-1)} \cup Q_{t(j-1)}$ and $L_{tj} = O_{t(j+1)} - O_{tj}$. Thus, we only have to determine Q_{tj} to construct node (tj). The current P satisfies all the precedence relationships among blocks in schedules 1, 2, …, t–1 and the partial schedule from node 1 to node j–1 of schedule t. When we insert a grid-block (let it be grid-block u) into Q_{tj}, we have to find additional grid-blocks (let the set of these grid-blocks be denoted by "V") to be inserted into Q_{tj} together with grid-block u. A grid-block v in V satisfies the following conditions: (1) ($v \rightarrow u$)$\in P$, and (2) $v = \bigcup_{k=j+2}^{n_t} O_{tk} - \left(S_{tj} \cup Q_{tj}\right)$. Condition 2 means that grid-block v is included in neither S_{tj} nor Q_{tj} under construction. However, because $v = \bigcup_{k=j+2}^{n_t} O_{tk}$, grid-block v must be reserved for future occupation. This means that, if grid-block v is not reserved at this stage, the precedence relationship ($u \rightarrow v$) results. From conditions 1 and 2, if grid-block v is not reserved at this stage, two conflicting precedence relationships result, which implies a deadlock. Q_{tj} is constructed in a way of satisfying the precedence relationship among grid-blocks not only in schedules 1, 2, …, t–1, but also in the partial schedule from node 1 to node $j-1$ of schedule t. For example, when an AGV travels onto a bidirectional spur, because the AGV must go into the end of the spur and then come back from the end, we can expect conflicts among multiple grid-blocks in the precedent relationship. In this case, all the grid-blocks in the spur must be reserved simultaneously. This way of reservation prevents other AGVs from entering the spur until all the grid-blocks in the spur are released.

The set of precedence relationships includes records, which represent a precedence relationship between two grid-blocks and consist of the ID number of

Table 3 P after the reservation schedule for route s (Fig. 4) is constructed ($v \rightarrow u$)

Route ID	v	u	Case
s	50	41	1
s	41	32	1
s	50	32	1
s	32	23	1
s	41	23	1
s	50	23	1
s	23	14	1
s	32	14	1
s	41	14	1
s	50	14	2
s	14	13	1
s	23	13	1
s	32	13	1
s	41	13	2
s	50	13	2
s	13	5	1
s	14	5	1
s	23	5	1
s	32	5	1
s	41	5	2
s	50	5	2
s	5	4	1
s	13	4	1
s	14	4	1
s	23	4	1
s	32	4	1
s	41	4	2
s	50	4	2
s	4	12	2
s	5	12	2
s	13	12	1
s	14	12	1
s	23	12	2
s	32	12	2
s	41	12	2
s	50	12	2

the corresponding route, a grid-block with higher precedence, and a grid-block with lower precedence. When a vehicle completes its route, all of the records with that route ID are deleted from *P*.

Suppose that $t-1$ vehicles have already started their routes and a reservation schedule has been made for the *t*th AGV. This reservation schedule can be constructed as follows:

Step 0: j=0.
Step 1: j=j+1. If $j>n_t$, then stop. Otherwise, set $S_{tj} = S_{t(j-1)} + Q_{t(j-1)} - L_{t(j-1)}$, $Q_{tj} = O_{t(j+1)} - S_{tj}$, and $L_{tj} = O_{tj} - O_{t(j+1)}$. If $Q_{tj}=\varnothing$, then go to the beginning of this step. Otherwise, for all the pairs, (u, v), such that $u \in S_{tj}$ and $v \in Q_{tj}$ (case 1), insert (t, u→v) into P. Add the precedence relationships resulting from the transitive property and the route ID of t into P (case 2). Set $U=Q_{tj}$. Go to step 2.
Step 2: If $U=\varnothing$, then go to step 1. Otherwise, arbitrarily select *u* from *U* and check whether any v exists such that $v \in V$ and $(s, v \rightarrow u) \in P$ for some s, $1 \leq s \leq t$, where $V = \bigcup_{k=j+2}^{n_t} O_{tk} - (S_{tj} \cup Q_{tj})$. Let the set of v satisfying the above conditions be V*. If $V^*=\varnothing$, then go to the beginning of this step. Otherwise, for all $v^* \in V^*$, set $U = U + \{v*\}$ and $Q_{tj} = Q_{tj} + \{v*\}$. For all $r \in S_{tj}$, insert (t, r→v*) into P (case 3). Add the precedence relationships resulting from the transitive property and the route ID of t to P (case 2). Set $U=U-\{u\}$ and go to the beginning of this step.

Table 4 An illustration of the reservation scheduling for route *t*

Stage (*j*)	S_{tj}	Q_{tj}	New entries *v*→*u*	L_{tj}	*u*	*V*	*V**
1	{49}	{40}	49→40	∅	40	{31,22,13,14,23,24}	∅
2	{40,49}	{31}	40→31, 49→31	∅	31	{22,13,14,23,24}	∅
3	{31,40,49}	{22}	31→22, 40→22 49→22	∅	22	{13,14,23,24}	∅
4	{22,31,40,49}	∅		{49}			
5	{22,31,40}	{13}	22→13, 31→13 40→13, 49→13	∅	13	{14, 23, 24}	{14, 23}
5	{22,31,40}	{13, 14, 23}	22→14, 31→14 40→14, 49→14 22→23, 31→23 40→23, 49→23	∅	14, 23	{24}	∅
6	{13,14,22,23,31,40}	∅		{40}			
7	{13,14,22,23,31}	∅		{13}			
8	{14,22,23,31}	∅		{14}			
9	{22,23,31}	∅		{31}			
10	{22,23}	{24}	22→24, 23→24, 31→24, 40→24, 49→24	∅	24	∅	∅
11	{22,23,24}	∅		∅			

Table 3 shows *P* after the reservation schedule for route *s* (Fig. 4) has been constructed.

Grid-blocks v^* in V^* in step 2 are those that will produce a cycle in the reservation graph unless they are reserved together with grid-block *u*. Because we assume that the reservation schedule for route *s* was constructed first, there is no possibility of a deadlock. Thus, the reservation schedule can be constructed by reserving only the grid-blocks required to travel route *s* and releasing the grid-blocks that an AGV actually releases while travelling its route as shown in Table 2.

Under the condition that a vehicle on route *s* has already begun its route, when a new vehicle with route *t* tries to begin its route, a new reservation schedule is constructed, as shown in Table 4, and new precedence relationships are added to *P*, as shown in Table 5. Note that, when grid-block 13 is reserved for an AGV, grid-blocks 14 and 23 must be reserved at the same time to prevent a deadlock with an AGV travelling on route *s*.

4 A simulation study to evaluate the performance of *DFRS*

A simulation study was conducted to evaluate the performance of the reservation scheduling algorithm used in this paper. The container terminal is assumed to have three QCs and seven yard blocks, as shown in Fig. 1. Each AGV lane allows travel

Table 5 The precedence table after constructing the reservation schedule for route *t*

Route ID	*v*	*u*	Case
t	49	40	1
t	40	31	1
t	49	31	1
t	31	22	1
t	40	22	1
t	49	22	1
t	22	13	1
t	22	14	3
t	22	23	3
t	31	13	1
t	31	14	3
t	31	23	3
t	40	13	1
t	40	14	3
t	40	23	3
t	49	13	2
t	49	14	2
t	49	23	2
t	22	24	1
t	23	24	1
t	31	24	2
t	40	24	2
t	49	24	2

in a predetermined direction. When an AGV travels in a straight lane, its speed is 6.5 m/s, whereas, when an AGV turns, its speed is reduced to 2 m/s. The simulation program was developed using Visual C++.

Because no previous algorithm was found to resolve deadlocks in this situation, we compared the algorithm in this paper with a simple reservation method (SM). SM reserves a specified number (α) of grid-blocks, which will be occupied in the future. However, when an AGV is traveling toward a transfer point in front of a block, approaches a turning point from a running lane to go to the transfer point, and arrives at the position where a specified number (β) of grid-blocks remain before the turning point, SM reserves all the grid-blocks through to the transfer position. When the AGV leaves the transfer point, SM reserves all the grid-blocks from the transfer point to the last grid-block on which the AGV will turn. α and β will be called the number of grid-blocks for advanced reservation when traveling in a straight lane and turning, respectively. After an AGV arrived at its destination, its next destination was generated randomly. The simulation program was run on a Pentium IV processor with 1.7 GHz and 512 MB of memory.

We could not find a simple method to determine the values of α and β to guarantee the deadlock-free operation of AGVs in a given operating situation. Thus, simulation studies were conducted to find the minimum values of α and β in which AGVs can travel without deadlocks. While increasing the values of α and β one by one, the simulation was conducted until the values of α and β for which no deadlock was observed were found. Table 6 shows the percentage of simulation runs during which deadlocks occurred before the number of trips reached 300. For each combination of α and β, the simulation was conducted 20 times. When α=3 and β=7, no deadlock was found during the simulation run. Thus, as a reference rule, the rule of SM 3/7 was used in the following comparisons.

To compare performances of *DFRS*, which is the deadlock-free reservation scheduling method in this study, with those of SM 3/7, a simulation study was conducted. Delivery orders were randomly generated between positions of QCs and transfer points of blocks. During the simulation study, the number of AGVs was changed from six to 16. The simulation run was repeated ten times for each condition. Figure 8 compares the average speed of the AGVs in SM 3/7 and *DFRS*. The simulations using *DFRS* showed that the AGVs traveled, on average, 9.3% faster than those on SM 3/7 because there were fewer blockages on *DFRS*.

Figure 9 compares the average percent of the time that an AGV was blocked by other AGVs during a trip from its starting position to its destination, which we call "blocked time." For both reservation methods, the average percent of the blocked time increased as the number of AGVs increased. However, the average percent of the blocked time was much lower for *DFRS* than for SM 3/7.

Table 6 The number of reserved grid-blocks in SM and the percentage of simulations with deadlocks

No of grid-blocks for advanced reservation during running/turning (α/β)		SM 3/7 (%)	SM 2/2 (%)	SM 3/3 (%)	SM 4/4 (%)	SM 5/5 (%)
No of AGVs	8	0	50	50	50	70
	16	0	70	70	70	80

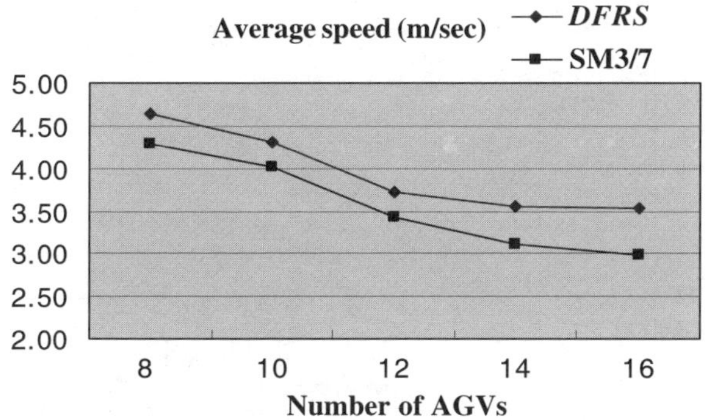

Fig. 8 Comparison of the average speed of AGVs

The yard space needed for AGV movement is another important resource that must be utilized efficiently. The amount of space used by an AGV can be determined by multiplying the amount of space an AGV reserves by the duration of the reservation. The average reservation space per unit time was evaluated by summing the amount of space used by all the vehicles, and then dividing the sum of travel times of all the vehicles. Figure 10 compares the average amount of space per unit time that *DFRS* and SM 3/7 reserved. Figure 10 shows that, as a whole, SM 3/7 used more than twice the space that *DFRS* used.

5 Partitioning traveling area to increase the utilization of space

In the previous section, *DFRS* considered all of the precedence relationships between all of the grid-blocks for all moving vehicles when scheduling a reservation for a newly dispatched vehicle. However, suppose that two different vehicles pass through the same area at very different times, so that there is no possibility for the two vehicles to be in the area at the same time. Then, when

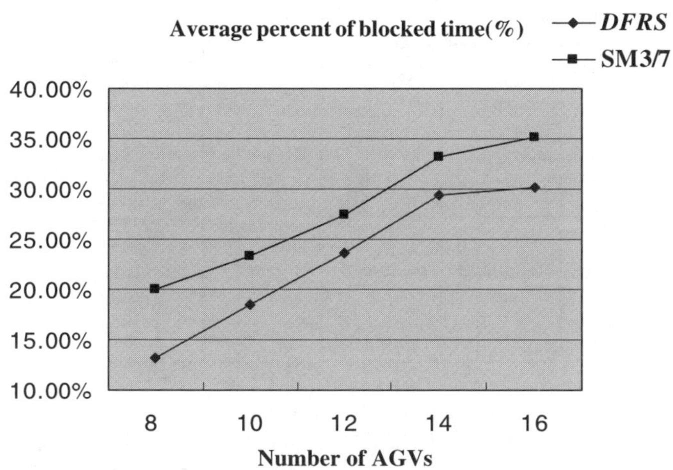

Fig. 9 Comparison of the average percent of blocked time of AGVs

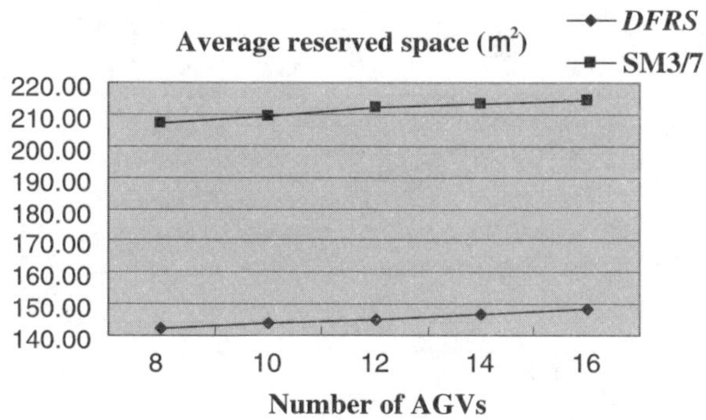

Fig. 10 Comparison of the average reserved space per unit time

constructing the schedule for one vehicle, it is not necessary to consider the precedence relationships between the grid-blocks reserved for the route of the other vehicle. In this sense, *DFRS* reserves more grid-blocks than is necessary to prevent deadlocks. That is, the method in this is a little restrictive because the time is not considered in the set of the precedence relationship (*P*). However, considering the time for constructing *P* will make the algorithm too complicated. To resolve this problem, a method to partition the apron area for AGV travelling into multiple modules is proposed in this section. One set of precedence relationships is maintained for each module and the set is updated only when a vehicle enters or departs that module. Thus, the set of the precedence relationships for a module considers the reservation schedules only for vehicles passing through that module. This method should result in improved utilization of space and higher vehicle

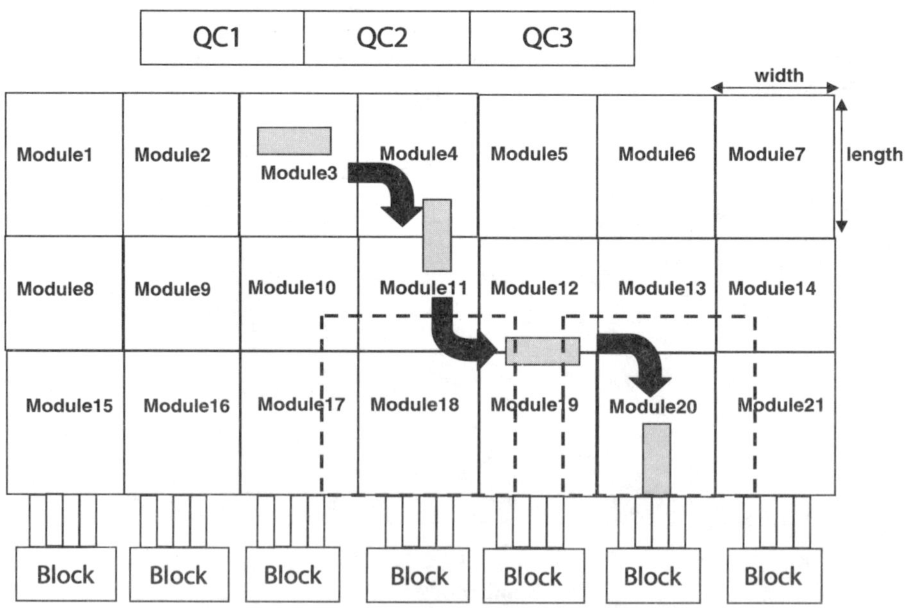

Fig. 11 Partitioning of the traveling area into modules

speeds. This method is a compromise between the complexity of considering the time for constructing *P* and the restrictiveness of the algorithm in Section 3.

Figure 11 illustrates an apron area for AGV travelling, which consists of 21 modules. The width of a module equals the width of a yard block, which was 42 m in the assumed layout. The lengths of a module under QCs and in front of each yard block equal the total width of AGV lanes under QCs (36 m) and that in front of a yard block (20 m), respectively. However, adjacent modules overlap each other, as shown in Fig. 12. The overlapping areas were used to prevent deadlocks on girds near the borders between two adjacent modules. Before a vehicle reaches position *A*, the reservation schedule of the vehicle included grid-blocks only in module 1, and this schedule only considered the value of *P* for module 1. However, after the vehicle entered the overlapping area and before it reached position *C*, the reservation schedule of the vehicle included grid-blocks in both modules 1 and 2, and the schedule considered not only the value of *P* for module 1 but also that for module 2.

When a vehicle enters a module, a new reservation schedule for the vehicle is constructed considering the value of *P* for the module, and updates the value of *P* for the module at the same time. When a vehicle departs a module, all the precedence relationships with the ID number of the corresponding route are deleted from *P*. The procedure to construct the reservation schedule is similar to that in the previous section, except that a *P* is maintained for each module. The deadlock-prevention algorithm in this section is called "modified *DFRS*."

The modified *DFRS* algorithm in this section was compared with *DFRS* in the previous section by a simulation study. The various conditions for the experiment were the same as in the experiment of the previous section. Figure 13 compares the average velocity of vehicles for different numbers of AGVs assigned to each QC, when the two deadlock prevention algorithms were used. The average speed of the vehicles was higher when the modified *DFRS* was used than when *DFRS* was used.

Figure 14 compares the average percent of blocked time during the travel from a starting position to an arrival position for the two deadlock prevention algorithms. The average blocked time was lower when the modified *DFRS* was used than when *DFRS* was used. Figure 15 compares the average reserved area for both deadlock prevention algorithms, which shows the modified *DFRS* utilizes the space more efficiently than *DFRS*.

Figure 16 shows the average computational time required to schedule a trip for an AGV when *DFRS* and modified *DFRS* are used. The computational time increased as the number of AGVs increased. When *DFRS* was used, as the number

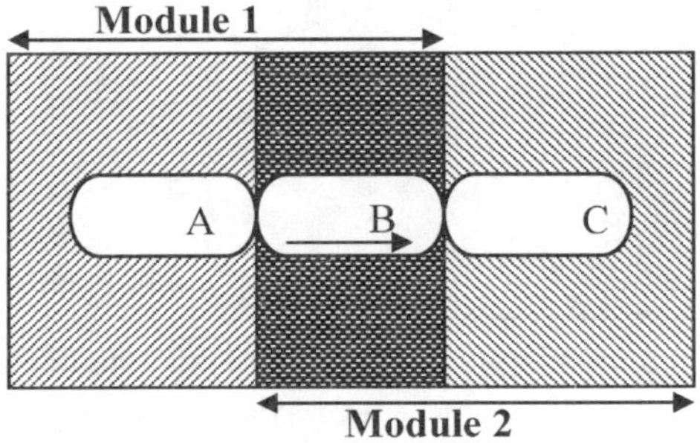

Fig. 12 An AGV crossing a border between modules

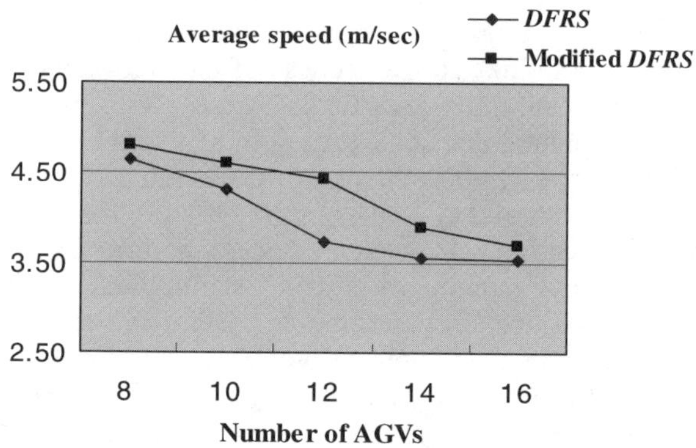

Fig. 13 The average speed of AGVs

of vehicles increased, the average computational time for scheduling a route increased significantly, as shown in Fig. 16. The computational time averaged 2.8 s for 16 AGVs, which is a long time for use in real time. The computational time was about 1 s when the modified *DFRS* was used. The value for the modified *DFRS* represents the average computational time for scheduling a route within a module. The computational time for the modified *DFRS* seems to be much shorter than that for *DFRS*.

The second simulation study was done by using a more realistic model. As in the first simulation study, the layout in Fig. 1 was assumed. However, unlike the first simulation study in which the delivery orders were issued randomly between QCs and blocks, the second simulation model described all the detailed operations in container terminals. Based on the arrival data of four vessels, containers for receiving, delivering, unloading, and loading were generated. The simulation was conducted for the unloading and loading operations of four vessels, with 1,480 unloading moves and 1,280 loading moves in total. For locating outbound containers, it was attempted to cluster containers with similar attributes in positions close to each other. For locating inbound containers, congestions in the transfer positions of the yard blocks were considered as having the highest priority. For dispatching vehicles, balancing loads between different yard blocks was emphasized.

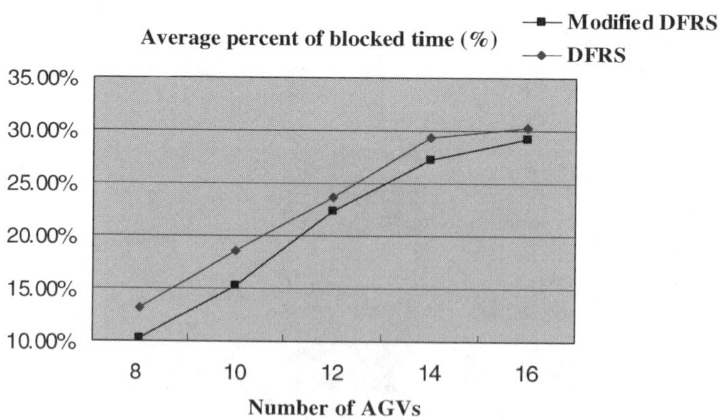

Fig. 14 The average percent of blocked time of vehicles

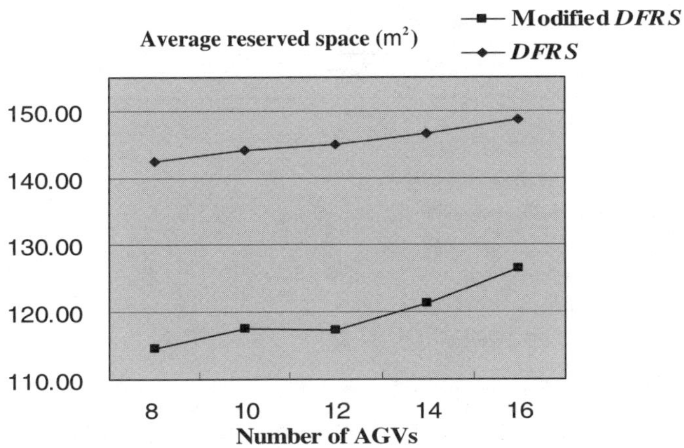

Fig. 15 The average reserved space

Table 7 summarizes the results of the experiment. The average speed of vehicles for different numbers of vehicles per QC was between 3.7 and 4.7 m/s. The average number of grid-blocks, which are reserved for preventing the deadlocks, divided by the total number of grid-blocks reserved was about 0.28–0.64%. This result implies that the number of grid-blocks to be reserved for the purpose of preventing deadlocks is very small compared with the number of grid-blocks to be reserved for preventing collisions among vehicles. As long as the number of AGVs assigned to a QC increased from 8 to 14, the average computational time to construct a reservation schedule stayed below 1 s. Although the productivity of the unloading and loading operations not only depends on the reservation scheduling algorithm but also on other operation rules for AGVs, including the dispatching rule and the routing rules, as the number of AGVs increased, the productivity of the loading operation did not show much change, while the productivity of the unloading operation increased slowly. Instead, as the number of vehicles assigned to a QC increased, the average service time of road trucks, which is the duration of stay of trucks at yard blocks for the service, decreased significantly.

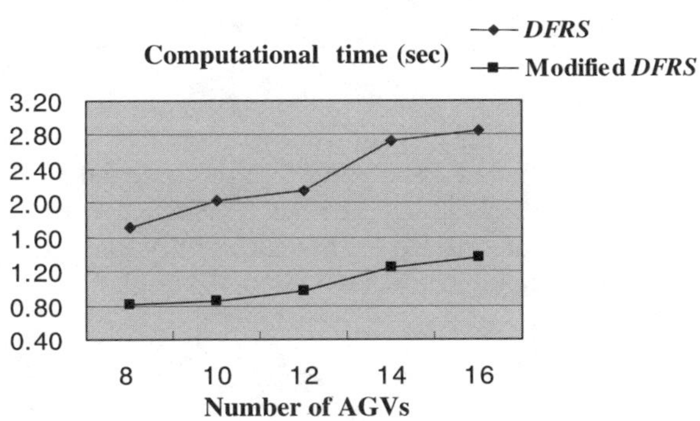

Fig. 16 The average computational time to schedule a trip

Table 7 Results of the second simulation study

	No of AGVs per QC			
	8	10	12	14
Average speed (m/s)	4.79	4.55	3.92	3.70
Average percent of blocked time (%)	13.2	20.3	25.6	30.1
Average percent of grid-blocks reserved for preventing deadlocks (%)	0.28	0.49	0.64	0.64
Average computational time for constructing a schedule (s)	0.84	0.86	0.88	0.90
Average no of loadings per hour per QC	25	27	27	26
Average no of unloadings per hour per QC	26.5	28.5	29	29
Average truck service time (s)	480	455	430	400

The second simulation study, which was done by using a more realistic model than that for the first simulation study, showed more promising results for the reservation scheduling method in this paper to be used for preventing deadlocks in real situations.

6 Conclusion

This paper addresses the operational problem of AGVs in container terminals. The number of AGVs used in container terminals is usually large (more than 50 or 60) and AGVs are larger than those used in manufacturing systems. Thus, controlling AGV traffic has been an important issue. This paper proposes deadlock detection and prevention algorithms for AGVs. It was assumed that vehicles reserve grid-blocks in advance to prevent collisions and deadlocks among AGVs.

A graphic representation method, called the "reservation graph," was proposed to express a reservation schedule in such a form that the possibility of a deadlock can be easily detected. A method to detect possible deadlocks by using the reservation graph was suggested. Also, a *DFRS* was proposed, which used a precedence relationship between the grid-blocks. The *DFRS* schedules the reservation for an AGV in a way such that the schedule does not violate the precedence relationships that have already been developed for other schedules.

A simulation was conducted to evaluate the *DFRS* in this study. The *DFRS* was compared with a heuristic rule for reservation scheduling (SM 3/7), in which a fixed number of grid-blocks are reserved whenever a new reservation is necessary. It was found that *DFRS* outperformed SM 3/7 in the average speed of AGVs, the average time AGVs were blocked per trip, and the amount of area reserved for the routes.

DFRS was modified to utilize space more efficiently. The modified *DFRS* partitioned the apron area for AGV traveling into multiple, smaller-sized modules and constructed a reservation schedule whenever a vehicle entered a module. When a reservation schedule was constructed for a vehicle, only the priority relationships for the grid-blocks in that module were considered. A simulation study showed that the modified *DFRS* had higher travel speeds, shorter blocked times, and better space utilization than the original *DFRS*.

DFRS was tested in a simulation environment that was similar to real ship operations in practice. The simulation study showed that the modified *DFRS* is satisfactory from the viewpoint of the average speed of vehicles, the space utilization, and the computational time, which implies that the algorithm has potentials to be used in practice.

Acknowledgement This research was accomplished as a part of the project "Development of Intelligent Port and Logistics System for Super-Large Container Ships," which was sponsored by the Ministry of Maritime and Fishery in Korea.

References

Aho AV, Hopcroft JE, Ullman JD (1974) The design and analysis of computer algorithms. Addison-Wesley, Reading, pp 189–195

Banaszak ZA, Krogh BH (1990) Deadlock avoidance in flexible manufacturing systems with concurrently competing process flows. IEEE Trans Robot Autom 6(6):724–734

Cho HB, Kumaran TK, Wysk RA (1995) Graph theoretic deadlock detection and resolution for flexible manufacturing systems. IEEE Trans Robot Autom 11(3):413–421

Duinkerken MB, Evers JJM, Ottjes JA (1999) TRACES: traffic control engineering system. Proceedings of the 31st Summer Computer Simulation Conference, pp 461–465

Evers JJM, Koppers SAJ (1996) Automated guided vehicle traffic control at a container terminal. Transp Res Part A Policy Pract 30(1):21–34

Fanti MP (2002) Event-based controller to avoid deadlock and collisions in zone-control AGVS. Int J Prod Res 40(6):1453–1478

Fanti MP, Maione B, Mascolo S, Turchiano B (1997) Event-based feedback control for deadlock avoidance in flexible production systems. IEEE Trans Robot Autom 13(3):347–363

Grunow M, Günther H-O, Lehmann M (2004) Dispatching multi-load AGVs in highly automated seaport container terminals. OR Spectr 26(2):211–236

Kim CW, Tanchoco JMA (1991) Conflict-free shortest-time bi-directional AGV routing. Int J Prod Res 29(12):2377–2391

Kim CW, Tanchoco JMA (1993) Operational control of bi-directional AGV system. Int J Prod Res 31(9):2123–2138

Lee CC, Lin JT (1995) Deadlock prediction and avoidance based on Petri nets for zone control Automated Guided Vehicle Systems. Int J Prod Res 33(12):3239–3265

Möhring RH, Köhler E, Gawrilow E, Stenzel B (2004) Conflict-free real-time AGV routing. Preprint 026-2004. Technical University Berlin, Institute of Mathematics

Rajeeva LM, Wee HG, Ng WC, Teo CP, Yang NS (2003) Cyclic dead-lock prediction and avoidance for zone-controlled AGV system. Int J Prod Econ 83:309–324

Reveliotis SA (2000) Conflict resolution in AGV system. IIE Trans 32:647–659

Viswanadham N, Narahari Y, Johnson TL (1990) Deadlock prediction and deadlock avoidance in flexible manufacturing systems using petri net models. IEEE Trans Robot Autom 6(6):713–723

Vis IFA, Harika I (2004) Comparison of vehicle types at an automated container terminal. OR Spectr 26(1):117–143

Wu NQ (1999) Necessary and sufficient conditions for deadlock-free operation in flexible manufacturing systems using a colored petri net model. IEEE Trans Syst Man Cybern Part C Appl Rev 29:182–204

Wu NQ, Zhou MC (2000) Resource-oriented petri nets for deadlock avoidance in automated manufacturing. Proceedings of 2000 IEEE International Conference on Robotics and Automation, San Francisco, pp 3377–3382

Yang CH, Choi YS, Ha TY (2004) Simulation-based performance evaluation of transport vehicles at automated container terminals. OR Spectr 26(2):149–170

Yeh MS, Yeh WC (1998) Deadlock prediction and avoidance for zone-control AGVs. Int J Prod Res 36(10):2879–2889

Part 3:

Cargo systems

Dirk C. Mattfeld · Holger Orth

The allocation of storage space for transshipment in vehicle distribution

Abstract We address the planning of transportation and storage capacity over time. In intermodal transshipment terminals, finished vehicles are assigned to yard locations for intermediate storage. The evolutionary algorithm proposed evolves a period-oriented capacity utilization strategy. This capacity utilization strategy then controls a construction heuristic which assigns vehicle movements to periods and vehicles to storage locations. It is aimed at efficient operations and at a balanced distribution of vehicle movements over the periods of the planning horizon.

Keywords Vehicle transshipment · Multi-period model · Task assignment · Storage space allocation · Construction heuristic · Evolutionary algorithm

1 Finished vehicle transshipment

Automobile manufacturers aim at strategic competitive advantages by distributing their activities around the globe (Spatz and Nunnenkamp 2002). The division of automobile production entails an increased volume of vehicles shipped by means of worldwide transportation networks. These networks are typically run by logistics service providers consigned with the transportation, transshipment, and storage of vehicles (Rodrigue 1999). Service providers aim at economies of scale due to a consolidation of transport volume incurred on behalf of different vehicle manufacturers (Tyan et al. 2003).

This has led to the emergence of inter-modal terminals at seaports, handling enormous volumes of vehicles. Carriers operate at a liner schedule and call at vehicle terminals at fixed points in time. Since the time for an oversea transport

D. C. Mattfeld (✉)
Technical University Braunschweig, Abt-Jerusalem-Str. 4, 38106 Braunschweig, Germany
E-mail: d.mattfeld@tu-braunschweig.de

H. Orth
Orth IT-Dienstleistungen, Marienthaler Str. 30, 20535 Hamburg, Germany
E-mail: holger@orth.de

may take several weeks, the number of vehicles to be transshipped is pretty much known in advance. Vehicles are transported in charges of a few dozens up to hundreds or even thousands in rare cases. Vehicles of a charge have the same destination and undergo an identical treatment concerning transportation and transshipment.

The inter-modal split at a port necessitates an intermediate storage of vehicles at a terminal. Import vehicles arrive by car-carrier and usually undergo an inspection procedure before they are forwarded to the hinterland. This requires an intermediate storage of vehicles. Export vehicles arrive via rail or feeder ship. They typically stay at the terminal yard for the purpose of consolidation before they are loaded onto car-carriers for oversea transport.

For each charge of vehicles, two relocations are considered: the storage from a transfer point (quay, rail ramp, and truck yard) into a storage location and the retrieval from a storage location to a transfer point. Since vehicles demand ground storage, the extent of the storage yard required is remarkable. Vehicle movements are performed by driving personnel, leading to a noticeable manpower demand. The movement of a charge of vehicles is performed by a typically small group of drivers. After every driver of a group has moved one vehicle, the group is picked up by a taxi.

For drivers, the avoidance of damage is of top priority. Nevertheless, efficient operations have to be ensured, which seemingly contradicts the principle of safety and reliability. In order to achieve safe and reliable operations, it is aimed at balancing the manpower demand over the planning horizon given in terms of periods, i.e., two to three working shifts per day in the interval of 1 to 2 weeks. To provide efficient operations, as a secondary goal, the sum of working hours needed over the entire planning horizon is minimized.

To achieve these goals, two decisions are to be taken. First, storage locations for the intermediate storage of vehicle charges are to be chosen. Second, a period for the execution of a relocation is to be determined subject to a typically narrow time-window given by the customer, i.e., the vehicle carrier.

The transport distance within the terminal is a crucial issue for planning. In order to minimize the sum of working hours for the driving personnel, the location with the smallest sum of storage and retrieval distance will be chosen. However, one may accept a longer driving distance for storage operation in periods of small to modest manpower utilization. This decision turns out advantageous if it yields short retrieval distance for vehicles in a forthcoming period of congested manpower utilization.

By performing storage operations early and retrieval operations late, the average number of vehicles stored in the terminal becomes minimal. A small utilization of the storage yard allows the greatest choice among storage locations, supporting efficient operations that way. By assigning an earlier period of storage and/or a later period of retrieval, the vehicles unnecessarily occupy storage space. However, this option can support the balancing of the manpower demand over the planning horizon. Thus, a decision is to be taken among conflicting options.

In this paper, a methodological support for terminal management is addressed in order to provide both safe and efficient operations for real world-sized vehicle transshipment problems. In Section 2, we review some related models. In Section 3, we present a mathematical formulation of the problem. In Section 4, we develop a construction heuristic and an evolutionary algorithm for the control of the construction heuristic. In Section 5, we perform a thorough computational investigation. Finally, we conclude.

2 Space allocation problems

The application at hand can be modeled as a space allocation problem, which considers the allocation of storage capacity to inventory with respect to transportation effort (Kusiak 2000). McKendall and Jaramillo (2006) report on the resource assignments of project activities in terms of a dynamic space allocation problem. The authors focus on the assignment of resources to locations over time with respect to minimizing the sum of the distances.

Literature concerned with space allocation in container transshipment stresses the particularities of equipment handling (Steenken et al. 2004). The stacking of containers decreases the storage space needed and, hence, also decreases the mean transportation effort required. To the opposite, handling work increases in case of stacking and, therefore, we observe a trade-off between the consumption of storage space and the handling work required (Taleb-Ibrahimi and Castilho 1993). The authors aim at calculating the minimum space required for a given transshipment rate. Furthermore, the minimal handling costs can be obtained for a given storage space. Both figures address strategic/tactic decisions only.

Preston and Kozan (2001) suggest optimizing the allocation of storage space at an operational level such that setup times (synonymous for handling work) and transport time become minimal. The authors use a genetic algorithm in order to generate a sequence of geo-coordinates, at which containers are to be placed. Stacking of containers is penalized by additional handling times, whereas unnecessary detouring during container placements is penalized by additional transport effort. The approach does not consider multiple periods of transshipment.

Time is incorporated implicitly by applying suitable rules for space allocation, i.e., based on the duration of stay of containers in a yard (Kim and Park 2003). Bish et al. (2001) make timing explicit by means of a (single period) scheduling model. The authors minimize the time needed to unload a container ship under the constraints of limited availability of storage areas and transport vehicles. The model developed assigns container to storage areas and vehicles to containers. Handling work is not considered in this rather simple model. Hartmann (2004) suggests a general model to schedule operations at straddle carriers, automated guided vehicles, stacking cranes, etc. A dispatching rule and an evolutionary algorithm are proposed to solve these problems.

Zhang et al. (2003) propose two successive MIP formulations to be solved for a multi-period problem. First, the workload is balanced over the storage areas available before, second, the transportation effort is minimized. Workload can be balanced between various periods by means of space allocation decisions. This multi-period assignment problem is applied on the basis of a rolling time horizon by currently adjusting the solution to forthcoming changing conditions. Murty et al. (2005) propose to incorporate dynamic load attributes into space allocation decisions.

However, space allocation approaches for container storage are fundamentally different from those for vehicle storage at transshipment terminals. Handling work is assumed constant and, therefore, just transport effort is to be minimized in storage space allocation models (see Mattfeld (2003) for an earlier work on this subject).

Moreover, the balancing of workload is of first importance. Different from container transshipment, where the buffering of containers at a marshaling area can be used to balance workload over time, in vehicle transshipment such buffer facilities do not exist. In order to avoid damages to vehicles, the number of vehicle movements is to be kept at the absolute minimum.

In the space allocation for container transshipment, the control of individual containers in a stochastic environment receives particular importance. In vehicle transshipment, an identical treatment applies to typically hundreds of vehicles grouped in a charge. Therefore, a charge can serve as an aggregate planning object for optimization in a deterministic multi-period model (Mattfeld 2006).

3 Problem modeling

We consider a vehicle transshipment terminal consisting of a network of spatially distributed storage locations of finite capacity interconnected by travel ways. This network is extended by transfer points (i.e., quays, rail ramps, and truck loading areas) depicting the origin and destination of vehicles. The manpower demand required by the transport of vehicles is determined as a function of volume and distance covered.

Services offered to customers comprise the transshipment of a charge of vehicles from one transfer point to another. Customers do not necessarily insist on the transshipment within a certain period but allow time-windows for both relocation types, the storage into the terminal, and its corresponding retrieval.

Central to our approach is the notion of a *task* (Mattfeld and Kopfer 2003). A task comprises the relocation of a charge of identical (assumed) vehicles, which are treated as entity for planning. The vehicles belonging to a task are supposed to be transported from an origin to a destination in a given, typically narrow, time-window. We differentiate between "storage tasks" entering vehicles to the terminal and "retrieval tasks" performing the vehicle dispatch from the terminal.

In order to model that the storage and retrieval may fall asunder, we consider the transshipment of a charge of vehicles as a pair of storage and retrieval tasks coupled by a precedence constraint. If intermediate storage is unavoidable, a storage location of sufficient capacity is chosen.

If the time-windows for a pair of storage and retrieval tasks overlap, both tasks are assigned to the same period and the vehicles are transshipped directly. This allows a routing of vehicles through the terminal without taking the capacity constraints of storage locations into account. Since storage space is a scarce resource, a direct transshipment is performed whenever possible. We model this case as a single transshipment task with a time-window covering the overlapping periods of the storage and retrieval tasks.

The assignment of tasks to periods and the allocation of storage space for each task have to be performed over the periods of the planning horizon. Thereby, the deviation of the manpower demand over the periods is to be minimized while keeping the overall manpower demand reasonably small. In the following, we describe the problem resources before we discuss decision variables and related consistency conditions. Then, we turn to a description of the objective function and the constraints involved.

3.1 Resources and variables

3.1.1 Storage locations and transfer points

We consider a terminal consisting of the set $F = \{1, \dots, m\}$ of storage locations and transfer points. For location $i \in F$, $H_i \in \{\mathtt{I}, \mathtt{E}\}$ denotes whether i is an internal storage location ($H_i = \mathtt{I}$) with capacity K_i or an external transfer point ($H_i = \mathtt{E}$) without storage capacity. Let K^{min} be the smallest storage location size to be utilized, then

$$0 \le B_i \le K_i \quad \forall i \in F \qquad \text{with } K_i \begin{cases} \ge K^{min}, & \textit{if } H_i = \mathrm{I} \\ = 0, & \textit{if } H_i = \mathrm{E} \end{cases}$$

holds, with B_i being the initial inventory of location i.

3.1.2 Productivity measure

The production coefficient $D_{i_1 i_2}$ determines the driving time required for the relocation of a single vehicle between location i_1 and location i_2. In this way, $D_{i_1 i_2}$ provides a manpower-oriented measure for the distance between i_1 and i_2. $D_{i_1 i_2} > 0 \ \forall i_1, i_2 \in F$ with $i_1 \neq i_2$ holds.

3.1.3 Tasks

A set $A = \{1, \dots, n\}$ of tasks is to be assigned to the periods of the planning horizon. A task $j \in A$ is labeled with $j_{(Y_j)}$ as one of three relocation types $Y_j \in \{\mathtt{S}, \mathtt{R}, \mathtt{T}\}$, namely *storage*, *retrieval*, and *transshipment*.

A task j consists of volume L_j given in terms of the number of vehicles to be relocated. A minimal task volume L^{min} limits tasks to useful transshipment operations, thus $L_j \ge L^{min} \ \forall j \in A$. $L^{min} \le K^{min}$ holds; otherwise, storage location i with $K_i = K^{min}$ does not provide capacity for the smallest possible task volume and is therefore not usable in any way.

Each task j consists of an origin $q_j \in F$ and a destination $z_j \in F$. Newly arriving vehicles typically have to be moved from their externally given transfer point Q_j to a storage location. The destination of a retrieval task is also a given transfer point Z_j. Note that capital letters Q_j and Z_j denote prescribed external transfer points, whereas lower case letters indicate that the destination of a storage task is not determined in advance.

The destination of a storage task z_j can be freely chosen. However, the destination of a storage task z_j and the origin of the corresponding retrieval task q_j are identical. To model this relation, we tie retrieval task j to its preceding storage task

V_j. Thus, for a retrieval task j, $q_j = z_{V_j}$ warrants that the internal inventory system is kept consistent.

A direct transshipment is carried out from a given origin Q_j to a given destination Z_j and, therefore, no choice can be taken. In summary, the following constraints restrict the choice of locations:

$$\begin{aligned} &\forall j \in A \text{ with } Y_j = \text{S hold}: \quad q_j = Q_j\,, \\ &\forall j \in A \text{ with } Y_j = \text{R hold}: \quad q_j = z_{V_j} \quad \text{and} \quad z_j = Z_j\,, \\ &\forall j \in A \text{ with } Y_j = \text{T hold}: \quad q_j = Q_j \quad \text{and} \quad z_j = Z_j\,. \end{aligned} \tag{1}$$

3.1.4 Discrete time model

Time is modeled as discrete periods. The transshipment of vehicles is planned up to T periods in advance; $t \in \{0, 1, \ldots, T\}$ addresses the period to consider.

3.1.5 Task-to-period assignment

For each task j, a period of processing $s_j \in \{1, \ldots, T\}$ has to be determined within the planning horizon. The period s_j of processing task j is constrained by a time-window $[EET_j, LET_j]$ between the earliest execution time EET_j given and the latest permissible execution time LET_j. Moreover, the processing of retrieval task j is restricted to periods after the processing of its corresponding storage task V_j, i.e., $s_{V_j} + 1 \leq s_j$ holds. The temporal constraints implied by the planning horizon, the time-windows, and the precedence relations existing are depicted by $\max\{1, EET_j, s_{V_j} + 1\} \leq s_j \leq \min\{LET_j, T\}$ for all j considered.

3.1.6 Inventory holding

The storage yard will be utilized by vehicles at the beginning of the first period considered $t = 1$. The initial inventory B_i of a dedicated location i depicts the vehicles stored at the fictitious period $t = 0$. In order to retrieve the initially stored vehicles, retrieval tasks without corresponding storage tasks may exist. For a retrieval task j, this case is labeled with $s_{V_j} := 0$.

The auxiliary variables $l_{t,i}$ determine the inventory of storage location i in period t starting from its initial inventory B_i. The variables $l_{t,i}$ of location i are tied in accordance to t by a dynamic inventory balance equation[1]:

$$\begin{aligned} &l_{0,i} = B_i && \forall i \in F, \\ &l_{t,i} = l_{t-1,i} + \sum_{j \,\middle|\, \substack{Y_j = S \\ s_j = t \\ z_j = i}} L_j - \sum_{j \,\middle|\, \substack{Y_j = R \\ s_j = t \\ q_j = i}} L_j && \forall i \in F,\ t = 1, \ldots, T. \end{aligned} \tag{2}$$

[1]In the following, the notation "$a|b$" is used as abbreviation of "a with condition b".

Because capacity limitations have to be kept, $l_{t,i} \leq K_i$ holds for every i and t. Since the terminal management will aim at a high utilization of the storage resource, $l_{t,i} \leq K_i$ turns out to be a very important constraint.

3.1.7 Manpower planning

Moving one vehicle from q_j to z_j requires $D_{q_j z_j}$ units of driving time. For the relocation of L_j vehicles from q_j to z_j, time $p_j := L_j \cdot D_{q_j z_j}$ is required. The driving time required in period t is determined by

$$p_t = \sum_{j|s_j=t} p_j = \sum_{j|s_j=t} L_j D_{q_j z_j} \,. \tag{3}$$

The driving time p_t is achieved by summing up all the time required for the storage, retrieval, and transshipment tasks carried out in period t. Due to the prescriptions with respect to origins and destinations given in Eq. 1, the terms of p_t can be distinguished as follows:

$$p_t \overset{(3)}{=} \sum_{j|s_j=t} L_j D_{q_j z_j} \overset{(1)}{=} \sum_{j \left| \substack{Y_j = S \\ s_j = t} \right.} L_j D_{Q_j z_j} + \sum_{j \left| \substack{Y_j = R \\ s_j = t} \right.} L_j D_{z_{V_j} Z_j} + \sum_{j \left| \substack{Y_j = T \\ s_j = t} \right.} L_j D_{Q_j Z_j} \,. \tag{4}$$

The mean of all p_t over the planning horizon is given by $\bar{p} = T^{-1} \sum_{t=1}^{T} p_t$. Provided with a perfectly balanced workload over the periods, p_t equals $\bar{p}$. The maximal driving time is bounded by p^{max}; therefore, the feasible interval of p_t is integrated into the model for all t: $p_t \leq p^{max}$.

3.2 Goals of operation

An evenly balanced workload over the T periods of the planning horizon is aimed at. Thus, we look for a way to minimize the sum of deviations of the manpower demand p_t in period t to the mean demand $\bar{p} = 1/T \sum_{t=1}^{T} p_t$:

$$\min \sum_{t=1}^{T} |p_t - \bar{p}| \,. \tag{5}$$

In periods with $p_t < \bar{p}$, a raise of p_t towards $\bar{p}$ might be achieved by uselessly allocating remote storage locations. The long distances to these remote locations will enhance p_t by larger driving times needed. In order to prevent this, we take only positive deviations of p_t to $\bar{p}$ into account:

$$\min \sum_{t=1}^{T} \max\{p_t - \bar{p}\,, 0\} \,. \tag{6}$$

Next to the balancing of workload, a minimization of the total driving time is pursued. Notice that Eq. 6 does not aim at reducing $\bar{p}$ because it will yield small values as long as $\bar{p}$ is large enough. To integrate the goal of minimizing driving time, the term $\min \sum_{t=1}^{T} p_t$ is added to Eq. 6. Consider that

$$\frac{1}{T}\sum_{t=1}^{T} p_t = \bar{p} \quad \Rightarrow \quad \sum_{t=1}^{T} p_t = T \cdot \bar{p} = \sum_{t=1}^{T} \bar{p}.$$

Thus, $\min \sum_{t=1}^{T} p_t$ corresponds to $\min \sum_{t=1}^{T} \bar{p}$ leading to the following function:

$$\min \sum_{t=1}^{T} \max\{p_t - \bar{p}, 0\} + \sum_{t=1}^{T} \bar{p} \quad = \quad \min \sum_{t=1}^{T} \max\{p_t, \bar{p}\}. \tag{7}$$

The linear integration of the minimizing and balancing function terms achieves a minimization of $\bar{p}$ and of the deviation of p_t from $\bar{p}$ at the same time.

3.3 Objective function and constraints

The function deduced in Eq. 7 is minimized by aligning p_t to $\bar{p}$ at the smallest achievable level. To achieve this goal, the objective function is set to

$$\min \sum_{t=1}^{T} \max\{p_t, \bar{p}\}. \tag{8}$$

The constraints described in Section 3.1 apply:

$$\max\{1, EET_j, s_{V_j} + 1\} \leq s_j \leq \min\{LET_j, T\} \quad \forall j \in A, \tag{9}$$

$$l_{t,i} \leq K_i \qquad \forall i \in F,\ t = 1, \ldots, T, \tag{10}$$

$$p_t \leq p^{max} \qquad t = 1, \ldots, T. \tag{11}$$

Tasks $j \in A$ are fully determined by the attributes expressing how many vehicles are relocated from where, whereto, and when, i.e., (L_j, q_j, z_j, s_j). From Eq. 4 we see that Eq. 8 is determined by fixation of variables s_j and z_j. The former variables determine a sequence of task operations whereas the latter variables determine the allocation of storage space.

4 Algorithmic approach

The model proposed above requires a vast number of binary variables. Many intricate constraints apply, which will prohibit the generation of optimal solutions for reasonably sized real-world problems. In particular, dynamic inventory constraints as given in Eq. 2 will prevent the formulation of an efficient exact algorithm (compare Neumann and Schwindt (1999); Laborie (2001) for earlier work in this direction).

We propose a construction heuristic to be controlled on a period-based level instead of focusing on the detailed level of the individual task. We formulate a reasonable default heuristic ("greedy strategy") and then allow variations from this heuristic. For this purpose, parameters indexed by period $t \in 1, \ldots, T$ control the assignment of tasks to periods and the allocation of storage space. An evolutionary algorithm is used to develop efficient "adaptive strategies" by setting the parameters of the construction heuristic in a suitable way.

4.1 Overview

In the following, we propose a procedure controlled by a set of parameters allowing for a variation from a default "greedy strategy": performing storages into the yard at the latest permissible period and retrievals in the earliest permissible period. Choose a storage location of sufficient capacity such that the distance of storage and retrieval is minimal. A variation from this "greedy procedure" is defined in two passes (illustrated in Fig. 1).

A greedy assigning of tasks to periods will lead to an unbalanced volume as sketched in the uppermost histogram. In order to balance the volume, we filter the task-to-period assignment by means of parameter α_t. This filtering may lead to a more balanced distribution of volume as it is sketched in the second histogram from the top. On the left-hand side of Fig. 1, we depict this filtering in analogy to the "funnel model" proposed by Wiendahl (1987) for the load-oriented production control.

In the second pass, we consider the distribution of manpower instead of volume. A balanced distribution of volume does not lead to a balanced distribution

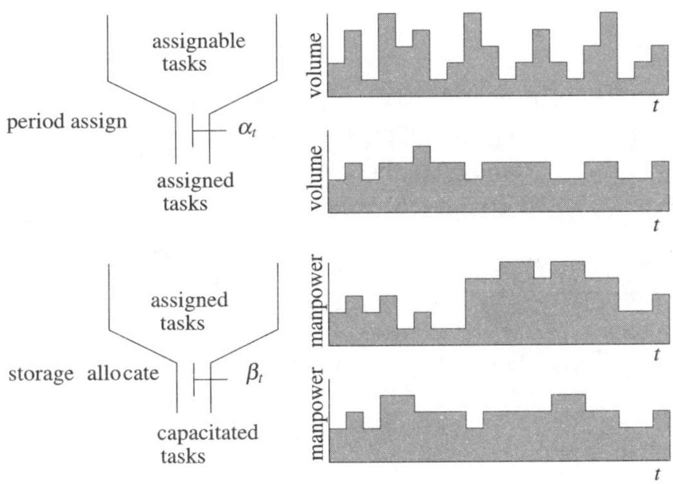

Fig. 1 Scheme of the two-pass construction heuristic

of manpower in every case (compare the third histogram from the top of Fig. 1). An uneven manpower distribution may be caused by a utilization of storage areas of varying productivity coefficients.

This time, we filter with respect to the storage capacity by means of the parameter β_t modifying the selection of storage locations. We end up with "capacitated tasks" in a hopefully well-balanced distribution of manpower as depicted in the lowermost histogram.

Note that variations of α and β can be used in order to achieve the same effect on the manpower utilization. Therefore, it depends on the problem at hand whether a modification to α or β should be applied. However, the period assignment of a task controlled by α has to be executed before the space allocation controlled by β can be applied.

4.2 Task-to-period assignment

In a first pass, the tasks are assigned to periods. Parameter α_t provides a way to shift task volumes between periods in order to achieve balanced transshipment volumes. Thereby, α_t decides upon the number of optionally executable tasks to be processed in period t. A large value of α_t causes the execution of almost all assignable tasks, whereas a small value of α_t tends to defer assignable tasks to forthcoming periods.

Set $\mathcal{S}_t$ contains all tasks assignable in period t, whereas set $\mathcal{R}_t \subseteq \mathcal{S}_t$ contains all tasks actually assigned to period t ($\mathcal{S}_0 = \mathcal{R}_0 = \emptyset$). The assignment of tasks to periods is performed for each consecutive period $t = 1, \ldots, T$ separately in three steps.

1. A fraction $\hat{L}_t$ of the optionally assignable volume in $\mathcal{S}_t$ is specified by $\alpha_t \in [0, 1]$. The total volume of all tasks assigned to period t is limited by $\hat{L}_t$:

$$\sum_{j \in \mathcal{R}_t} L_j \leq \hat{L}_t \quad \text{with} \quad \hat{L}_t := \lfloor \alpha_t \sum_{j \in \mathcal{S}_t} L_j \rfloor. \tag{12}$$

   In this way the manpower demand of period t is controlled indirectly by the vehicle volume $\hat{L}_t$. As only exception, we consider tasks in $\mathcal{S}_t$ with $LET = t$ and volume $\sum_{j|LET_j=t} L_j$ exceeding $\hat{L}_t$. In order to allow for the complete assignment of this volume to t, constraint (12) is relaxed:

$$\sum_{j \in \mathcal{R}_t} L_j \leq \max \left\{ \hat{L}_t, \sum_{j|LET_j=t} L_j \right\}.$$

2. Optionally processable tasks for $\mathcal{R}_t$ are determined from $\mathcal{S}_t$, which are $\emptyset$ for $\alpha_t = 0$ and $\mathcal{S}_t$ for $\alpha_t = 1$. For $\alpha_t \in (0, 1)$, a subset of $\mathcal{S}_t$ is selected by means of a utility function $\Phi_{t,j}$ for all $j \in \mathcal{S}_t$, i.e.,

$$\mathcal{R}_t := \Phi_{t,j}(\mathcal{S}_t).$$

3. $\mathcal{S}_{t+1}$ is built from the backlog of tasks considered in $\mathcal{S}_t$ but not yet assigned in $\mathcal{R}_t$ and then updated by tasks with an earliest processing time $t+1$, such that

$$\mathcal{S}_{t+1} := \mathcal{S}_t \setminus \mathcal{R}_t \cup \{j \in A | EET_j = t+1\}$$

holds and $\mathcal{S}_{t+1}$ contains the assignable tasks for period $t+1$.

The specification of function $\Phi_{t,j}$ in step 2 remains as an open issue. $\Phi_{t,j}$ selects tasks for $\mathcal{R}_t \subseteq \mathcal{S}_t$ with respect to period t and the maximal volume of transshipment $\hat{L}_t$. All tasks with $LET_j = t$ have to be assigned to t and, therefore, have to enter the set of tasks assigned to period t, i.e., $\{j \in \mathcal{S}_t | LET_j = t\} \subseteq \Phi_{t,j}(\mathcal{S}_t)$.

We model the selection problem as a knapsack model, which maximizes its total utility with respect to the constrained knapsack volume $\hat{L}_t$. For that purpose, the utility of tasks has to be determined. To this end, a priority $\phi_{t,j}$ is assigned to each feasible period of processing t for task j. Its value is determined by two factors: firstly, by the impact on the capacity of the storage system and, secondly, by the pressure of time caused by the time-windows imposed.

4.2.1 Impact on the storage system

The three types of tasks Y_j are classified into priority classes in accordance with their impact on the storage system. A transshipment does not alter the inventory since no storage into or retrieval from a storage area takes place. Its priority class is therefore set to *default*, which is used to initialize the classification process. A storage task burdens the storage system with additional vehicles and is, therefore, classified subordinate compared to a pure transshipment. We accordingly assign the priority class *low*. A retrieval task disburdens the storage system and should be preferred for that reason over a transshipment. We accordingly set its priority class to *high*.

With respect to the impact imposed on the storage system, selecting $\mathcal{R}_t$ from $\mathcal{S}_t$ favors retrieval tasks over transshipment tasks and transshipment tasks over storage tasks. Since $\mathcal{R}_t$ represents the faction α_t of $\mathcal{S}_t$, storage tasks tend to be deferred to $\mathcal{S}_{t+1}$, whereas retrieval tasks tend to be processed at the beginning of the time-window.

4.2.2 Urgency of processing

The urgency of processing becomes greater with the progression of time regardless of the type of task to perform. The deferment of a task increases the urgency of its processing in the next period because the time-window remaining becomes smaller with each deferment. If the processing of a task is deferred to its LET period, it has to be executed unconditionally. The priority class *maximal* is introduced for tasks with $LET = t$. With progression of time, each type of task may pass through two priority classes (refer to Table 1).

Table 1 Scheme of priority classes

Type of task Y_j	Priority class
Tasks in *LET*	Maximal
Retrieval up to $LET - 1$	High
Transshipment up to $LET - 1$	Default
Storage up to $LET - 1$	Low

Following the above scheme, the priority of storage and transshipment tasks is raised drastically in the last possible period such that execution is warranted. To achieve a more balanced distribution, in the following, we propose to raise the priority in a stepwise fashion. The importance of urgency of processing is favored over the impact on the storage system in order to allow a smoother raise of priority towards *maximal* at *LET*. Table 2 depicts the class scheme obtained.

Since the priority class of tasks is raised with elapsing time, even tasks of low priority become labeled for execution after waiting for a certain span of time.

4.2.3 Order of priority classes

By combining the impact on storage resources and the task urgency, a priority class has been assigned to every task in every period (refer Table 2). To build an order, the sequence of execution within a priority class has to be determined.

Provided that different types of tasks share a priority class in S_t, the impact on the storage system decides upon the order of execution. The rank order induced is taken as the basis for the assignment of priorities $\phi_{t,j}$. Nine scenarios may occur, which are distinguished as depicted in Table 3.

Priority $\phi_{t,j}$ is interpreted as the utility gained through the relocation of a vehicle of task j in period t. The utility $\Phi_{t,j}$ of task j with volume L_j in period t is

Table 2 Assignment of tasks to priority classes in accordance with urgency

Type of task Y_j	Period s_j	Priority class
Retrieval	Up to $LET - 1$	High
	In *LET*	Maximal
Transshipment	Up to $LET - 2$	Default
	In $LET - 1$	High
	In *LET*	Maximal
Storage	Up to $LET - 3$	Low
	In $LET - 2$	Default
	In $LET - 1$	High
	In *LET*	Maximal

determined by $\Phi_{t,j} := \phi_{t,j} \cdot L_j$. The utility values $\Phi_{t,j}$ of tasks are in $\mathbb{N}$. The knapsack problem can now be formulated with decision variable x_j such that $\mathcal{R}_t = \{j \in \mathcal{S}_t | x_j = 1\}$ constitutes a solution to the problem:

$$\max \sum_{j \in \mathcal{S}_t} x_j \Phi_{t,j}, \tag{13}$$

$$\sum_{j \in \mathcal{S}_t} x_j L_j \leq \max \left\{ \hat{L}_t, \sum_{j|LET_j = t} L_j \right\}, \tag{14}$$

$$x_j \in \{0, 1\} \qquad \forall j \in \mathcal{S}_t. \tag{15}$$

Equation 13 maximizes the utility while Eq. 14 depicts the capacity constraint. Finally, Eq. 15 expresses the integer condition of the problem.

This MIP has been implemented using the software package **lp_solve** version 3.2. Solving a knapsack problem entailed from constructing a solution for one of the test problems defined in Section 4 takes just a fraction of a second only. As the construction heuristic requires to solve $T = 50$ knapsack problems, solving each problem to optimality may become computationally prohibitive when used as a base heuristic inside an evolutionary algorithm (refer to Section 4.4. For the computational investigation performed, the mixed integer formulation of the knapsack problem is therefore replaced by a greedy heuristic task selection scheme with almost no loss of solution quality.

4.3 Allocating storage space

This second pass allocates storage space for the tasks involved. Parameter β_t allows shifting of the manpower demand between periods by controlling the way storage locations are chosen. Deciding upon the manpower demand of a storage task V_j, the manpower demand of its corresponding retrieval task j is determined. Storage and retrieval of a transshipment are performed in different periods t_1 and t_2 ; β_{t_1} and β_{t_2} determine the relative importance of saving manpower in t_1 with respect to t_2 and

Table 3 Rank order of scenarios

Priority class	Task type Y_j with s_j	$\phi_{t,j}$
Maximal	Retrieval in *LET*	9
	Transshipment in *LET*	8
	Storage in *LET*	7
High	Retrieval up to $LET - 1$	6
	Transshipment up to $LET - 1$	5
	Storage in $LET - 1$	4
Default	Transshipment up to $LET - 2$	3
	Storage in $LET - 2$	2
Low	Storage up to $LET - 3$	1

vice versa. This way, parameter β_t controls whether transports to nearby or remote locations are preferably carried out in a period. The steps to perform are described in the following:

1. The dynamic inventory balance equation (2) do not track inventory fluctuations within a period. In order to comply with the consecutive space allocation of the heuristic procedure with the model proposed in Section 3, for each period, retrieval tasks are processed before storage tasks. Thus, storage space is always freed before being reused.
2. To provide a (contingently feasible) solution in every case, a storage location i^* with unlimited capacity K_{i^*} and extremely large production coefficient D_{i^*i} with respect to all other locations $i \in F$ is provided in $F^* := F \cup \{i^*\}$. The production coefficient of i^* is set ten times larger than the mean of coefficients observed in the storage system. In order to process storage task $j_{(\mathrm{S})}$, internal storage locations of sufficient capacity $G_{j_{(\mathrm{S})}} := \{i \in F^* | K_i - l_{s_{j_{(\mathrm{S})}},i} \geq L_{j_{(\mathrm{S})}}\}$ are identified.
3. A storage location for storage task $j_{(\mathrm{S})}$ with corresponding retrieval task $j_{(\mathrm{R})}$ is determined by
$$z_{j_{(S)}} = i | \min_{i \in G_{j_{(\mathrm{S})}}} \{\beta_{s_{j_{(\mathrm{S})}}} D_{Q_{j_{(\mathrm{S})}},i} + \beta_{s_{j_{(\mathrm{R})}}} D_{i,Z_{j_{(\mathrm{R})}}}\}, \quad \beta_t \in [0, 1] .$$
Whenever $\beta_{s_{j_{(\mathrm{S})}}} = \beta_{s_{j_{(\mathrm{R})}}}$ for the period $s_{j_{(\mathrm{S})}}$ and the period $s_{j_{(\mathrm{R})}}$, the efforts for storage and retrieval contribute at the same rate and the location of sufficient capacity with the smallest overall distance available is consequently chosen. In case of a comparably larger $\beta_{s_{j_{(\mathrm{S})}}}$, the storage effort is favored over the retrieval effort. As a consequence, a nearby location will be chosen at the expense of a higher transportation effort for the corresponding retrieval task (and vice versa). Figure 2 provides an example of the storage allocation scheme considering a storage task with origin a to be executed in $t = 1$ and its corresponding retrieval task with destination d to be executed in $t = 2$. In comparison to $t = 1$, manpower capacity is more constrained in $t = 2$ expressed with $\beta_1 = 0.4$ and $\beta_2 = 0.8$, respectively. Two storage locations, b or c, of sufficient capacity can be chosen for intermediate storage. For the example, due to the setting of β_t, location c is selected with $\min_{i\in\{b,c\}} \{\beta_1 \cdot D_{a,i} + \beta_2 \cdot D_{i,d}\}$. In this case, min $\{(0.4 \cdot 4 + 0.8 \cdot 1), (0.4 \cdot 1 + 0.8 \cdot 3)\} = \min\{2.4 , 2.8\} = 2.4$ favors the alternative with the larger sum of productivity coefficients because of the higher relevance of manpower productivity in $t = 2$.
4. Finally, inventory is tracked by freeing storage space $l_{t,i} := l_{t,i} - L_j$ in case of a retrieval task and allocating space $l_{t,i} := l_{t,i} + L_j$ in case of a storage task.

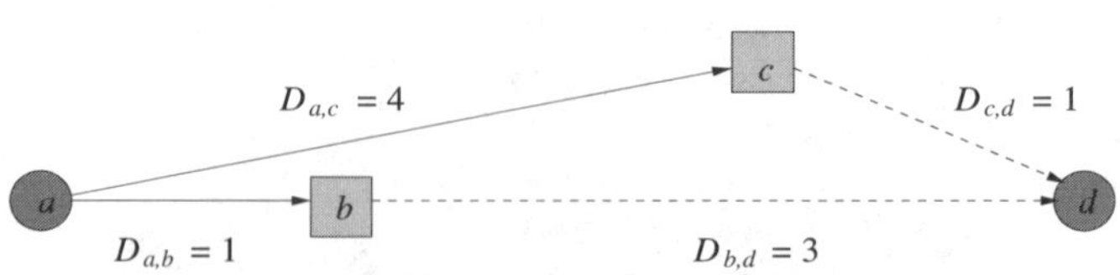

Fig. 2 Example of selecting a location

4.4 An evolutionary algorithm approach

In the remainder of this section, we propose an integrative setting of α_t and β_t by means of an evolutionary algorithm (EA). EAs are iterative stochastic search methods based on the principles of natural evolution. General introductions to EAs can be found, e.g., in Goldberg (1989) and Michalewicz (1996).

EAs can be applied independently from specific properties of the optimization model. For parameter optimization problems, EAs evolve suitable parameter sets with regard to the objective function value. The EA proposed in this research adapts the capacity utilization strategy by means of the parameters α_t and β_t. The construction heuristic proposed is integrated as a base heuristic in order to evaluate the strategies evolved.

In order to evolve suitable strategies, a pretty standard EA design is taken from the "Evolvable Objects" software library (Keijzer et al. 2001). To be specific, the **esea** program provided with the library in version 0.9.3 is adopted for the adaptation of a storage utilization strategy. Without performing any parameter optimization in advance, the program has been configured with commonly accepted parameter values:

- A real coded vector has been chosen for representation purposes consisting of α_t and β_t of all $t = 1, \ldots, T$ periods. All vector elements cover the domain $[0, 1]$.
- Mutations are applied at a rate of $p_m = 0.2$ and alter a vector element by $\epsilon = 0.03$ (see Bäck et al. 2000).
- A standard hypercube crossover produces superior results and is therefore applied at a rate of $p_c = 0.8$ (see Booker et al. 2000).
- A population size of 200 individuals has been shown to work sufficiently well. Four hundred offspring are produced and are subject to (μ, λ) replacement (see Deb 2000).
- A generational reproduction model is run for 200 generations. The $400 \cdot 200 = 80,000$ evaluations performed require approximately 30 s on a Pentium IV 2.6 GHz.

5 Computational investigation

In this section, we first introduce problem instances used for benchmarking. On the basis of these instances, we compare results achieved by the "greedy strategy" and the "adaptive strategy" and discuss the impacts for terminal operations management.

5.1 Setting up problem instances

To investigate the algorithmic approach to the problem at hand, problem instances are needed. An instance should be characterized by a few parameters meaningful to the problem. To meet this requirement, we refrain from using real-world problems and generate artificial problem instances instead. The detailed way of generating test problems is described in the Appendix.

For the problem instances, a storage system $F = \{1, \ldots, 40\}$ is used with 10 storage locations and 30 transfer points. The capacity K_i for a storage location is drawn from a uniform distribution in $[100, 1900]$. The total capacity of the storage system is summed up to $\hat{K} = 13529$.

All problem instances generated consist of $T = 50$ periods with an intended duration of storage of $\Delta = 8.0$ time units. The task volumes L_j are generated from a uniform distribution $[50, 250]$ with mean $\bar{L} = 150$. In accordance with Eq. 16, the mean inter-arrival time $\bar{\Lambda}$ is determined by $(150/13529) \cdot 8.0 = 0.088$, such that, per period, on average $0.088^{-1} = 11.3$ storage tasks are generated. As retrieval tasks are produced reactively at the same rate, a problem consists of approximately $22.6 \cdot 50 = 1,130$ tasks, diminished by the number of direct transshipments.

For these problems, we prescribe different inventory levels and time-windows of tasks in order to validate our algorithmic approach. We vary the overall inventory level $\Gamma \in \{0.8, 0.9, 1.0\}$ to produce modestly to heavily utilized storage systems. The mean extension of time-windows (given as a fraction of $\Delta = 8.0$) is prescribed by $\Omega \in \{0.000, 0.250, 0.500\}$. For each combination of Γ and Ω, we generate problems by varying the random seed $\Sigma \in \{1, \ldots, 50\}$ responsible for the arrival process, the task volumes, and the choice of storage locations; thus, each of the 450 test problems is uniquely referred to by (Γ, Ω, Σ).

Table 4 shows the mean values observed over $\Sigma = 50$ instances of an attribute combination, i.e., the intended inventory level Γ and the extension of time-windows Ω. Column $\bar{p}$ refers to the mean manpower demand given in terms of driving time. The standard deviation $v := T^{-1}\sqrt{\sum_t (p_t - \bar{p})^2}$ provides a measure

Table 4 Results obtained for the "greedy strategy" and the "adaptive strategy"

Ω	Γ = 0.8			Γ = 0.9			Γ = 1.0		
	$\bar{p}$	v	$\bar{l}$	$\bar{p}$	v	$\bar{l}$	$\bar{p}$	v	$\bar{l}$
				Greedy strategy					
0.000	6591.5	1304.3	0.74 (50	6849.7	1554.6	0.84 (50	7304.5	1912.7	0.91 (44
0.250	6284.2	1314.8	0.52 (50	6436.1	1399.3	0.61 (50	6537.2	1606.2	0.69 (50
0.500	6114.9	1418.2	0.37 (50	6245.3	1499.2	0.45 (50	6344.3	1702.8	0.52 (50
				Adaptive strategy					
0.000	6572.2	899.8	0.74 (50	6763.8	1133.6	0.84 (50	7127.5	1571.2	0.91 (50
0.250	6346.2	575.2	0.71 (50	6592.6	797.2	0.82 (50	6950.2	1675.5	0.87 (41
0.500	6149.0	580.3	0.64 (50	6330.7	700.2	0.75 (50	6547.4	1058.8	0.84 (50

Each figure depicts the mean over 50 different problem instances solved for a combination of Γ and Ω. Mean manpower demand $\bar{p}$, deviation of manpower demand v, and mean inventory level observed $\bar{l}$. The number of runs obtaining a feasible solution is given in braces

for the balance of manpower over the periods considered. The mean overall inventory level observed is given by

$$\bar{l} = \frac{1}{T}\sum_{t=1}^{T}\left(\frac{1}{\hat{K}}\sum_{i\in F} l_{t,i}\right) = \frac{1}{T\cdot\hat{K}}\sum_{t=1}^{T}\sum_{i\in F} l_{t,i}\,.$$

As the problems are generated by means of a simulation, the observed inventory level $\bar{l}$ is slightly lower than intended by Γ even for $\Omega = 0.0$. Another reason for $\bar{l} < \Gamma$ is the direct transshipment in the case of overlapping time-windows.

Although each problem instance is solvable, a heuristic algorithm may not find a feasible solution with respect to capacity constraints of the storage system. For these cases, we provide an additional location i^* which may be used at extremely high costs (compare Section 4.3). If no feasible solution can be obtained, it is omitted. The superscript at column $\bar{l}$ denotes the number of feasible solutions contributing to the figure.

5.2 Discussion of results

Let us start with a discussion of $\bar{l}$ with respect to variations of Ω. As time-windows expand with increasing Ω, direct transshipment become more frequent and the utilization of the storage yard $\bar{l}$ consequently decreases. This is particularly true for the "greedy strategy", whereas for the "adaptive strategy" $\bar{l}$ decreases only gradually.

Recall that the latest-in-earliest-out rule for the execution of storage and retrieval tasks forced by the "greedy strategy" ensures a minimal utilization of the storage yard. The "adaptive strategy" applies variations to the above rule in contingently performing an earlier storage and a later retrieval. This leads to increase of the overall inventory level $\bar{l}$.

If applied in an intelligent way, the "adaptive strategy" will lead to a positive effect on the manpower deviation v. At the same time, we can expect a deterioration of the mean manpower demand $\bar{p}$ because a large $\bar{l}$ potentially restricts the choice of suitable storage locations.

By comparing the "greedy strategy" and the "adaptive strategy", we observe a slight increase of $\bar{p}$ and a strong decrease of v for $\Omega > 0$. For modest inventory levels $\Gamma \in \{0.8, 0.9\}$, the improvements in manpower balancing are worth the marginal losses concerning the overall manpower demand.

With $\Gamma = 1.0$ and $\Omega = 0.25$, only for 41 of 50 problem instances has a feasible solution been found. An increase of $\bar{l}$ in the face of a heavily utilized storage system is obviously not helpful at all. If $\bar{l}$ is further decreased $(\Omega = 0.5)$, the "adaptive strategy" starts to work again in producing a superior manpower balance v.

Finally, we consider the absence of time-windows $(\Omega = 0)$. Variations with respect to the period of task execution cannot be applied and, therefore, the inventory level $\bar{l}$ is fixed. For these cases, we observe improvements for both, $\bar{p}$ as well as v. It is obvious that an intelligent choice of storage locations with respect to the current workload of the system can gain significant improvements.

With $\Gamma = 1.0$, the "greedy strategy" generates a feasible solution for 44 out of 50 problem instances. Vice versa, a variation of the choice of storage locations due to parameter β_t in the "adaptive strategy" is successful for all 50 instances.

In summary, the two goals, (a) manpower balancing and (b) manpower minimization, do not necessarily conflict. The "adaptive strategy" can improve transshipment operations to some extent by means of a load-oriented control of the allocation of storage space.

Further significant improvements with regard to manpower balancing can be obtained making use of time-windows. However, the negative side effect of the additional inventory burden prevails in the presence of a high inventory level. Thus, extending the duration of stay becomes a valid option for modestly utilized storage yards only.

6 Conclusion

We have proposed a multi-period-capacitated transshipment problem to improve the efficiency of operations for inter-modal vehicle terminals. For this problem, we have developed an optimization model.

We have proposed an evolutionary algorithm evolving a capacity utilization strategy with respect to the periods considered. The decisions concerning the individual tasks are performed by a construction heuristic parameterized by a capacity utilization strategy.

We have chosen a period-oriented control because of the large-size vehicle transshipment problems tend to have in practice. We have investigated whether improvements over a reasonable greedy strategy can be gained. The results obtained render the approach powerful.

Further work will go into two directions. First, the applicability of the approach is to be evaluated and extended for its use in container terminals. Second, an implementation of the proposed algorithm as an out-of-the-box support for terminal management needs further development.

Appendix

The generation of solvable test problems is a challenge. In the following, we present a way of producing solvable test problems, i.e., problems for which a feasible solution exists. We commence by firstly generating a storage system and then proceed with the stepwise generation of a task sequence.

Storage system generation

The storage system consists of m locations, where the ratio between the number of internal and external locations is externally given. The storage capacities of the locations are determined with respect to K^{min}:

$$K_i \begin{cases} \text{follows a discrete uniform distribution in } [K^{min},\ 2\bar{K} - K^{min}], & \text{if } H_i = \mathtt{I} \\ = 0, & \text{if } H_i = \mathtt{E}, \end{cases}$$

where $\bar{K}$ denotes the expectation value of K_i to be determined. The total capacity of the storage system is given by $\hat{K}$ and is determined by summing the capacities of all locations:

$$\hat{K} = \sum_{i \in F} K_i.$$

The coordinates of the storage locations are distributed to a standard normal, while the coordinates of transfer points are generated from a uniform distribution in $[-3, 3]$. Since a variable drawn from the standard normal distribution falls into this interval with probability 0.997, storage locations tend to be located in the median of the transfer points considered.

The storage locations are accessible at reasonable costs from many transfer points leading to a competition for central locations. This characteristic is often found for transshipment terminals, where interfaces to the seaside and the hinterland are located at the terminals' periphery. Figure 3 provides a bird's eye view to the locations and distances of a simulated terminal used later on for the computational study.

After the assignment of coordinates of all $i \in F$, a projection of the interval $[-3, 3]$ onto a distance measured is performed. The distances observed are taken as production coefficient of the manpower demand $D_{i_1 i_2}$, i.e., the driving time needed to transport one vehicle of L_j of task j from location i_1 to location i_2.

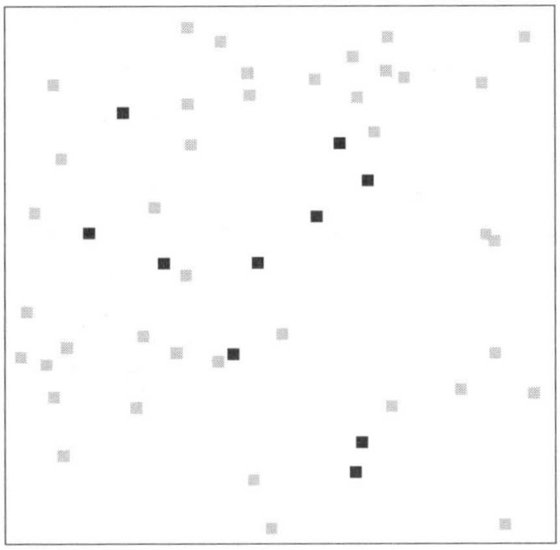

Fig. 3 Distribution of storage locations (*black*) and transfer points (*gray*)

Generating storage tasks

Tasks are produced by simulating a transshipment scenario for a number of periods. In this simulation, storage tasks are assigned to periods with increasing time. The inter-arrival time of storage tasks is consecutively drawn from an exponential distribution with mean $\bar{\Lambda}$. The corresponding task volumes L_j are generated from a discrete uniform distribution in $[L^{min}, 2\bar{L} - L^{min}]$ with mean $\bar{L}$.

Assuming that the total capacity of the storage system $\hat{K}$ is fully utilized over the entire time horizon, the mean inter-arrival time of storage tasks is

$$\bar{\Lambda} = \frac{\bar{L}}{\hat{K}} \cdot \Delta \,, \tag{16}$$

with Δ being the externally given and intended mean duration of storage. In the simulation run, a storage task $j_{(\mathrm{S})}$ is assigned to period $s_{j_{(\mathrm{S})}} = t$ determined by rounding its arrival time up to the next largest period number. After $j_{(\mathrm{S})}$ is assigned to a period, a storage location i of sufficient capacity is chosen at random, where $j_{(\mathrm{S})}$ is placed with volume $L_{j_{(\mathrm{S})}}$.

Generating retrieval tasks

Retrieval tasks are generated in dependency of the arrival process of storage tasks. Whenever the vehicles of a storage task $j_{(\mathrm{S})}$ cannot be stored anymore because of capacity constraints, a storage location i is selected at random where $j_{(\mathrm{S})}$ could be placed if the location were empty, i.e., $L_{j_{(\mathrm{S})}} \leq K_i$.

To place storage task $j_{(\mathrm{S})}$, capacity is freed from storage location i by generating retrieval task $j_{(\mathrm{R})}$, removing its volume from i. The volume $L_{j_{(\mathrm{R})}}$ of retrieval task $j_{(\mathrm{R})}$ equals the volume of its already processed storage counterpart, i.e.,

$$L_j = L_{V_j} \quad \forall j \in A | Y_j = \mathrm{R} \,.$$

Retrieval tasks are contingently generated for location i and assigned to period t until the initiating storage task $j_{(\mathrm{S})}$ can be placed in i. Finally, volume $L_{j_{(\mathrm{S})}}$ is stored in i.

Figure 4 depicts the scheme of generating tasks. Λ prescribes the growth rate of the accumulated inventory, i.e., the angle of the arrival process. The capacity of the storage system $\hat{K}$ as well as the intended duration of stay Δ tie the retrieval process to the storage process. In the long run, the capacity limitation of the storage system entails the generation of retrieval tasks at the same rate as observed for storage tasks.

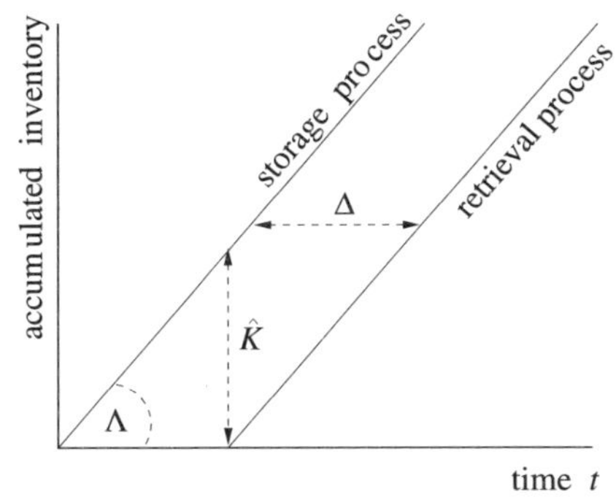

Fig. 4 Scheme of task generation

Altering the inventory level

Since retrieval tasks are generated reactively, the storage system is almost fully utilized. Therefore, its overall inventory level is close to 1.0 denoting a storage system with $\hat{K}$ units stored, whereas 0.0 denotes an empty storage yard. To produce test differing problem instances, we lower the overall inventory level with hindsight. Once a retrieval task j has been assigned to period s_j, we can safely pre-draw j for a number of periods without risking violation of inventory constraints, i.e.,

$$s_j := s_{V_j} + \lfloor (s_j - s_{V_j}) \cdot \Gamma \rfloor \quad \forall j \in A | Y_j = \mathrm{R}$$

with $0 \leq \Gamma \leq 1$ being the intended inventory level.

Deriving a dynamic problem

In the initial phase of the simulation, merely storage locations are filled. This phase ends with the execution of the first retrieval task $j_{(\mathrm{R})}$ in period $s_{j_{(\mathrm{R})}}$. To discard this initial phase from the problem instance, time $t = 1$ of the problem is determined by $s_{j_{(\mathrm{R})}}$. The inventory level at the beginning of period 1 is stored as the initial inventory level B_i for each internal location $i \in F$ of the problem instance.

All retrieval tasks j with $s_{V_j} < 1$ are assigned the storage location chosen for V_j as their prescribed origin Q_j. In this way, these retrieval tasks completely withdraw the initial inventory level B_i over the course of the simulation.

In order to depict T periods later on in the problem, the simulation has to be performed for at least $s_{j_{(\mathrm{R})}} + T$ periods. All tasks j with $s_j > T$ are discarded from consideration; merely for the corresponding storage tasks V_j with $s_{V_j} < T$ is the destination Z_j of retrieval task j kept to allow the determination of suitable storage locations for V_j.

In this way, the set A of the problem instance only consists of tasks whose period of performance s_j in the simulation falls into the planning horizon, $A = \{j \mid 1 \leq s_j \leq T\}$.

Generating transfer points

External locations are chosen at random for all the tasks $j \in A$ generated. Transfer points are determined as origin Q_j for storage tasks and as destinations Z_j for retrieval tasks.

Generating time-windows

To ensure that the simulation storage assignments performed in the simulation are incorporated in the problem to be constructed, simulated times s_j have to fall into the interval $[EET_j, LET_j]$. This is achieved by extending the time-window starting from s_j. By extending a time-window beyond $t = T$, tasks may not be considered in the planning horizon anymore. Therefore, if required,

$$\begin{aligned} EET_j &:= s_j && \forall j \in A, \\ LET_j &:= \min\{s_j + w_j\,, T\} && \forall j \in A \end{aligned}$$

with w_j discrete uniformly distributed in $[0, (2 \cdot \Delta \cdot \Omega)]$. Thus, the extension of time-windows of tasks is specified in proportion $0 \leq \Omega \leq 1$ of the intended duration of stay Δ. The existence of time-windows allows the execution of storage tasks later than performed in the simulation run. This gives the opportunity to further decrease the overall inventory level as a matter of optimization.

Whenever the time-windows of retrieval j and its corresponding storage V_j overlap, i.e., $LET_{V_j} \geq EET_j$ holds, the affected number L_j of vehicles are transshipped directly. The pair of j and V_j is replaced by a single transshipment $j_{(\mathrm{T})}$. Next to the volume $L_{j_{(\mathrm{T})}} := L_j$, the following parameters apply for $j_{(\mathrm{T})}$:

$$\begin{aligned} EET_{j_{(\mathrm{T})}} &:= EET_j\,, & Q_{j_{(\mathrm{T})}} &:= Q_{V_j}\,, \\ LET_{j_{(\mathrm{T})}} &:= LET_{V_j}\,, & Z_{j_{(\mathrm{T})}} &:= Z_j\,. \end{aligned}$$

By generating transshipments, the set A of tasks is modified. Despite the modification of A, the simulation still warrants a feasible solution because the transshipment merely depicts a simultaneous processing of storage and corresponding retrieval. With regard to Eq. 3, p_t for all T periods of the simulation are calculated. Capacity p^{max} is subsequently determined as the largest p_t observed in the simulation:

$$p^{max} := \max_{t=1,\ldots,T} p_t\,.$$

In this way, the feasibility with respect to Eq. 11 is kept for every period of the planning horizon.

Finally, the information generated by the simulation about the period assignment and storage location of tasks is discarded from the problem instance before it is used for benchmark purposes.

References

Bäck T, Fogel DB, Whitley D, Angeline P (2000) Mutation operators. In: Bäck T, Fogel DB, Michalewicz T (eds) Evolutionary computation 1: basic algorithms and operators, chapter 32. IOS, pp 237–255

Bish E, Leong T-Y, Li C-L, Ng J, Simchi-Levi D (2001) Analysis of a new vehicle scheduling and location problem. Nav Res Logist 48:363–385

Booker L, Fogel DB, Whitley D, Angeline P, Eiben A (2000) Recombination. In: Bäck T, Fogel DB, Michalewicz T (eds) Evolutionary computation 1: basic algorithms and operators, chapter 33. IOS, pp 256–269

Deb K (2000) Introduction to selection. In: Bäck T, Fogel DB, Michalewicz T (eds) Evolutionary computation 1: basic algorithms and operators, chapter 22. IOS, pp 166–171

Goldberg D (1989) Genetic algorithms. Addison–Wesley

Hartmann S (2004) A general framework for scheduling equipment and manpower at container terminals. OR Spectrum 26:51–74

Keijzer M, Merelo J, Romero G, Schoenauer M (2001) Evolvable objects: a general purpose evolutionary computation library. In: 5th Conference Internationale sur l' Evolution Artificielle. Universite de Bourgogne, France (http://eodev.sourceforge.net)

Kim KH, Park KT (2003) A note on a dynamic space-allocation method for outbound containers. Eur J Oper Res 148:92–101

Kusiak A (2000) Computational intelligence in design and manufacture. Wiley, pp 412–427

Laborie P (2001) Algorithms for propagating resource constraints in AI planning and scheduling: existing approaches and new results. In: Cesta A, Borrajo D (eds) Proceedings of the sixth European conference on planning, pp 205–216

Mattfeld D (2003) Mid-term planning of transshipment tasks. In: Spengler T, Voß S, Kopfer H (eds) Logistikmanagement 2003. Springer, Berlin Heidelberg New York, pp 411–424

Mattfeld D (2006) The management of transshipment terminals. Springer, Berlin Heidelberg New York

Mattfeld D, Kopfer H (2003) Terminal operations management in vehicle transshipment. Transp Res Part A Policy Pract 37:435–452

McKendall A, Jaramillo J (2006) A tabu search heuristic for the dynamic space allocation problem. Comput Oper Res 33:768–789

Michalewicz Z (1996) Genetic algorithms + data structures = evolution programs, 3rd edn. Springer, Berlin Heidelberg New York

Murty K, Liu J, Wan Y, Linn R (2005) A decision support system for operations in a container terminal. Decis Support Syst 39:309–332

Neumann K, Schwindt C (1999) Project scheduling with inventory constraints. Technical report WIOR-572, University of Karlsruhe

Preston P, Kozan E (2001) An approach to determine storage locations of containers at seaport terminals. Comput Oper Res 28:983–995

Rodrigue J (1999) Globalization and the synchronization of transport terminals. J Transp Geogr 7:255–261

Spatz J, Nunnenkamp P (2002) Globalization of the automobile industry—traditional locations under pressure? Aussenwirtschaft 57(4):469–493

Steenken D, Voß S, Stahlbock R (2004) Container terminal operation and operations research. OR Spectrum 26(1):3–49

Taleb-Ibrahimi M, Castilho Bd (1993) Storage space vs handling work in container terminals. Transp Res Part B Methodol 27(1):13–32

Tyan J, Wang F-K, Du T (2003) An evaluation of freight consolidation policies in global third party logistics. Omega 31:55–62

Wiendahl H-P (1987) Belastungsorientierte Fertigungssteuerung. Hanser Verlag, München

Zhang C, Liu J, Wan Y, Murty K, Linn R (2003) Storage space allocation in container terminals. Transp Res Part B Methodol 37:883–903

Huei Chuen Huang · Chulung Lee · Zhiyong Xu

The workload balancing problem at air cargo terminals

Abstract We consider a large air cargo handling facility composed of two identical cargo terminals. In order to improve the operational efficiency, the workload must be balanced between the terminals. Thus, we must assign each airline served by the facility to one of the terminals such that (ideally): (1) each terminal has the same total workload, and (2) the workload at each terminal is distributed evenly along the timeline. Complicating the problem is that cargo loads are difficult to predict (stochastic). We develop a stochastic mixed integer linear program model in which a weighted sum of the balance measures is minimized. We employ sample average approximation for the stochastic program and develop an accelerated Benders decomposition algorithm to reduce the computational time. The proposed model can also be applied to partially reassign the airlines for the operational schedule changes. The computational results show that a small number of reassignments are often sufficient to rebalance the workload. The simulation results based on data from a large international airport show that the proposed algorithms efficiently balance the workload and the cargo service time is consistently reduced.

Keywords Air cargo terminal · Workload balancing · Stochastic mixed integer linear program · Benders decomposition · Simulation

---

H. C. Huang
Department of Industrial and Systems Engineering, National University of Singapore,
10 Kent Ridge Crescent 119260, Singapore, Singapore
E-mail: isehhc@nus.edu.sg

C. Lee (✉)
Department of Industrial, Systems and Information Engineering, Korea University, Anam-Dong, Seongbuk-Gu, Seoul 136-713, Korea
E-mail: leecu@korea.ac.kr

Z. Xu
Singapore Airlines, 04-N, SIAEC Hangar, 31 Airline Road 819831, Singapore, Singapore
E-mail: Zhiyong_Xu@singaporeair.com.sg

1 Introduction

The efficiency of airport operations has received considerable research interest but most of the works focus on passenger terminals (see, for example, Baron 1969; De Neufville and Rusconi-Clerici 1978; Wirasinghe and Bandara 1990; Haghani and Chen 1998). However, as the revenue from cargo transportation has substantially increased in the last decade, competitions have become fierce and airlines and airports strive to streamline their cargo handling operations. Air cargo terminals at an airport are considered a type of warehouses. Although there is a large literature on traditional warehouse operations (see, for example, Ashayeri and Gelders 1985; Cormier and Gunn 1992; Rouwenhorst et al. 2000), very few have discussed the operations at air cargo terminals.

We observed the operations of an air cargo terminal operator at a leading international airport that handles high cargo volume from many airlines. This operator has four equal-sized terminals, two of which are dedicated to import operations and two to export operations. To maintain accountability and tractability, import (export) cargo belonging to one airline is always handled by just one of the two terminals even though the terminals are located adjacent to each other.

Due to the time-sensitive nature of air cargo handling, it is imperative that cargo must be processed swiftly. Consequently, there is often a surge of manpower and equipment demand when a large number of flights are scheduled to depart or arrive within a short time interval. The cargo volume to handle is difficult to forecast, varying from day to day and week to week, which makes it difficult to schedule a regular workforce to meet this highly stochastic workload with the minimum cost. In practice, no matter how experienced the planners are or how efficient the scheduling method is, a baffling problem is frequently encountered: Manpower shortages are often worsened by the fact that completely idle time arises from time to time during working hours. This dilemma is termed *the self-contradiction of hands shortage and idleness*. The chief cause of this problem is the unbalanced workload distribution because the quality and cost of manpower schedule hinges very much on the workload. At air cargo terminals, the workload distribution depends on flight schedules, which vary in intensity throughout the day; however, the manpower schedule is comprised of a regular 8-h shift for each employee. Under these circumstances, when the workload surges, employees may be overloaded and when it slows, they may become underutilized or even idle. In summary, the relatively regular manpower schedule does not match the irregular workload distribution. On the other hand, balancing the workload between the two import (export) terminals by assigning the proper set of airlines to each should help to improve this situation.

In this paper, for a given number of airlines, each of which may provide a number of flights, we assign each airline to one of the two import (export) terminals, T1 and T2, for the objective of balancing the workload between the terminals along the time line. The operator's workload can be measured by the total number of pallets/containers that must be handled. While the workload from each flight is highly stochastic, an empirical distribution for the workload can be obtained from the historical data. The flight schedules are given by the airlines and are assumed to be fixed.

A clear definition for the measure of balance (imbalance) is necessary for the study of workload balancing. Naturally, *variance* or *deviation* is a good measure; however, it is hard to employ due to its nonlinear nature. Instead, several different linear measures have been used. These measures include the maximal workloads, which is broadly employed (e.g., Berrada and Stecke 1986) the sum of the overloads and underloads (Moreno and Ding 1989), the functions of the maximal deviations (Shanker and Tzen 1985), and so forth. Regarding the effectiveness, Guerro et al. (1999) stated that the simplest and the most effective measure is the "*difference between max and min.*" A comprehensive comparison of balance measures was presented by Kumar and Shanker (2001) and they concluded that the "min [*average pairwise difference*]" is the best objective for workload balancing. Although all these studies were conducted in the areas of manufacturing and production, they provide a good reference to define a measure of workload balance in the context of cargo terminals.

In this paper, two of the above-mentioned balance measures are employed. The first is the "the maximal workloads," the second is what we call "sum pairwise difference," which is derived from "minimization of variance." After we obtain the solution using these two linear measures, we also compute the variance as an additional indicator of workload balancing.

In the rest of the paper, a stochastic mixed integer linear program (S-MILP) model is developed in Section 2 in view of the stochastic nature of the workload. In the model, a weekly schedule with time points wrapping around from the end to the beginning is considered. This is similar to the way that is practiced in the airline scheduling; as in all these problems, the work demand is driven by the flight schedules, which repeat weekly. To compute the workload, the time span of 1 week is further divided into equal time periods. Sample average approximation (SAA) is employed to transform the stochastic program to a solvable deterministic model, which is then solved by decomposition-based algorithms developed in Section 3. In Section 4, an extended model called partial reassignment is presented, followed by the computational results in Section 5. The performance of the optimal assignment is examined in Section 6 with both numerical and simulation experiments.

2 Mathematical formulations

The following notations are used:

Sets and indexes:

J:	Set of all time periods, indexed by j
I:	Set of all airlines, indexed by i

Parameters:

a_{ij}:	Workload of airline i at period j
A_j:	Total workload at period j. $A_j = \sum_{i \in I} a_{ij}$
l_i:	Total workload for airline i. $l_i = \sum_{j \in J} a_{ij}$
L:	Total workload for all airlines. $L = \sum_{i \in I} \sum_{j \in J} a_{ij}$
w_j^1, w_j^2:	Total workload at period j assigned to T1 and T2, respectively

Decision variables:

x_i:	(0, 1); 1 if airline i is assigned to terminal T1, 0 if otherwise (airline i is assigned to terminal T2)

2.1 Deterministic model

An instance of the problem can be modeled as a deterministic linear program as follows:

$$[\text{OP}] \quad \text{Minimize } C \cdot z + \sum_{j \in J} y_j \tag{1}$$

$$s.t. \quad z \geq \sum_{i \in I} a_{ij}\, x_i \quad \textit{for all } j \tag{2a}$$

$$z \geq A_j - \sum_{i \in I} a_{ij}\, x_i \quad \textit{for all } j \tag{2b}$$

$$y_j \geq 2 \cdot \sum_{i \in I} a_{ij}\, x_i - A_j \quad \textit{for all } j \tag{3a}$$

$$y_j \geq A_j - 2 \cdot \sum_{i \in I} a_{ij}\, x_i \quad \textit{for all } j \tag{3b}$$

$$\begin{aligned} &(0.5 - \tau)L \leq \sum_{j \in J} w_j^1 \leq (0.5 + \tau)L \\ &x \in \{0, 1\} y, z \geq 0 \end{aligned} \tag{4}$$

The objective in Eq. 1 is to minimize a weighted sum of two objectives: “maximal workload” and sum of “pairwise difference,” represented by auxiliary decision variables z and y_j, respectively.

$$z = \max\left\{w_j^1, w_j^2 \text{: } \textit{for all } j\right\} \tag{5}$$

$$y_j = \left|w_j^1 - w_j^2\right| \; \textit{for all } j \tag{6}$$

The overall workload is required to spread as evenly as possible between the terminals at all periods because the two terminals are identical in terms of size and equipment, and equipment and/or space are very often in shortage (causing delays in cargo processing). Thus, evenly distributing the workloads between the terminals improves overall resource utilization. Secondly, we minimize the maximum workload among time and terminals. The labor cost of the facilities is determined by the workload at the peak period, thus, it is important to minimize the peak period workload. In addition, this will ensure that the highly promised service level can be achieved even during the busiest time. Although these two requirements are not completely conflicting, they may not always be achieved simultaneously. Therefore, a weighted sum of these two measures with the weight parameter C is employed. C is a constant and user-defined parameter (depending on users' cost parameter values), which would affect the optimal solution. An insight to choose an appropriate C value can be obtained from the computational results provided in Section 5.1.

Constraints 2a, 2b, 3a, and 3b are to build the linear program model for z and y_j. Constraint 4 restricts the total workloads of the two terminals to be of similar size. Lower and upper bounds of total workload for each terminal must be $0.5+\tau$ of the total workload:

$$(0.5-\tau)L \leq \sum_{j\in J} w_j^1 \leq (0.5+\tau)L \tag{7}$$

For example, when τ=0.05, 45 and 55% of the total workload are lower and upper bounds. It may not be possible to satisfy this constraint for some data sets. However, for the data sets used in this study, which we obtained from real terminal operations, infeasibility has not occurred. In case of infeasibility, we can adjust the parameter.

[OP] is a difficult mixed integer program to solve because the solution for the underling LP is:

$$x_i = 1/2, \quad \textit{for all } i \tag{8}$$

which makes it difficult to obtain a good branch-and-bound starting basis. Because the two terminals are identical, by preassigning the largest airline to either terminal, that is,

$$x_1 = 1 \tag{9}$$

the computational time can be reduced by nearly 50%.

2.2 Model reformulation

In [OP], the maximum workload z is bounded by a pair of constraints, Eqs. 2a and 2b at all periods. Similarly, the absolute value of difference y is constrained by Eqs. 3a and 3b. Each pair of these constraints can be combined by the following transformations.

To transform the constraints in Eqs. 3a and 3b, auxiliary variables p_j and q_j are introduced:

$$y_j - p_j = 2 \cdot \sum_{i \in I} a_{ij}\, x_i - A_j \quad p_j \geq 0 \tag{10a}$$

$$y_j - q_j = A_j - 2 \cdot \sum_{i \in I} a_{ij}\, x_i \quad q_j \geq 0 \tag{10b}$$

Equations 11 and 12 can be obtained from the transformation of Eqs. 10a and 10b, which are equivalent to Eqs. 3a and 3b.

$$p_j + 4\sum_{i \in I} a_{ij}\, x_i \geq 2A_j \tag{11}$$

$$y_j = p_j + 2\sum_{i \in I} a_{ij}\, x_i - A_j \tag{12}$$

Because all the constraints for z and y are computed from the terms $\sum_{i \in I} a_{ij}x_i$, A_j, and their differences, a new set of constraints that consist of z and y can be derived from the two pairs of constraints.

For the constraint in Eq. 2a, when we multiply both sides by 2 and subtract A_j from both sides, we obtain:

$$2z - A_j \geq 2\sum_{i \in I} a_{ij}\, x_i - A_j \tag{13a}$$

Similarly, for the constraint in Eq. 2b, multiply by 2 and then subtract A_j:

$$2z - A_j \geq A_j - 2\sum_{i \in I} a_{ij}x_i \tag{13b}$$

The left hand side of Eqs. 13a and 13b are exactly the same with Eqs. 3a and 3b. Thus, we can tighten the bounds for z by introducing a new set of constraints as follows:

$$2z - A_j \geq y_j \quad \textit{for all } j \tag{13}$$

In addition, z must be minimized in the objective function. Therefore, Eq. 13 is equivalent with Eqs. 2a and 2b. Substituting y with Eq. 12, the number of constraints is reduced by half. The reformulation is denoted by [OP'] where a constant term is removed from the objective function.

$$
\begin{aligned}
\text{Minimize} \quad & 2\sum_{i\in I} l_i\, x_i + \left(C \cdot z + \sum_{j\in J} p_j \right) \\
s.\,t. \quad & z \geq p_j/2 + \sum_{i\in I} a_{ij}\, x_i && \textit{for all } j \\
& p_j + 4\sum_{i\in I} a_{ij}\, x_i \geq 2A_j && \textit{for all } j \qquad \text{(OP')} \\
& 0.45L \leq \sum_{i\in I} l_i\, x_i \leq 0.55L \\
& z, p \in R^+\; x \in \{0,1\} x_1 = 1
\end{aligned}
$$

2.3 Stochastic model and sample average approximation

In [OP], the average workload is used for a_{ij}. It is doubtful whether the optimal solution provides the robust performance when the actual workload varies over time. To model this stochastic behavior, a_{ij} is considered a random variable following a probability distribution obtained from the historical data. Define P to be a collection of possible instances of the workload vector (a_{ij}). The assignment problem can be formulated as follows:

$$
\text{Min}_{x\in\Theta}\left\{ f(x) := E_{\omega^k\in P} h\left(x, \omega^k\right) \right\} \tag{14}
$$

where $\Theta := \{x | x \in \{0,1\}, x_1 = 1\}$, ω^k stands for an instance of the workload vector, and k is an integer number representing the kth particular instance.

$$
\begin{aligned}
h\left(x, \omega^k\right) = \min_{y^k, z^k} \quad & C z^k + \sum_{j\in J} y_j^k \\
s.t. \quad & z^k \geq \sum_{i\in I} a_{ij}^k x_i && \textit{for all } j \\
& z^k \geq A_j^k - \sum_{i\in I} a_{ij}^k x_i && \textit{for all } j \\
& y_j^k \geq 2 \cdot \sum_{i\in I} a_{ij}^k x_i - A_j^k && \textit{for all } j \qquad (15) \\
& y_j^k \geq A_j^k - 2 \cdot \sum_{i\in I} a_{ij}^k x_i && \textit{for all } j \\
& 0.45L^k \leq \sum_{i\in I} l_i^k x_i \leq 0.55L^k \\
& y^k, z^k \geq 0
\end{aligned}
$$

It is very difficult if not impossible to solve this optimization problem. First, for a given decision variable x, computing the objective function value requires computing the expected value of a linear programming value function $h(x, \omega)$. The exact expectation involves the probability distributions of multiple dimensions and there is no closed form to express this. Secondly, even if the expectation can be computed numerically for a given assignment, it is computationally intractable to find the best assignment among all the possible assignments.

To overcome this problem, sampling-based algorithms may be employed (Shapiro and Homem-de-Mello 1998; Verweij et al. 2003; Kleywegt et al. 2001). The basic idea is simple: A random sample is obtained and the expectation is approximated by the corresponding sample average value. The idea of using SAAs for solving stochastic programs is a natural one and has been employed by various researchers over the years.

Assume that $\{\omega^1, \omega^2, \ldots \omega^N\}$ is a random sample where N is defined by the sample size. The objective becomes:

$$\mathrm{Min}_{x\in\Theta}\left\{\widehat{f}_N(x) := N^{-1}\sum_{k=1}^{N} h\left(x, \omega^k\right)\right\} \tag{16}$$

Combining Eqs. 15 and 16, the assignment problem can be modeled deterministically for a given sample. To distinguish this from [OP], we call it the *stochastic problem* SP].

$$\begin{array}{lll}
\text{Minimize} & \frac{1}{N}\left(C\sum_{k=1}^{N} z^k + \sum_{k=1}^{N}\sum_{j\in J} y_j^k\right) & \\
s.t. & z^k \geq \sum_{i\in I} a_{ij}^k\, x_i & \textit{for all } j, k \\
 & z^k \geq A_j^k - \sum_{i\in I} a_{ij}^k\, x_i & \textit{for all } j, k \\
 & y_j^k \geq 2\cdot\sum_{i\in I} a_{ij}^k\, x_i - A_j^k & \textit{for all } j, k \\
 & y_j^k \geq A_j^k - 2\cdot\sum_{i\in I} a_{ij}^k\, x_i & \textit{for all } j, k \\
 & 0.45L^k \leq \sum_{i\in I} l_i^k\, x_i \leq 0.55L^k & \textit{for all } j, k \\
 & x \in \{0,1\}\ x_1 = 1\ y^k, z^k \geq 0 &
\end{array} \tag{SP'}$$

It was shown (Shapiro, 2001) that the solutions of Eq. 16 converge to the optimal solution of Eq. 14 with the probability of one as N goes to infinity. In addition, Shapiro and Homem-de-Mello (2001) and Kleywegt et al. (2001) proved that the convergence rate is exponentially fast if the involved probabilistic

Table 1 Optimality gap (the gap between lower and upper bounds) with different sample sizes

Sample size	Lower bound	Upper bound	Relative gap (%)
10	6,782.8	6,871.69	1.3105
20	6,800.66	6,874.76	1.0896
30	6,845.39	6,879.99	0.5054
40	6,835.71	6,858.91	0.3394
50	6,833.12	6,856.54	0.3427
60	6,852.14	6,857.32	0.0756
70	6,849.11	6,854.03	0.0718

distributions are discrete. It was also suggested that a fairly good approximate solution to the problem in Eq. 14 can be obtained by solving the SAA problem in Eq. 16 with a modest sample size. In practice, the SAA method computes the solutions of Eq. 16 with independently distributed samples for any given sample size. Santoso et al. (2003) developed a method to quantitatively estimate the gap between the optimal solution and the SAA approximation. In their approach, input data with the same sample size N is randomly generated. Each of the problems is considered an independent deterministic problem and is solved. The statistical lower and upper bounds are obtained accordingly. By increasing the sample size N, the gap between the lower and the upper bounds is reduced. Using this approach, we solve the problem [SP] for the data set ImD. The computational results are reported in Table 1, which shows that the gap between these bounds decreases fast as the sample size increases. When the sample size is greater than 30, the gap becomes less than 1%, and when the size is greater than 60, the gap is less than 0.1%. Based on these observations, we propose to use N=100 to solve [SP].

Using the same method of reformulating [OP] to [OP'], the [SP] can be reformulated as follows:

$$\begin{aligned}
\text{Minimize} \quad & \frac{1}{N}\sum_{k=1}^{N}\left(2\sum_{i\in I} l_i^k\, x_i + \left(C\cdot z^k + \sum_{j\in J} p_j^k\right)\right) & & \\
s.t. \quad & z^k \geq p_j^k\Big/2 + \sum_{i\in I} a_{ij}^k\, x_i & \textit{for all } j,k & \\
& p_j^k + 4\sum_{i\in I} a_{ij}^k\, x_i \geq 2A_j^k & \textit{for all } j,k & \qquad \text{(SP)}\\
& 0.45L^k \leq \sum_{i\in I} l_i^k\, x_i \leq 0.55L^k & \textit{for all } k & \\
& z,\ p \geq 0 \quad x\in\{0,1\}\ x_1 = 1 & &
\end{aligned}$$

The reformulation reduces the problem complexity and thus is able to reduce the computational time as well. To illustrate the reduction in the computational time, a computational experiment is conducted and the results are reported in Table 2.

The results show that the new formulation significantly reduces the computational time, especially when the sample size is not larger than 50. The time reduction ranges from 40.6 to 53.4% of the computational time under the original formulation. When the sample size becomes large, e.g., 100, as proposed in Table 1, the computation takes longer time and the reformulation only reduces 8.7% of the computational time. In the following section, the Benders decomposition algorithm is developed to solve [SP]

Table 2 The computation time in seconds under formulation [SP] and [SP']

Sample size	Formulation [SP]	Formulation [SP′]	Time reduction as (%)
1	1.39	0.78	43.9
5	23.05	10.75	53.4
20	349.11	175.89	49.6
50	1,851.16	1,099.36	40.6
100	7,771.02	7,095.41	8.7

with large sample size. Furthermore, an acceleration technique is developed to further shorten the computational time.

3 Benders decomposition

The problem [SP] can be considered a two-stage program. The first stage is to determine the airline assignment represented by the binary decision variable x and the second stage is to obtain the optimal balance measures y and z based on the predetermined airline assignment and parameter values. The second stage problem is trivial because the optimal solution can be easily obtained by

$$z^k_{opt} = \max\left(\sum_{i\in I} a^k_{ij}\, x_i,\ A^k_j - \sum_{i\in I} a^k_{ij} x_i \quad for\ all\ j\right) \tag{17}$$

$$y^k_{j_{opt}} = \left|2\cdot\sum_{i\in I} a^k_{ij}\, x_i - A^k_{ij}\right| \quad for\ all\ j \tag{18}$$

As the number of airlines or the sample size increases, it is very difficult to solve [SP]. To solve this type of two-stage large-scale problems, a commonly used technique is to employ the Benders decomposition algorithms (e.g., Benders 1962; Higle and Sen 1991; Birge and Louveaux 1997). We therefore develop the Benders decomposition algorithm for [SP]. The solution procedures are provided as follows.

Step 0 Set lower and upper bounds, $lb=-\infty$ and $ub=+\infty$, respectively. Set the iteration index $i=0$. Let $\tilde{x}$ denote the current solution.

Step 1 Solve the relaxed master problem.

$$\begin{aligned} &\text{Minimize} && \xi \\ &s.\ t. && 0.45L^k \le \sum_{i\in I} l^k_i\, x_i \le 0.55L^k \quad for\ all\ k \\ & && [\text{Benders Cuts}] \\ & && \xi \in R\ \ x \in \{0,1\}\ \ x_1 = 1 \end{aligned} \tag{MP}$$

Let the optimal solution be $\hat{\xi}$ and $\hat{x}$. Update lower bound $lb = \hat{\xi}$.

Step 2 For k=1, 2,…, N, subproblems for the corresponding $\widehat{x}$ are as follows:

$$\begin{aligned}
&\text{Minimize } \theta^k = Cz^k + \sum_{j \in J} y_j^k \\
&s.\,t.\; z^k \geq \sum_{i \in I} a_{ij}^k \widehat{x}_i && \textit{for all } j : \pi_j^{1k} \\
&z^k \geq A_j^k - \sum_{i \in I} a_{ij}^k \widehat{x}_i && \textit{for all } j : \mu_j^{1k} \\
&y_j^k \geq 2 \cdot \sum_{i \in I} a_{ij}^k \widehat{x}_i - A_j^k && \textit{for all } j : \pi_j^{2k} \\
&y_j^k \geq A_j^k - 2 \cdot \sum_{i \in I} a_{ij}^k \widehat{x}_i && \textit{for all } j : \mu_j^{2k} \\
&z, y \geq 0
\end{aligned} \qquad (\text{Sub}^k)$$

where $\pi_j^{1k}, \mu_j^{1k}, \pi_j^{2k}$, and μ_j^{2k} are the corresponding dual variables for [Sub^k]. In the subproblems [Sub^k], k=1, 2,…, N are always feasible and the optimal solutions are given by Eqs. 17 and 18. Suppose that the corresponding optimal dual variable values are $\widehat{\pi}_j^{1k}, \widehat{\mu}_j^{1k}, \widehat{\pi}_j^{2k}, \textit{ and } \widehat{\mu}_j^{2k}$ and the objective function value is $\widehat{\theta}^k$. Let

$$\widehat{\theta} = \frac{1}{N} \sum_{k=1}^{N} \widehat{\theta}^k.$$

If $ub > \widehat{\theta}$, update the upper bound $ub = \widehat{\theta}$ and update the current solution $\widetilde{x} = \widehat{x}$.

Step 3 If $ub - lb \leq \varepsilon$ where $\varepsilon \geq 0$ is a prespecified optimality tolerance, stop. $\widetilde{x}$ is the optimal solution and ub is the optimal objective function value; otherwise, proceed to step 4.

Step 4 A Benders optimality cut is generated as follows:

$$\begin{aligned}
N \cdot \xi \geq \sum_{k=1}^{N} \Big(&\sum_{j=1}^{J} (\pi_j^{1k} \cdot \sum_{i \in I} a_{ij}^k\, x_i) + \sum_{j=1}^{J} (\mu_j^{1k} \cdot (A_j^k - \sum_{i \in I} a_{ij}^k\, x_i)) \\
&+ \sum_{j=1}^{J} (\pi_j^{2k} \cdot (2 \cdot \sum_{i \in I} a_{ij}^k\, x_i - A_j^k)) + \sum_{j=1}^{J} (\mu_j^{2k} \cdot (A_j^k - 2 \cdot \sum a_{ij}^k\, x_i))\Big)
\end{aligned}$$

We add this to [MP]. Update the iteration index i to i+1 and go to step 1.

3.1 Accelerating convergence: a new feasibility cut

The Benders decomposition algorithm converges to the optimal solution. However, convergence might require a large number of iterations for real problems. For [SP], we observe that the convergence rate is extremely slow (see Table 5). Thus, we accelerate convergence by adding some auxiliary linking variables of the two stages to the first stage problem; from these, a new set of Benders feasibility cuts are generated in each iteration in place of the earlier optimality cuts. The new formulation is given in [SP2]. New decision variables $\bar{z}$ and $\bar{y}_j$ are defined and [SP] is reformulated as follows:

$$[\text{SP2}] \quad \text{Minimize } C \cdot \bar{z} + \sum_{j \in J} \bar{y}_j \tag{SP2}$$

$$\text{s. t. } \bar{z} = \frac{1}{N} \sum_{k=1}^{N} z^k \tag{SP2.1}$$

$$\bar{y}_j = \frac{1}{N} \sum_{k=1}^{N} y_j^k \; \textit{for all } j \tag{SP2.2}$$

$$z^k \geq \sum_{i \in I} a_{ij}^k \, x_i \; \textit{for all } j,\ k \tag{SP2.3}$$

$$z^k \geq A_j^k - \sum_{i \in I} a_{ij}^k x_i \; \textit{for all } j,\ k \tag{SP2.4a}$$

$$y_j^k \geq 2 \cdot \sum_{i \in l} a_{ij}^k x_i - A_j^k \; \textit{for all } j,\ k \tag{SP2.4b}$$

$$y_j^k \geq A_j^k - 2 \cdot \sum_{i \in l} a_{ij}^k \, x_i \; \textit{for all } j,\ k \tag{SP2.5a}$$

$$0.45L^k \leq \sum_{i \in l} l_i^k \, x_i \leq 0.55L^k \; \textit{for all } k \tag{SP2.5b}$$

$$\bar{y},\ y,\ z,\ \bar{z} \geq 0\ x \in \{0,\ 1\}\ x_1 = 1 \tag{SP2.6}$$

From Eq. SP2.4a,

$$\sum_{k=1}^{N} z^k \geq \sum_{k=1}^{N} \sum_{i \in I} a_{ij}^k x_i \ \ \textit{for all } j \tag{SP2.4a'}$$

Dividing both sides by N and switching the sequence of the summation on the right hand side, we obtain:

$$\frac{1}{N} \sum_{k=1}^{N} z^k \geq \sum_{i \in I} \left(\frac{1}{N} \sum_{k=1}^{N} a_{ij}^k \right) \cdot x_i \ \ \textit{for all } j$$

Let $\bar{a}_{ij} = \frac{1}{N} \sum_{k=1}^{N} a_{ij}^k$ and $\bar{A}_j = \sum_{i \in I} \bar{a}_{ij}$; we have the following inequalities for $\bar{z}$:

$$\bar{z} \geq \sum_{i \in I} \bar{a}_{ij}\ x_i \ \textit{for all } j \tag{SP2.4a*}$$

Similarly, for Eqs. SP2.4b, SP2.5a, and SP2.5b:

$$\bar{z} \geq \bar{A}_j - \sum_{i \in I} \bar{a}_{ij} x_i \ \ \textit{for all } j \tag{SP2.4b*}$$

$$\bar{y}_j \geq 2 \cdot \sum_{i \in I} \bar{a}_{ij} x_i - \bar{A}_j \ \ \textit{for all } j \tag{SP2.5a*}$$

$$\bar{y}_j \geq \bar{A}_j - 2 \cdot \sum_{i \in I} \bar{a}_{ij} x_i \ \ \textit{for all } j \tag{SP2.5b*}$$

Adding Eqs. SP2.4a*, SP2.4b*, SP2.5a*, and SP2.5b* into [SP2], we obtain an equivalent new formulation [SP3].

$$\text{Minimize } C \cdot \bar{z} + \sum_{j \in J} \bar{y}_j \tag{SP3}$$

$$s.t. \bar{z} \geq \sum_{i \in I} \bar{a}_{ij}\ x_i \ \ \textit{for all } j \tag{SP3.1}$$

$$\bar{z} \geq \overline{A}_j - \sum_{i \in I} \overline{a}_{ij}\, x_i \;\; \textit{for all } j \tag{SP3.2a}$$

$$\bar{y}_j \geq 2 \cdot \sum_{i \in I} \overline{a}_{ij}\, x_i - \overline{A}_j \;\; \textit{for all } j \tag{SP3.2b}$$

$$\bar{y}_j \geq \overline{A}_j - 2 \cdot \sum_{i \in I} \overline{a}_{ij}\, x_i \;\; \textit{for all } j \tag{SP3.3a}$$

$$\bar{z} = \frac{1}{N} \sum_{k=1}^{N} z^k \tag{SP3.3b}$$

$$\bar{y}_j = \frac{1}{N} \sum_{k=1}^{N} y_j^k \;\; \textit{for all } j \tag{SP3.4}$$

$$z^k \geq \sum_{i \in I} a_{ij}^k x_i \quad \textit{for all } j, k \tag{SP3.5}$$

$$z^k \geq A_j^k - \sum_{i \in I} a_{ij}^k\, x_i \;\; \textit{for all } j,\ k \tag{SP3.6a}$$

$$y_j^k \geq 2 \cdot \sum_{i \in I} a_{ij}^k\, x_i - A_{ij}^k \;\; \textit{for all} \quad j,\ k \tag{SP3.6b}$$

$$y_j^k \geq A_{ij}^k - 2 \cdot \sum_{i \in I} a_{ij}^k\, x_i \;\; \textit{for all} \quad j,\ k \tag{SP3.7a}$$

$$0.45L^k \le \sum_{i\in I} l_i^k x_i \le 0.55L^k \quad \textit{for all } k \tag{SP3.7b}$$

$$\bar{y},\ y,\quad z,\ \bar{z} \ge 0\, x \in \{0, 1\}\, x_1 = 1 \tag{SP3.8}$$

For this new formulation, the decision variables x, $\bar{z}$, and $\bar{y}$ are optimized in the first-stage master relaxed problem [MP*].

$$\begin{aligned}
\text{Minimize}\quad & C \cdot \bar{z} + \sum_{j\in J} \bar{y}_j \\
s.t.\quad & \bar{z} \ge \sum_{i\in I} \bar{a}_{ij}\, x_i && \textit{for all } j \\
& \bar{z} \ge \bar{A}_j - \sum_{i\in I} \bar{a}_{ij}\, x_i && \textit{for all } j \\
& \bar{y}_j \ge 2 \cdot \sum_{i\in I} \bar{a}_{ij}\, x_i - \bar{A}_j && \textit{for all } j \\
& \bar{y}_j \ge \bar{A}_j - 2 \cdot \sum_{i\in I} \bar{a}_{ij}\, x_i && \textit{for all } j \\
& 0.45L^k \le \sum_{i\in I} l_i^k\, x_i \le 0.55L^k && \textit{for all } k \\
& \bar{y},\ \bar{z} \ge 0\ x \in \{0, 1\}\ x_1 = 1 \\
& \text{[Benders Feasibility Cuts]}
\end{aligned} \tag{MP*}$$

We describe next why feasibility cuts instead of optimality cuts are necessary for this new formulation. Let the optimal solution of [MP*] be $\widehat{x}$, $\widehat{\bar{z}}$, and $\widehat{\bar{y}}$. Then, the second-stage subproblem becomes:

$$\textit{Minimize}\quad 0 \tag{Sub*}$$

$$s.t.\quad \sum_{k=1}^{N} z^k = N \cdot \widehat{\bar{z}}\ \ (\alpha) \tag{Sub*.1}$$

Table 3 Data sets for the computational experiments

Instance	Number of flights	Number of airlines	Duration of each period (h)	Number of periods
ExD	252	17	1	168
ImD	752	50	2	84

Table 4 Workload variances and values of C

C	0, 10, 20	40, 50	70, 84, 100, 200, 1,000	$10e^4$	$10e^6$
Maximum workload	76.4	75.6	75.5	75.0	75.0
Sum of difference	255.3	275.5	282.1	316.3	448.7
Workload variance	61.08	60.97	60.75	62.00	63.69

$$\sum_{k=1}^{N} y_j^k = N \cdot \hat{\bar{y}}_j \ \ for\ all\ j\ \ (\beta_j) \tag{Sub*.2}$$

$$z^k \geq \sum_{i \in I} a_{ij}^k \hat{x}_i \ \ for;\ \ all\ j, k\ \ \left(\gamma_j^k\right) \tag{Sub*.3}$$

$$z^k \geq A_j^k - \sum_{i \in I} a_{ij}^k \hat{x}_i \ \ for\ all\ j,\ k\ \ \left(\lambda_j^k\right) \tag{Sub*.4a}$$

$$y_j^k \geq 2 \cdot \sum_{i \in I} a_{ij}^k \hat{x}_i - A_j^k \ \ for\ all\ j,\ k\ \ \left(\pi_j^k\right) \tag{Sub*.4b}$$

$$y_j^k \geq A_j^k - 2 \cdot \sum_{i \in I} a_{ij}^k \hat{x}_i \ \ for\ all\ j,\ k\ \ \left(\mu_j^k\right) \tag{Sub*.5a}$$

Table 5 Number of iterations for the Benders decomposition-based algorithms

Sample size	General Benders decomposition	Accelerated Benders decomposition	Reduction (%)
1	261	1	99.62
5	289	77	73.36
20	269	95	64.68
50	267	86	67.79
100	251	80	68.13
200	252	84	66.67

Table 6 Computational time in seconds for different algorithms and models

Sample size	MIP optimizer	General Benders decomposition	Reduction (%)	Accelerated Benders decompositions	Reduction (%)[a]
1	1	67	−6,600	0.95	5.00 (98.58)
5	23	111	−382.61	185	−704.35 (−66.67)
20	349	185	46.99	320	8.31 (−72.97)
50	1,851	370	80.01	378	79.58 (−2.16)
100	7,771	620	92.02	527	93.22 (15.00)
200	36,254	1,216	96.65	1,114	96.93 (8.39)

[a] The number in parentheses is the relative reduction between the general and accelerated Benders decomposition

$$y,\ z \geq 0 \qquad \text{(Sub*.5b)}$$

where the Greek alphabets in the brackets are the dual variables for the corresponding constraints.

In [MP*], $\hat{\bar{z}}$ represents the maximum value of the average workload over the periods, but the corresponding z represents the absolute maximum value of the workload over all the scenarios and the equations (Eq. Sub*.2) cannot be satisfied unless the solution is optimal. Thus, a Benders feasibility cut is generated by the following procedure.Consider the dual of [Sub*]:

$$\begin{aligned}
\text{Maximize}\quad & \alpha N \cdot \hat{\bar{z}} + \sum_{j\in J} \beta_j N \cdot \hat{\bar{y}}_j + \sum_{k=1}^{N}\sum_{j\in J} \gamma_j^k \sum_{i\in I} a_{ij}^k \hat{x}i + \sum_{k=1}^{N}\sum_{j\in J} \lambda_j^k \left(A_j^k - \sum_{i\in I} a_{ij}^k \hat{x}i \right) \\
& + \sum_{k=1}^{N}\sum_{j\in J} \pi_j^k \left(2\sum_{i\in I} a_{ij}^k \hat{x}i - A_j^k \right) + \sum_{k=1}^{N}\sum_{j\in J} \mu_j^k \left(A_j^k - 2\sum_{i\in I} a_{ij}^k \hat{x}i \right) \\
s.t.\left(z^k\right)\quad & \alpha + \sum_{j\in J} \gamma_j^k + \sum_{j\in J} \lambda_j^k \leq 0 \quad \mathit{for\ all}\ k \\
& \left(y_j^k\right) \quad \beta_j + \pi_j^k + \mu_j^k \leq 0 \quad \mathit{for\ all}\ j,\ k \\
& \alpha,\ \beta\,\text{unrestricted}\,\gamma, \lambda,\ \pi,\ \mu \geq 0
\end{aligned} \qquad \text{(DP)}$$

Table 7 Case 1 partial reassignment: terminate the services of the second biggest airline

P	Airlines reassigned		CPU time in seconds	Gap in objective value (%)
	#	Percentage (%)		
0	0	0.00	0.9	19.19
5	2	4.08	4.6	4.19
7	3	6.12	5.7	2.66
10	4	8.16	11	1.98
30	14	28.57	102	0.14
50	23	46.94	187	0.00
100	23	46.94	199	0.00

Table 8 Case 2 partial reassignment: terminate the services of the second smallest airline

P	Airlines reassigned		CPU time in seconds	Gap in objective value (%)
	#	Percentage (%)		
0	0	0.00	0.67	1.71
5	2	4.08	6.9	1.30
7	3	6.12	11	1.11
10	4	8.16	18	0.90
30	13	26.53	138	0.20
50	24	48.98	248	0
100	24	48.98	262	0

If [Sub*] is infeasible, then the objective function of [DP] is unbounded. Hence, we add a feasibility cut:

$$0 \geq \alpha N \cdot \hat{\bar{z}} + \sum_{j \in J} \beta_j N \cdot \hat{\bar{y}}_j + \sum_{k=1}^{N} \sum_{j \in J} \gamma_j^k \sum_{i \in I} a_{ij}^k \widehat{x}i + \sum_{k=1}^{N} \sum_{j \in J} \lambda_j^k \left(A_j^k - \sum_{i \in I} a_{ij}^k \widehat{x}i \right)$$
$$+ \sum_{k=1}^{N} \sum_{j \in J} \pi_j^k \left(2 \sum_{i \in I} a_{ij}^k \widehat{x}i - A_j^k \right) + \sum_{k=1}^{N} \sum_{j \in J} \mu_j^k \left(A_j^k - 2 \sum_{i \in I} a_{ij}^k \widehat{x}i \right)$$

It is notable that although the master problem [MP*] is more difficult to solve than the original [MP] due to the increased number of variables and constraints, convergence is accelerated considerably and the total computational time is significantly reduced (see Table 6).

4 Partial reassignment

Once airlines are assigned to the terminals, relocating the services to another terminal may require costly readjustments. While airports often experience operational (or seasonal) schedule changes by airlines, these changes may not justify the relocation of the large number of airline services. With the slightest change in services, the workload imbalance may be propagated. To avoid such imbalance, the seemingly best approach is to reshuffle the airlines and to reoptimize the workload-balancing problem, but this is not practical because the airline assignment should not be changed too often. Thus, we introduce a new constraint that restricts the number of airlines to be relocated in the model [SP].

Table 9 Imbalance measures under the current and the optimal assignments

	Current assignment		Optimal solutions		Reduction (%)
	T1	T2	T1	T2	
Maximum workload	31.8	47.6	35.85	35.85	24.68
Variance	32.65	39.17	26.73	26.22	26.27
Sum of difference	1,736.8		258.8		85.10

Table 10 Simulation results for the scenario of inadequate capability

Assignment terminal	Current assignment		Optimal assignment	
	T1	T2	T1	T2
Average cycle time (min)	6.66	31.22	14.43	8.08
Overall average (min)	21.55		11.40	
Reduction (%)	47.10			

Assume that the current assignment is x^0 obtained from solving [SP]. After changes in airline services, such as the addition or deletion of flights, the balancing problem is solved again without reshuffling all the airlines. Let the new solution be $\widetilde{x}$, then the new assignment should satisfy

$$\sum_{i \in I'} \left| \widetilde{x}_i - x_i^0 \right| \leq \frac{P}{100} \cdot \left| x^0 \right|, \tag{19}$$

where the set I' is the set of airlines that remains from the previous assignment, the parameter P is defined by the percentage of the total number of airlines whose original assignment can be changed, and $P \in [0,100]$. For the two extreme points, P=0 means no permission to change any airline assignment and P=100 allows the overall reoptimization. The absolute term $\left| \widetilde{x}_i - x_i^0 \right|$ causes the model to be nonlinear. However, we note that x^0 and $\widetilde{x}$ are both binary variables and x^0 is given. If $x_i^0 = 1$, then $\left| \widetilde{x}_i - x_i^0 \right| = \left| \widetilde{x}_i - 1 \right| = 1 - \widetilde{x}_i$; otherwise, if $x_i^0 = 0$, then $\left| \widetilde{x}_i - x_i^0 \right| = \left| \widetilde{x}_i - 0 \right| = \widetilde{x}_i$. Therefore, the model is still a mixed integer linear program after adding the constraint.

5 Computational results

In this section, computational experiments are conducted to examine the performance of the proposed solution procedure with the data from an international airport cargo terminal. To obtain a_{ij}, the following three steps of preprocessing are taken: (1) The total cargo load estimation for each flight, (2) arrival (request) time distribution of export (import) cargo, and (3) Summation of the cargo load for all the flights that belong to an airline. Table 3 lists the two data sets used for these computational experiments. ImD is a comprehensive set of data, which contains 50 airlines from two Import Terminals. ExD is an independent data set from one Export Terminals, which contains 17 airlines. In the computational experiments, to compare the performance of different algorithms and formulations, the small data set ExD is

Table 11 Simulation results for the scenario of adequate capability

Assignment terminal	Current assignment		Optimal assignment	
	T1	T2	T1	T2
Average cycle time (min)	6.85	6.83	6.73	6.77
Overall average (min)	6.84		6.75	
Reduction (%)	1.34			

employed for the sake of computational (experimental) time. Wherever the computation is to validate the results for different assignments, the complete data set ImD is employed. The computer codes were written in Microsoft Visual C++ with ILOG CPLEX8.1 and Concert Technology 1.3 and executed in a DELL Pentium IV computer with 2.4 GHz CPU and 512 MB RAM in the Microsoft Windows XP environment.

5.1 Effect of *C*

An experiment is conducted using the data set ImD to investigate the impact of *C* on the optimal solution. [SP] is solved with different *C* values ranging from 0 to $10e^6$, and the two objective function values are obtained accordingly. With the optimal solution obtained, the variance of workload among the periods is computed as shown in Table 4.

The results show that the optimal solution may change with the change of *C* but not all the time. C=0 and C=$10e^6$ are the two extreme cases that consider only one objective. The best solution is obtained when both objectives are considered. Thus, we propose the *C* value that equals to the total number of periods (e.g., 168 for ExD) where both objectives are considered with equal importance.

5.2 Results of benders decomposition algorithms

A computational study is carried out using the data set ExD for different sample sizes. Table 5 shows the number of iterations for both the general and the accelerated Benders decomposition algorithms. The reduction in the number of iterations is provided by the percentage in the fourth column. Table 6 reports the computational time of the two algorithms with respect to the default MIP solver (in CPLEX). The relative time reductions by these algorithms are also computed with respect to the computational time of the default solver.

The results show that the general Benders decomposition algorithm reduces the computation time considerably when the sample size is greater than 50. The reduction is more than 90% when the sample size is 100, which is considered large enough (see Section 2.3). It reaches 96% when the sample size becomes 200. In addition, the accelerated Benders decomposition reduces the number of iterations by more than 60% compared with the general Benders decomposition. As discussed in Section 3, the complexity of a decomposed subproblem increases in the accelerated algorithm, but the convergence is significantly improved. Despite the tradeoff between these two, the solution time is still reduced as the sample size increases: When the sample size is 100, this time reduction is 15%.

When the sample size is very small (less than 10), the default MIP solver is the most efficient. As the sample size increases, the Benders decomposition algorithms become more efficient, and especially with the sample size being large enough, the accelerated Benders decomposition is the most efficient algorithm.

5.3 Partial reassignment

Two different scenarios are considered to illustrate the partial reassignment:

Case 1 From the ImD data set, the services of the second biggest airline are terminated.

Case 2 From the same data set, the services of the second smallest airline are terminated.

Different P values that represent the different number of airlines to be reassigned may affect the resulting workloads of the terminals. While the increase of P values may improve the solution quality, this significantly increases the computational time as well. Thus, the impact of P on the objective function value is examined. As shown in Tables 7 and 8, the "gap in the objective value" is measured by the difference between the optimal objective function values from the partial assignment and the full assignment models. The number of airlines whose assignment *is actually changed* is reported in the second column.

Changes in airline services may cause the imbalance of the terminal workloads. The imbalance becomes greater with the changes in the services of bigger airlines (19.19 vs 1.71%). As the P value increases, the imbalances are drastically reduced. With P=10, the gap between the objective function values of the optimal and the partial reassignments becomes 1.98% only and in both cases, as P becomes 30 or 40, the partial reassignment provides a near-optimal solution with only half the computational time.

6 Evaluation of the optimal airline assignments

The optimal airline assignment is obtained by solving [SP] with N=100, which is large enough (see Section 2.3). The performance of the optimal solution vs the current assignment is examined under a real workload scenario. The two balance measures as defined in the model, the maximum workload $\left(\max\left\{w_j^1, w_j^2 : forall j\right\}\right)$, and the sum of difference $\left(\sum_{j\in J}\left|w_j^1 - w_j^2\right|\right)$ are computed. In addition, the variance of workload among the periods at each terminal is also computed. Table 9 shows these results. The maximum workload is reduced by 24.68% and the variance reduction is 26.27%. The reduction in the sum of difference is 85.10%. These results show that the optimal solution significantly reduces the workload imbalance.

To illustrate the benefit of the workload balancing at a cargo terminal, simulation experiments are conducted using a simulator developed by Leong (2004). This simulator visualizes the simulation of cargo handling process in an import terminal. The cycle time per cargo retrieval trip is measured as a performance indicator. This simulator was built on a simplified assumption: The cargo handling capability is constant over all the time. It is the most ideal resource schedule for an air cargo terminal because the manpower schedule is comprised of a continually regular 8-h shift for each employee, which is just one of the motivations of this research that we introduced in Section 1.

The experiment simulates a 5-week operation with the actual flight schedule. The result from the first week is not analyzed (warm-up period). For the actual data set ImD, the simulation run was exploded with unserved cargoes due to the large deviation of workload that cannot be coped with a constant limited capability. In such a scenario, the balanced assignment reduces the average cycle time by 47.10% compared with the current assignment. Table 10 shows the results.

As a comparison, the cargo loads was reduced by 20% with the removal of the services for the second and third biggest airlines to obtain a steady state performance statistics. After the reduction, the workloads were accomplished even during its peak periods and the steady-state statistics are obtained as shown in Table 11.

The result shows that the proposed optimal airline assignments reduce the retrieval cycle time by 1.34%. The improvement is marginal due to the assumptions that each terminal has a constantly adequate capability over time. This result explains again that workload balancing is marginally meaningful under the condition of adequate resource capability.

In reality, however, the main resource, manpower, is always sought to be minimized to maintain low cost. Therefore, the comparison of these two scenarios commendably justified that workload balancing is able to contribute significant efficiency improvement under the condition of inadequate resource capability, which is exactly a reality that many air cargo terminals are facing today.

7 Conclusions

An S-MILP approach for the workload-balancing problem at air cargo terminals was developed. The services for airlines are assigned to two identical terminals so that the stochastic workload is balanced between terminals. Two measures of the workload imbalance are introduced and minimized. SAA is employed to transform the stochastic program to a solvable deterministic model. The model is further simplified by a reformulation, which significantly reduces the computational time especially for the problems with small sample size. The Benders decomposition algorithm is then developed to solve the problems with large sample size. In addition, new Benders cuts are developed to accelerate the Benders decomposition algorithm, which further reduces the computational time. Two actual examples from an international air cargo terminal are employed in our computational experiments. The results show that the proposed algorithms provide the optimal solution efficiently. For practical applications when operational or seasonal schedule changes for some airlines occur, partial reassignment is proposed to promptly reoptimize the airline assignments. The computational results show that the partial reassignment approach is efficient and practical. Finally, a simulation experiment shows that the proposed optimal airline assignment significantly reduces the cycle time of the terminal operations by balancing the workloads between the terminals.

Acknowledgement This research is partially supported by the National University of Singapore research grant R-266-000-025-112.

References

Ashayeri J, Gelders LF (1985) Warehouse design optimization. Eur J Oper Res 21:285–294
Baron P (1969) A simulation analysis for airport terminal operations. Transp Res 3:481–491
Benders JF (1962) Partitioning procedures for solving mixed-variables programming problems. Numer Math 4:238–252
Birge JR, Louveaux FV (1997) Introduction to stochastic programming. Springer, Berlin Heidelberg New York
Berrada M, Stecke KE (1986) A branch and bound approach for machine load balancing in flexible manufacturing systems. Manage Sci 32(10):1316
Cormier G, Gunn EA (1992) A review of warehouse models. Eur J Oper Res 58:3–13
De Neufville R, Rusconi-Clerici I (1978) Designing airport terminals for transfer passengers. J Transp Eng 104:775–787
Guerro F, Lozano S, Koltai TJ (1999) Machine loading and part type selection in flexible manufacturing systems. Int J Prod Res 37:1303–1317
Haghani A, Chen M (1998) Optimizing gate assignments at airport terminals. Transp Res A 32 (6):437–454
Higle JL, Sen S (1991) Stochastic decomposition: an algorithm for two-stage linear programs with recourse. Math Oper Res 16:650–669
Kleywegt AJ, Shapiro A, Homem-De-Mello T (2001). The sample average approximation method for stochastic discrete optimization. SIAM J Optim 12:479–502
Kumar N, Shanker K (2001) Comparing the effectiveness of workload balancing objectives in FMS loading. Int J Prod Res 39(5):843–871
Leong CH (2004) A simulation model of an air cargo terminal. Master dissertation, Department of Industrial and Systems Engineering, National University of Singapore
Moreno AA, Ding F-Y (1989) Goal oriented heuristics for the FMS loading (and part type selection) problems. Proceedings of the third ORSA/TIMS conference on flexible manufacturing systems, Cambridge, MA, pp 105–110
Rouwenhorst B, Reuter B, Stockrahm V, Houtum GJ, Mantel RJ, Zijm WHM (2000) Warehouse design and control: framework and literature review. Eur J Oper Res 122:515–533
Santoso T, Ahmed S, Goetschalckx M, Shapiro A (2003) A stochastic programming approach for supply chain network design under uncertainty. Logistics Institute, Georgia Institute of Technology
Shanker K, Tzen Y–JJ (1985) A loading and dispatching problem in a random flexible manufacturing system. Int J Prod Res 23:579–595
Shapiro A (2001). Monte Carlo simulation approach to stochastic programming. Proceedings of the 2001 winter simulation conference, 1999
Shapiro A, Homem-de-Mello T (1998) A simulation-based approach to stochastic programming with recourse. Math Program 81:301–325
Shapiro A, Homem-de-Mello T (2001) On the rate of convergence of optimal solutions of Monte Carlo approximations of stochastic programs. SIAM J Optim 11:70–86
Verweij B, Ahmed S, Kleywegt AJ, Nemhauser G, Shapiro A (2003) The sample average approximation method applied to stochastic routing problems: a computational study. Comput Optim Appl 24:289–333
Wirasinghe SE, Bandara S (1990) Airport gate position estimation for minimum total costs-approximate closed form solution. Transp Res B 24:287–297

D. Li · H.-C. Huang · A. D. Morton · E.-P. Chew

Simultaneous fleet assignment and cargo routing using benders decomposition

Abstract In this paper, we incorporate the cargo routing problem into fleet assignment to model the fleet assignment more accurately. An integrated model and a Benders decomposition-based approach are developed to simultaneously obtain the optimal assignment of fleet to legs and the routing of forecasted cargo demand over the network. Computational experiments show that this integrated approach converges very fast for all different test scenarios.

Keywords Airline planning · Fleet assignment · Benders decomposition

1 Introduction

Airline schedule planning consists of all planning activities that have to be carried out so that the schedule is operationally feasible and profitable. A set of mathematical models are often employed to support these activities. Normally, these activities are carried out in sequence and the corresponding models are solved individually. Such a sequential approach ignores the linkages between these planning activities and leaves scope for improvement.

Fleet assignment is a planning activity that aims to maximize profitability by optimally assigning fleets (types of airplanes) to legs (a nonstop flight from an

---

D. Li
Intel Corporation (Shanghai), Solver Solution Engineering Team, No 999 Ying Lun Road, Waigaoqiao Free Trade Zone, Pudong, 200131, Shanghai, People's Republic of China
E-mail: devin.li@intel.com

H.-C. Huang (✉) · E.-P. Chew
Department of Industrial & Systems Engineering, National University of Singapore, 1 Engineering Drive 2, Singapore, 117576, Singapore
E-mail: isehhc@nus.edu.sg, isecep@nus.edu.sg

A. D. Morton
Department of Operational Research, London School of Economics, Houghton Street, London, WC2A 2AE, UK
E-mail: a.morton@lse.ac.uk

origin to a destination with specific departure and arrival times). In most published literature, this step considers only the leg-based passenger flow. As all legs in a network are interdependent (Barnhart et al. 2002), failing to capture the network effect may result in not addressing the actual problem accurately. The incorporation of the itinerary-based passenger flow into fleet assignment has been studied by Barnhart et al. (2002), but they do not take into account the cargo flow. As the aircargo traffic has grown robustly relative to passenger traffic over the decade 1993–2003, and as industry forecasters expect this trend to continue (Boeing 2004), the cargo business is increasingly important for any airline which provides cargo capacity. The route of cargo is determined to a large extent by the cargo capacity of every leg, which depends on the fleet assignment decision. As a result, the fleet assignment has great influence on the cargo flow and, thus, on revenue. A traditional approach to the fleet assignment may cause great loss in the cargo revenue and, thus, the total profit of an airline. Incorporating the cargo routing into the fleet assignment can help the combination carrier to better balance its resources and the forecasted cargo demand.

Few references can be found in the operations research literature regarding the integration of planning activities in the airline industry. One of the early efforts in this direction is the work of Barnhart et al. (1998a), who proposed a string-based integrated model to simultaneously solve the fleet assignment and aircraft routing problems. A string was defined as a sequence of connected legs beginning and ending at a maintenance station, without violating the flow balance and the maintenance feasibility requirements. Another work by Barnhart et al. (1998b) describes the integrated approximate modeling of the fleet assignment and crew-pairing problems. Because the crew-pairing problem takes a very long time to be solved, an approximate duty-based model (Barnhart and Shenoi 1998) was used in place of it to maintain solvability. The integrated model was solved by an advanced sequential approach; thus, it was not really a simultaneous optimization. Desaulniers et al. (1997) presented an integration of the fleet assignment model and time windows. Two equivalent models were constructed. Column generation and a Dantzig–Wolfe decomposition approach were employed to solve the problem. Another example was given by Rexing et al. (2000), where the assignment of aircraft was considered together with the flight departure times to improve flight connection opportunities.

Recently, Cordeau et al. (2001) and Mercier et al. (2003) introduced an integrated model and a solution approach based on Benders decomposition to optimize the crew pairing and the aircraft routing at the same time. The linear relaxation of the combined model was split into a master problem and a subproblem by Benders decomposition. Then, both problems were solved by column generation. In the work of Cordeau et al., the aircraft routing formed the master problem and the crew-pairing problem formed the subproblem, while in the work of Mercier et al., the relationship was reversed. Other contributions to improve the crew-pairing solution by incorporating other planning problems are to be found in the works of Klabjan et al. (2002) and Cohn and Barnhart (2003).

Particularly interesting is the integrated model of fleet assignment and passenger mix introduced by Barnhart et al. (2002) to generate improved solutions. This itinerary-based fleet assignment model was able to capture network effects and estimate spill and recapture of passengers more accurately. Due to the size of the model, the problem was solved by a two-step approach. First, the problem's relaxed linear program (LP) was solved by column and row generation, and then an integer solution was obtained through branch and bound with no new columns generated.

In this paper, we propose a different integrated model that combines the fleet assignment with the cargo routing, instead of the passenger mix, to obtain better solutions. The passenger revenue is simply estimated linearly in our integrated model. Because of the nature of the cargo business, the cargo flow is modeled in a very different way from the key path approach used by Barnhart et al. (2002) in their work on the passenger mix model. Unlike passengers, cargo has no strong preference for a particular itinerary, as long as it can be delivered on time. There are also no available industry data to calculate the spill cost and the recapture rate for the cargo flow. Furthermore, for some airlines, cargo is allowed to transfer between different aircraft only at the hub, while there are no equivalent constraints for passengers.

Besides the integration model above, the main contributions of this paper also include the introduction of a Benders decomposition-based algorithm. A set of computational experiments are carried out and the results show that our algorithm converges very fast for all test scenarios.

The rest of this paper is organized as follows: "Mathematical formulation" introduces the notation used in this paper and presents the mathematical formulations. The solution approach based on the Benders decomposition algorithm is described in "Solution methodology." In "Computational experiments," the computational experiments and their results are reported. The conclusion and direction for future work are discussed in the final section.

2 Mathematical formulation

We based our models on the planning problems faced by a major Asia-Pacific combination carrier. It operates a weekly flight schedule through a passenger network with six different fleets and a freighter (airplane carrying only cargo) network with only one fleet. The entire network has only one hub.

We developed our formulation based on a 1-week wrapped-around timeline and an underlying graph that consists of a number of time–space networks indexed by the fleet types. This graph of fleeted time–space networks is commonly used in airline problems (Hane et al. 1995). In a time–space network for a specific fleet type, such as the one given in Fig. 1, a node is associated with a station and an

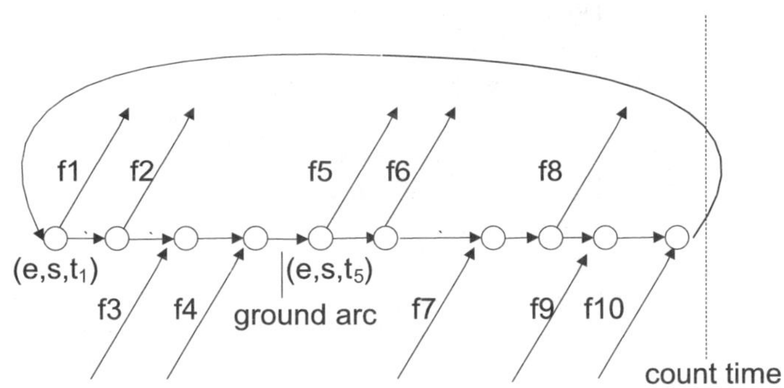

Fig. 1 A time–space network

arrival or departure event. The arcs in the network represent either the flight legs from one station to another or the ground arcs linking two nodes on the same station between two event time points.

Before describing the mathematical formulations, we first introduce the following notations:

Notation

L	The set of legs in the flight schedule indexed by i
E	The set of different fleet types indexed by e
S	The set of stations indexed by s
T	The set of arrival and departure event times at all stations in a week, indexed by t_j. The event at t_j occurs before the event at t_{j+1}. If $\|T\|=m$, t_m is the last event before the count time, and t_1 is the first event after the count time. The count time is a specific time point in a week used to count the number of aircraft in the schedule. To facilitate the counting, the count time is best chosen at a time when most of the aircraft are on the ground
N	The set of nodes in the fleeted time–space networks indexed by n or (e, s, t_j). Here, $e\in E$, $s\in S$, $t_j\in T$. To elaborate, for each fleet type e, there is an associated time–space network, with the node set $\{n=(e, s, t_j)\| s\in S, t_j\in T\}$
$O(ct)$	The set of legs that pass the count time
$I(n)$	The set of inbound legs to node n or (e, s, t_j)
$O(n)$	The set of outbound legs from node n or (e, s, t_j)
$c^e_{(0)_i}$	The operational cost, including the cost of fuel, handling, take-off, and landing
τ^e_i	The estimated passenger revenue of flying leg i with fleet e, which is a linear function of the seat number and the block time of this leg. The block time is the time from when the plane leaves the gate at the departure station to when it arrives at the gate of the arrival station
c^e_i	The cost of assigning fleet type e to leg i, which is the operational fleet assignment cost minus the estimated passenger revenue, written as $c^e_i = c^e_{(0)_i} - \tau^e_i$
Num^e	The number of aircraft in fleet type e
K	The set of commodities indexed by k, which is defined as a time-related origin–destination cargo demand pair. K is denoted by (o, t, D_o, T_o, T_w), where $o\in S$ is the cargo origin; $t\in S$ is the cargo destination; D_o is the day on which the cargo is ready for shipment at the origin, from Monday to Sunday; T_o is the ready time for shipment at the origin, either in the morning, at noon, or in the evening; and T_w is the time window for the cargo to reach the destination, 1 day (24 h) or 2 days (48 h)
$P_f(k)$	The set of feasible paths or itineraries for commodity k
d_i	Cargo capacity of leg $i\in L$, which is determined by the fleet type assigned to the leg

B^k	Demand for commodity k
r_p^k	Per unit revenue of flowing commodity k on path p
$\delta_i^p = \begin{cases} 1, \text{if } i \in p \\ 0, \text{otherwise} \end{cases}$	This is an incidence coefficient, i.e., it is 1 only if path p includes leg i

Decision variables

$$x_i^e = \begin{cases} 1, \text{if flight leg } i \text{ is assigned} \\ \text{to fleet type } e; \\ 0, \text{otherwise} \end{cases}$$

$g_{e,s,(t_j,t_{j+1})}$	The number of aircraft of fleet type e that are on the ground at station s immediately after event time t_j or just before event time t_{j+1}
y_p^k	The amount of commodity k on path p

2.1 Basic fleet assignment model

Given a flight schedule and a set of fleets, the fleet assignment is made to determine which fleet to assign to each leg, with the objective of minimizing the total assignment cost or maximizing the total fleet assignment contribution. The fleet assignment problem (FAM) has been well studied in airline problems (Abara 1989; Subramanian et al. 1994; Hane et al. 1995; Rushmeier and Kontogiorgis 1997).

The mathematical formulation of the basic fleet assignment model (Hane et al. 1995) is:

$$\min \sum_{i \in L} \sum_{e \in E} c_i^e x_i^e \tag{0}$$

subject to:

$$g_{e,s,(t_{j-1},t_j)} + \sum_{i \in I(n)} x_i^e = g_{e,s,(t_j,t_{j+1})} + \sum_{i \in O(n)} x_i^e, \forall n = (e, s, t_j) \in N, e \in E \tag{1}$$

$$\sum_{s \in S} g_{e,s,(t_m,t_1)} + \sum_{i \in O(ct)} x_i^e \leq Num^e, \forall e \in E \tag{2}$$

$$\begin{gathered} \sum_e x_i^e = 1, \forall i \in L \\ x_i^e \in \{0, 1\}, \forall i \in L, e \in E \\ g_{e,s,(t_j,t_{j+1})} \geq 0, \forall e \in E, s \in S, t_j, t_{j+1} \in T \end{gathered} \tag{3}$$

The objective Eq. (0) minimizes the total fleet assignment costs or maximizes the total passenger revenue minus operational costs. The constraints in Eq. (1)

ensure the conservation of flow balance on each fleeted time–space network. That is, aircraft going into a station s at a time point t_n must leave the same station at some other time later. The constraints in Eq. (2) ensure that the number of every fleet type in use does not exceed the total number available of that fleet type. The final set of constraints (Eq. 3) can be regarded as linking or side constraints imposed on all the time–space networks. They ensure that each flight is covered once, and only once, by a fleet type.

2.2 Cargo routing model

Given the forecasted cargo demand and a fleeted flight schedule in which all legs' fleet types are known, cargo routing maximizes the revenue of the commodity flow without exceeding either aircraft capacity or the time window. Similar to passenger routing, cargo routing with multiple origin–destination demand pairs can be modeled as a multicommodity network flow (MCNF) problem. MCNF has been studied extensively (see Ahuja et al. 1993 for a thorough description). For the solution methodology, Jones et al. (1993) studied the impact of three different formulations on the decomposition solution approach; namely, the node-arc, the path, and the tree formulations. Their results showed that the second formulation outperforms the other two in general cases. We adopt this path formulation in our approach too.

For the air cargo routing problem studied here, three additional side constraints have to be satisfied. First, the carrier allows cargo to be transferred from one aircraft to another only at the hub and at no other stations. Second, if transferring happens at the hub, the transit time between the two connected legs must be longer than the minimum cargo handling time. We supposed in this modeling that there was a common minimum cargo handling time for transfers between passenger aircraft, whatever the aircraft type. Third, if a cargo path does not pass through the hub, the legs covered must fly in one direction, while for a path transferring at the hub, the legs can change the flying direction only when leaving the hub. The single flying direction is ensured in our study by restricting the number of legs in a path. Explicitly modeling the time window constraints and the side constraints will result in a very complicated model. To capture these complex rules while maintaining the tractability, we apply a two-step modeling approach. Firstly, all feasible paths that satisfy the time window and the side constraints are generated for all commodities. Then, an MCNF path formulation with only the columns of feasible paths and rows of capacity and demand constraints is set up.

To view the cargo routing model (CRM) in a nutshell, we first simplify the presentation by ignoring the fleet types and the additional side constraints. In other words, we simply assume that there is only one time–space network defined by the flight legs, and the flow variables y along the paths are well defined as in MCNF. Then, the path-oriented CRM is:

$$\max \sum_{k \in K} \sum_{p \in P_f(k)} r_p^k y_p^k \tag{4}$$

subject to:

$$\sum_{k \in K} \sum_{p \in P_{\mathrm{f}}(k)} y_p^k \delta_i^p \leq d_i, \forall i \in L \tag{5}$$

$$\sum_{p \in P_{\mathrm{f}}(k)} y_p^k \leq B^k, \forall k \in K \tag{6}$$

$$y_p^k \geq 0, \forall p \in P_{\mathrm{f}}(k), k \in K \tag{7}$$

The objective (Eq. 4) maximizes the total cargo revenue. The capacity constraints in Eq. (5) restrict the total cargo flown on a leg i to its cargo capacity d_i. By flow the constraints in Eq. (6), the cargo actually shipped is less than or equal to the demand.

2.3 Integrated model

For combination carriers, the network usually can be divided into two subnetworks according to services, one for the passenger fleet flow and the other for the cargo fleet flow. Because the freighter cannot be assigned to passenger flights, and vice versa, the fleet assignment should be modeled separately for these two networks. On the contrary, cargo routing must take account of the capacity available on both freighter and passenger networks simultaneously, because cargo can also flow on a passenger flight, in the belly of the passenger aircraft. The feasible paths are thus constructed over the entire network.

The integrated formulation comprises the fleet assignment model and the CRM, but ignores the aircraft routing problem. The aircraft routing problem determines the sequence of flights, or routes, for each individual aircraft such that all legs are flown exactly once and each aircraft visits maintenance stations at regular intervals (Barnhart et al. 1998a). One may want to incorporate the aircraft routing model because one side constraint in the CRM requires cargo to be transferred to other aircraft only at the hub. Therefore, a truly integrated model would be able to identify different physical aircraft or tail numbers and involve three sequential problems. The first-stage problem would deal with the fleet assignment, the second-stage, with the aircraft routing, and the third-stage, with the cargo routing. Nevertheless, we argue that our two-stage integrated model approximates the actual three-stage problem very well for the carrier in question. This is because for most spoke stations, the frequency of flights is quite low, and so, to minimize the number of aircraft used at that station and to avoid having unnecessary aircraft on the ground overnight, the aircraft flying the arrival flight is usually assigned to the immediate departure flight. As a result, the same fleet types for the two connecting legs can be reasonably taken to imply the same physical aircraft. Although this is not the case for the hub, we do not care about the physical aircraft because cargo is allowed to transfer at the hub. Our two-stage integrated model, therefore, is a reasonable approximation to the actual problem.

To obtain the integrated model, we combine the passenger FAM, the freighter FAM, and the CRM together, and link them by multiplying the right-hand sides of the CRM capacity constraints with the fleet assignment variables. For a combination carrier that has only one type of freighter, the fleet assignment is not required for freighter flights. The integrated model thus includes only the passenger FAM and the CRM.

Because the fleet assignment and the cargo routing are determined simultaneously, the feasible paths are not fixed but altered whenever the fleet assignment changes. To accommodate this uncertainty, all potential feasible paths must be included in the integrated model. A direct path from the origin to the destination of a commodity is regarded as *potentially feasible* if it satisfies the time window constraint, the minimum transit time at the hub, and the requirement of the single flying direction. Let $P(k)$ denote the set of potential feasible paths for commodity k. The potential feasible path is generated based on the nonfleeted flight schedule and becomes valid as soon as all sequential pairs of flights, connected at a spoke, in this path are assigned with the same fleet type.

Let $L(pax)$ be the set of passenger legs, and $E(pax)$ be the set of passenger fleets. Let $L(frt)$ denote the set of freighter legs and $E(frt)$ be the set of freighter fleets (this set has only one element for the carrier in question). Then, we have $L=L(pax)\cup L(frt)$ and $E=E(pax)\cup E(frt)$. The cargo capacity of a fleet type is denoted by d^e, $e\in E$. To capture the feasible path correctly, we consider the fleeted time–space networks again. Let $N(pax)$ be the set of nodes in the passenger network. Instead of the constraints in Eq. (5), there will be a capacity constraint for each leg and fleet type combination. To define the flow variables y on the fleeted time–space networks, we divide the set $P(k)$ into two subsets. The set $P^{\mathrm{T}}(k)$ contains the paths transferring at the hub, and the other set, $P^{\mathrm{D}}(k)$, contains the rest of the paths.

Because every path p in the set $P^{\mathrm{T}}(k)$ may be formed from subpaths from two fleeted networks, we split it into two *subpaths*, $i(p)$ arriving at the hub and $o(p)$ departing from the hub (Fig. 2), each of which represents a feasible subpath on the fleeted networks. Accordingly, the set $P^{\mathrm{T}}(k)$ is divided into $PI^{\mathrm{T}}(k)$ and $PO^{\mathrm{T}}(k)$, which contain all $i(p)$ and $o(p)$, respectively, for commodity k. The flow variables are then defined for every path or subpath and every fleet combination. Let $y_p^{k,e}$, $y_{i(p)}^{k,e}, y_{o(p)}^{k,e}$ denote the decision variables corresponding to the path and subpath flown by fleet type e. Let r_p^k and $r_{i(p)}^k$ be the associated objective coefficients for

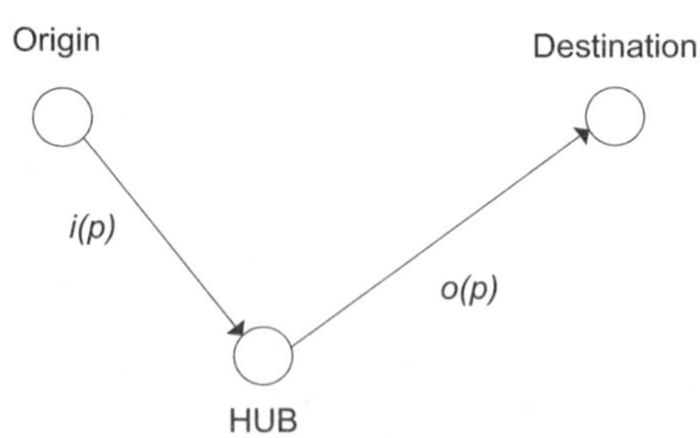

Fig. 2 A transferring path

$y_p^{k,e}$ and $y_{i(p)}^{k,e}$, respectively. Note that $r_{i(p)}^k$ denotes the unit revenue generated by the whole path p, and not by its subpath $i(p)$.

The integrated formulation is:

$$\max \sum_{k\in K}\left(\sum_{p\in P^{\mathrm{D}}(k)}\sum_{e\in E} r_p^k y_p^{k,e} + \sum_{\substack{i(p)\in PI^{\mathrm{T}}(k),\\ p\in P^{\mathrm{T}}(k)}}\sum_{e\in E} r_{i(p)}^k y_{i(p)}^{k,e}\right) - \sum_{i\in L(pax)}\sum_{e\in E(pax)} c_i^e x_i^e \tag{8}$$

subject to:

$$g_{e,s,(t_{j-1},t_j)} + \sum_{i\in I(n)} x_i^e = g_{e,s,(t_j,t_{j+1})} + \sum_{i\in O(n)} x_i^e, \forall n = (e,s,t_j) \in N(pax), e \in E(pax) \tag{9}$$

$$\sum_{s\in S} g_{e,s,(t_m,t_1)} + \sum_{i\in O(ct)} x_i^e \le Num^e, \forall e \in E(pax) \tag{10}$$

$$\sum_{e\in E(pax)} x_i^e = 1, \forall i \in L(pax) \tag{11}$$

$$x_i^e = \begin{matrix} 0, i \in L(pax), e \in E(frt) \\ 0, i \in L(frt), e \in E(pax) \\ 1, i \in L(frt), e \in E(frt) \end{matrix} \tag{12}$$

$$\sum_{k\in K}\sum_{p\in P^{\mathrm{D}}(k)} y_p^{k,e}\delta_i^p + \sum_{k\in K,}\sum_{\substack{i(p)\in PI^{\mathrm{T}}(k)\\ p\in P^{\mathrm{T}}(k)}} y_{i(p)}^{k,e}\delta_i^{i(p)} + \sum_{k\in K,}\sum_{\substack{o(p)\in PO^{\mathrm{T}}(k)\\ p\in P^{\mathrm{T}}(k)}} y_{o(p)}^{k,e}\delta_i^{o(p)} \le x_i^e d^e, \forall i \in L, e \in E \tag{13}$$

$$\sum_{p\in P^D(k)}\sum_{e\in E} y_p^{k,e} + \sum_{i(p)\in PI^T(k) p\in P^T(k)}\sum_{e\in E} y_{i(p)}^{k,e} \le B^k, \forall k \in K \tag{14}$$

$$\sum_{e \in E} y_{i(p)}^{k,e} = \sum_{e \in E} y_{o(p)}^{k,e}, \forall p \in P^T(k), k \in K \tag{15}$$

$$x_i^e \in \{0, 1\}, \forall i \in L, e \in E; g_{e,s,\left(t_j,t_{j+1}\right)} \geq 0, \forall e \in E(pax), s \in S, t_j, t_{j+1} \in T;$$
$$y_p^{k,e} \geq 0, \forall p \in P^{\mathrm{D}}(k), k \in K, e \in E;$$
$$y_{i(p)}^{k,e}, y_{o(p)}^{k,e} \geq 0, \forall i(p) \in PI^{\mathrm{T}}(k), o(p) \in PO^{\mathrm{T}}(k), p \in P^{\mathrm{T}}(k), k \in K, e \in E;$$

The objective (Eq. 8) maximizes the total cargo and passenger profit. The first part is cargo revenue contributed by the passenger network and the freighter network. The second part is the passenger profit. Because the freighter legs' assignment costs are constant (remember, there is only one freighter type), they do not appear in the objective function. We include only variables for subpath *i*(*p*), but not for *o*(*p*), in the objective function to avoid the double-counting of revenue. The first three sets of constraints (Eqs. 9, 10, 11) are the constraints of the passenger FAM. By Eq. (12), the condition that a freighter cannot be assigned to a passenger leg is enforced, and vice versa. Also, all the variables x_i^e , $i \in L(frt)$, and $e \in E(frt)$ are assigned to 1, as only one freighter type is available for the carrier in question. The rest of the constraints (Eqs. 13, 14, 15) are for the CRM. The set of inequations (13) are capacity constraints, and the set of inequations (Eq. 14) are the demand constraints. By Eq. (15), the flow consistency along a transferring path is ensured.

3 Solution methodology

One would expect that the real-life applications of problems (Eqs. 8, 9, 10, 11, 12, 13, 14, 15) are too large to be solved by general mixed-integer programming codes. The integrated model, however, naturally decomposes into two subproblems that are less difficult to solve. For any feasible solution to constraints (Eqs. 9, 10, 11, 12), problems (Eqs. 8, 9, 10, 11, 12, 13, 14, 15) reduce to a cargo routing problem. This observation motivates the development of the solution approach based on Benders decomposition (Benders 1962). Benders decomposition is a resource–directive decomposition method. It can be applied to linear programming, mixed-integer programming, and nonlinear programming problems. In a general sense, Benders decomposition can be viewed as a dual form of column generation (Bertsimas and Tsitsiklis 1997). It reformulates a problem by replacing part of the variables with constraints. As normally the number of constraints to replace the variables is enormous, the reformulated problem is solved in an iterative way, where active constraints are generated and added through the iteration. The iteration stops when optimality is reached. We will use this framework to reformulate our integrated problem in the next subsection, in which cargo flow variables are replaced by constraints involving only fleet assignment variables.

3.1 Benders reformulation

For the given value $\bar{x}$, $\bar{g}$ satisfying fleet assignment constraints (Eqs. 9, 10, 11, 12), the integrated model reduces to the following Benders *primal subproblem* that involves only cargo flow variables:

$$\max \sum_{k \in K} \left(\sum_{p \in P^D(k)} \sum_{e \in E} r_p^k y_p^{k,e} + \sum_{i(p) \in PI^T(k), p \in P^T(k)} \sum_{e \in E} r_{i(p)}^k y_{i(p)}^{k,e} \right) \tag{16}$$

subject to:

$$\sum_{k \in K} \sum_{p \in P^{\mathrm{D}}(k)} y_p^{k,e} \delta_i^p + \sum_{k \in K,} \sum_{\substack{i(p) \in PI^{\mathrm{T}}(k) \\ p \in P^{\mathrm{T}}(k)}} y_{i(p)}^{k,e} \delta_i^{i(p)} + \sum_{k \in K,} \sum_{\substack{o(p) \in PO^{\mathrm{T}}(k) \\ p \in P^{\mathrm{T}}(k)}} y_{o(p)}^{k,e} \delta_i^{o(p)} \le \bar{x}_i^e d^e, \forall i \in L, e \in E \tag{17}$$

$$\sum_{p \in P^D(k)} \sum_{e \in E} y_p^{k,e} + \sum_{i(p) \in PI^T(k) p \in P^T(k)} \sum_{e \in E} y_{i(p)}^{k,e} \le B^k, \forall k \in K \tag{18}$$

$$\sum_{e \in E} y_{i(p)}^{k,e} = \sum_{e \in E} y_{o(p)}^{k,e}, \forall p \in P^T(k), k \in K \tag{19}$$

$$y_p^{k,e} \ge 0, \forall p \in P^{\mathrm{D}}(k), k \in K, e \in E;$$
$$y_{i(p)}^{k,e}, y_{o(p)}^{k,e} \ge 0, \forall i(p) \in PI^{\mathrm{T}}(k), o(p) \in PO^{\mathrm{T}}(k), p \in P^{\mathrm{T}}(k), k \in K, e \in E;$$

After the passenger fleet assignment variables are fixed, the cargo capacity of every passenger leg is fixed accordingly. With the knowledge of the fleet assignment, the infeasible paths are excluded from the model. Because only one fleet type is assigned to a leg, at most, one of the flow variables $\{y_p^{k,e} : e \in E\}$ defined for a feasible path p is nonzero. For ease of presentation, we can combine all of these variables together and replace them with a single flow variable y_p^k, which is fleet independent. Similarly, because all but one of the capacity constraints defined for a leg has a zero right-hand side, we aggregate the constraints together to form a single capacity constraint. Given the fleet assignment, it is known to which fleet the cargo is transferred to at the hub. We can now combine the two subpaths into an origin–destination path, and therefore, the constraints in Eq. (19) are redundant. After these simplifications, the primal subproblem becomes equivalent to the simple CRM of the previous section.

$$\max \sum_{k \in K} \sum_{p \in P_{\mathrm{f}}(k)} r_p^k y_p^k \tag{20}$$

subject to:

$$\sum_{k\in K}\sum_{p\in P_{\mathrm{f}}(k)} y_p^k \delta_i^p \le \sum_{e\in E} \overline{x}_i^e d^e, \forall i \in L \tag{21}$$

$$\begin{aligned}\sum_{p\in P_{\mathrm{f}}(k)} y_p^k \le B^k, \forall k \in K \\ y_p^k \ge 0, \forall p \in P_{\mathrm{f}}(k), k \in K\end{aligned} \tag{22}$$

Let π_i and $i\in L$, and σ^k and $k\in K$, be the dual variables associated with constraints (Eqs. 21 and 22), respectively. The *dual subproblem* is written as:

$$\min \sum_{i\in L}\sum_{e\in E} (\overline{x}_i^e d^e)\pi_i + \sum_{k\in K} B^k \sigma^k \tag{23}$$

subject to:

$$\begin{aligned}\sum_{i\in L} \delta_i^p \pi_i + \sigma^k \ge r_p^k, \forall p \in P_{\mathrm{f}}(k), k \in K \\ \pi_i, i \in L, \sigma^k \ge 0, k \in K\end{aligned} \tag{24}$$

As the zero vector serves as a feasible solution to the primal subproblem, we note that it is always feasible and bounded by the objective value of its dual subproblem. The following constraint is called an optimality cut in Benders decomposition:

$$\eta \le \sum_{i\in L}\sum_{e\in E} (x_i^e d^e)\overline{\pi_i} + \sum_{k\in K} B^k \overline{\sigma}^k \tag{25}$$

Here, $(\overline{\pi},\overline{\sigma})$ is an extreme point of the dual polyhedron Q, defined by:

$$\begin{aligned}\sum_{i\in L} \delta_i^p \pi_i + \sigma^k \ge r_p^k, \forall p \in P_{\mathrm{f}}(k), k \in K \\ \pi_i, i \in L, \sigma^k \ge 0, k \in K\end{aligned}$$

Let $\mathbf{P_Q}$ be the set of extreme points of Q.

The integrated mixed-integer model (Eqs. 8, 9, 10, 11, 12, 13, 14, 15) can be reformulated as:

$$\max \eta - \sum_{i\in L(pax)}\sum_{e\in E(pax)} c_i^e x_i^e \tag{26}$$

subject to:

$$\eta \le \sum_{i\in L}\sum_{e\in E} (x_i^e d^e)\overline{\pi_i} + \sum_{k\in K} B^k \overline{\sigma}^k, \ \left((\overline{\pi},\overline{\sigma}) \in \mathrm{P_Q}\right) \tag{27}$$

$$g_{e,s,\left(t_{j-1},t_j\right)} + \sum_{i \in I(n)} x_i^e = g_{e,s,\left(t_j,t_{j+1}\right)} + \sum_{i \in O(n)} x_i^e, \forall n = \left(e, s, t_j\right) \in N(pax), e \in E(pax) \tag{28}$$

$$\sum_{s \in S} g_{e,s,(t_m,t_1)} + \sum_{i \in O(e)} x_i^e \leq Num^e, \forall e \in E(pax) \tag{29}$$

$$\sum_{e \in E(pax)} x_i^e = 1, \forall i \in L(pax) \tag{30}$$

$$x_i^e = \begin{cases} 0, i \in L(pax), e \in E(frt) \\ 0, i \in L(frt), e \in E(pax) \\ 1, i \in L(frt), e \in E(frt) \end{cases} \tag{31}$$

$$\begin{aligned} x_i^e &\in \{0, 1\}, \forall i \in L, e \in E \\ g_{e,s,\left(t_j,t_{j+1}\right)} &\geq 0, \forall e \in E(pax), s \in S, t_j, t_{j+1} \in T \end{aligned}$$

This model (Eqs. 26, 27, 28, 29, 30, 31) is called the Benders *master problem*. Compared with the original model, the Benders reformulation has fewer variables but contains more constraints, although most of these constraints are inactive in the optimal solution. Instead of enumerating them explicitly, we start with a relaxation of the master problem without Benders cuts and generate them on the fly by solving the subproblem. A *Benders relaxed master problem* is defined by Eqs. (26 and 28, 29, 30, 31), and a subset of Benders cuts by Eq. (27).

3.2 Solution algorithm

This algorithm is developed based on Benders decomposition. Each iteration solves a relaxed master problem to integral optimality, and the objective value becomes the updated upper bound, as not all constraints are included in the solving. A subproblem is set up and solved using the master problem solution. This constitutes a feasible solution to the integrated model, and its value can serve as a lower bound. Hence, the lower bound is updated from the best subproblem objective value obtained so far. If the relative difference between the upper and lower bounds is within a designated tolerance ζ, the algorithm stops and an optimal solution of the original problem is found. Otherwise, a Benders optimality cut is constructed from the dual subproblem solution and is added to the relaxed master problem. The same process is repeated until optimality is reached.

4 Computational experiments

To study the algorithm's performance, computational experiments were performed on a set of test scenarios based on available data about the airline. The weekly flight schedule of the airline contains 1,404 passenger legs and 201 freighter legs. Six passenger fleets and one freighter fleet are used to cover all flights. Besides this entire schedule, we generated four of its subschedules. The commodities and their demands are generated from the historical data. All cost coefficients are derived from the airline's annual report. The potential feasible paths for all commodities over the network (both passenger and freighter subnetworks) are generated for every test instance, as described in the previous section. Table 1 describes our five test data instances.

4.1 Computation results

The algorithm was implemented in C++. CPLEX8.1 was used to solve the master problems and the subproblems. The master problem was solved by the mixed-integer optimizer that uses the branch-and-cut method. The dual simplex algorithm with the steepest edge pricing was employed to solve the LP relaxation at each node in the branch-and-bound tree. The subproblem was solved by primal simplex optimizer. All experiments were performed on a computer with an 866-MHz CPU and 256 MB of RAM. The relative convergence tolerance ζ is set to 0.1%. Table 2 reports the computational results for all test instances

It is shown in Table 2 that the number of iterations is quite small and the convergence is very fast for every test instance. D2 spent only 1.3 s and three iterations to obtain the optimal solution. Even for the full instance, D5, the optimality was reached within 116.7 s by four iterations. This demonstrates the efficiency of the algorithm for our problem.

Also, we applied a sequential approach to plan the fleet assignment and the cargo routing. Compared to results of the integrated approach, the annual cargo traffic decreased 12.3% and the annual cargo and passenger profit decreased

Table 1 The characteristics of data instances

Instance	Total number of passenger legs $\|L(pax)\|$	Total number of freighter legs $\|L(frt)\|$	Total number of passenger fleets $\|E(pax)\|$	Total number of freighter types	Total number of commodities $\|K\|$	Total number of potential feasible paths $\sum_k \|p(k)\|$
D1	62	0	6	0	63,798	573
D2	102	0	6	0	63,798	2,051
D3	520	201	6	1	63,798	28,221
D4	884	151	6	1	63,798	73,880
D5(full)	1,404	201	6	1	63,798	173,285

Table 2 Computational results

Approach	Instance	Convergence CPU time (s)	Number of Benders iterations	Times per iteration (s)	Benders final relative gap (%)
Basic algorithm	D1	3.1	11	0.3	0.02
	D2	1.3	3	0.4	0.09
	D3	14.7	4	3.7	0.08
	D4	103.5	13	8.0	0.08
	D5	116.7	4	30.0	0.09
Pareto-optimal cut generation approach	D1	4.2	11	0.4	0.02
	D2	1.9	3	0.6	0.09
	D3	36.8	4	9.2	0.09
	D4	870.9	11	79.2	0.06
	D5	2,254.2	5	450.8	0.07
ε-Optimal approach	D1	5.4	18	0.3	N/A
	D2	1.4	3	0.5	N/A
	D3	14.8	5	3.0	N/A
	D4	Does not converge in 24 h			
	D5	565.0	14	40.4	N/A

12.7%. Although an ideal measure of the integrated approach should be the comparison with the actual fleet assignment planning and cargo routing in practice, these figures provide some indications on how important it is to take account of cargo routing when determining fleet assignment.

4.2 Comparison with two other Benders decomposition variants

This algorithm spent only a few minutes to obtain the best solution of our problem. As this is a planning problem, time is not normally a tight constraint and this is well within the bounds of acceptability. The application of the basic algorithm was thus successful for our problem.

In an attempt to further accelerate the convergence of the algorithm, we implemented two variants of Benders decomposition: the Pareto-optimal cut generation (Magnanti and Wong 1981) and the ε-optimal approach (Geoffrion and Graves 1974). Pareto-optimal cut generation exploits the fact that, in the standard Benders decomposition approach, there may be primal degeneracy in the dual subproblem, and so, multiple Benders cuts are possible: the Magnanti and Wong procedure provides a way to find strong cuts. The ε-optimal approach attempts to save computation time by solving the relaxed master problem to find an ε-improved solution rather than full optimality. This approach terminates and an optimal solution is obtained when it is infeasible to find an ε-improved solution for the relaxed master problem. Numerical results are reported in Table 2. It is shown that there were no improvements obtained by these efforts. The Pareto-optimal cut generation approach must solve one more linear program at each iteration, and the solution time increases very quickly with the problem size. For the ε-optimal approach, it took an extremely long time to prove the infeasibility of the last relaxed master problem.

5 Conclusion

To enhance the fleet assignment model, we proposed an integrated approach that incorporates the cargo routing into the fleet assignment. In contrast to passenger traffic, cargo has no strong preference on the specific itinerary, as long as its commitment is satisfied. There is also no available industry data to calculate the spill cost and the recapture rate for the cargo flow. Moreover, cargo is allowed to transfer between different aircraft only at the hub, while this requirement is not applicable to the passenger. The cargo flow is thus modeled in a way different from the passenger mix model. Computational results show that our Benders decomposition-based approach can generate the optimal solution within several minutes (very fast for planning problems) with an improved estimate of total profit, compared with that generated by the isolated fleet assignment.

It is worthwhile to restate that the passenger revenue is only linearly estimated in our integrated model. For combination carriers, passenger delivery is still their main source of profit; hence, the passenger flow problem should be properly modeled. Furthermore, our integrated model is an approximation of the three-stage problem that includes the aircraft routing. We believe that the integration of fleet assignment, cargo routing, and passenger mix, and the integration of fleet assignment, aircraft routing, cargo routing, and/or passenger mix are interesting problems to explore in the future. Moreover, the forecasted cargo demand is deterministic in our problem. This simplification may cause the solution to be less convincing because the demand is unknown at the time of planning. Therefore, another interesting research direction would be to explore ways to capture the demand uncertainty.

References

Abara J (1989) Applying integer linear programming to the fleet assignment problem. Interfaces 19(4):20–28

Ahuja RK, Magnanti TL, Orlin JB (1993) Network flows: theory, algorithms and applications. Prentice-Hall, Englewood Cliffs, pp 649–694

Barnhart C, Shenoi RG (1998) An approximate model and solution approach for the long-haul crew pairing problem. Transp Sci 32:221–231

Barnhart C, Boland NL, Clarke LW, Johnson EL, Nemhauser GL, Shenoi RG (1998a) Flight string models for aircraft fleeting and routing. Transp Sci 32:208–220

Barnhart C, Lu F, Shenoi R (1998b) Integrated airline schedule planning. In: Yu G (ed) Operations research in the airline industry. Kluwer, Boston

Barnhart C, Kniker TS, Lohatepanont M (2002) Itinerary-based airline fleet assignment. Transp Sci 36:199–217

Benders JF (1962) Partitioning procedures for solving mixed-variables programming problem. Numer Math 4:238–252

Bertsimas D, Tsitsiklis JN (1997) Introduction to linear optimization. Athena Scientific, Belmont

Boeing (2004) 2004–2005 World air cargo forecast. http://www.boeing.com/commercial/cargo/WACF_2004–2005.pdf

Cohn AM, Barnhart C (2003) Improving crew scheduling by incorporating key maintenance decisions. Oper Res 51:387–396
Cordeau JF, Stojkovic G, Soumis F, Desrosiers J (2001) Benders decomposition for simultaneous aircraft routing and crew scheduling. Transp Sci 35:375–388
Desaulniers G, Desrosiers J, Dumas Y, Solomon MM, Soumis F (1997) Daily aircraft routing and scheduling. Manage Sci 43:841–855
Geoffrion AM, Graves GW (1974) Multicommodity distribution system design by Benders decomposition. Manage Sci 20:822–844
Hane CA, Barnhart C, Johnson EL, Marsten RE, Nemhauser GL, Sigismondi G (1995) The fleet assignment problem—solving a large-scale integer-program. Math Program 70:211–232
Jones KL, Lustig IJ, Farvolden JM, Powell WB (1993) Multi-commodity network flows: the impact of formulation on de-composition. Math Program 62(1):95–117
Klabjan D, Johnson EL, Nemhauser GL, Gelman E, Ramaswamy S (2002) Airline crew scheduling with time windows and plane-count constraints. Transp Sci 36:337–348
Magnanti TL, Wong RT (1981) Accelerating Benders decomposition: algorithmic enhancement and model selection criteria. Oper Res 29:464–484
Mercier A, Cordeau J, Soumis F (2003) A computational study of Benders decomposition for the integrated aircraft routing and crew scheduling problem. GERARD technical report G-2003-48, 24 pp
Rexing B, Barhart C, Kniker T, Jarrah A, Krishnamurthy N (2000) Airline fleet assignment with time windows. Transp Sci 34(1):1–20
Rushmeier RA, Kontogiorgis SA (1997) Advances in the optimization of airline fleet assignment. Transp Sci 31(2):159–169
Subramanian R, Scheff RP, Quillinan JD, Wiper DS, Marsten RE (1994) Coldstart—fleet assignment at delta-airlines. Interfaces 24(1):104–120

P. Bartodziej · U. Derigs · M. Zils

O&D revenue management in cargo airlines—a mathematical programming approach

Abstract In this paper we present a mathematical programming based approach for revenue management in cargo airlines. The approach is based on a modified version of a multicommodity network flow model which has been developed in a decision support approach for schedule planning in cargo airlines. We think that using the same concept for planning and revenue management is essential for consistency of planning and operation. To meet the real-time requirements of revenue management special computational strategies for solving the large models are necessary.

Keywords Revenue management · Mathematical programming · Multi-commodity flow · Column generation · Simulation study

1 Introduction

Revenue management deals with the problem of effectively using perishable resources or products in industries or markets with high fixed cost and low margins, which are price-segmentable (see Cross 1997; McGill and van Ryzin 1999; Talluri and van Ryzin 2004). Revenue management is a complex problem which has to be supported by sophisticated forecasting methods/systems and optimization methods/systems.

Revenue management has its origin and has found broad application in the airline business, and here especially in passenger flight revenue management. Today almost all airline passenger revenues of over US $310 billion p.a. are actively managed through revenue management, and the industry claims to have generated as much as US $500 million in additional profits from the early 1980s onward (see Pompeo and Sapountzis 2002). Only recently, the concepts which have been developed for the passenger sector have been adequately modified and transferred

P. Bartodziej · U. Derigs (✉) · M. Zils
WINFORS, University of Cologne, Pohligstr. 1, 50969 Cologne, Germany
E-mail: derigs@informatik.uni-koeln.de

to the cargo sector. As demonstrated later in this paper, an immediate transfer of the concepts developed for passenger airlines to the cargo airlines is not feasible since the two kinds of business, although at first sight offering a similar kind of transportation service, face differences in product type and in production which have consequences for scheduling and capacity planning as well as revenue management.

While the practice of revenue management is well-established in the passenger airline industry, and there is a vast amount of literature on concepts, methods, and implementations, the revenue management concept is underdeveloped in the cargo industry and there is only a small number of publications focussing on this application.

Kasilingam (1996) and Billings et al. (2003) analyze the major differences between passenger and cargo revenue management. While our approach is assuming a pure cargo airline with certain and fixed capacities, Kasilingam (1996) is concerned with the integrated management of the available belly space after accommodating passengers which in addition to forecasting cargo market demand requires to forecast capacity available for cargo sale. Slager and Kapteijns (2004) describe the implementation of cargo revenue management at KLM, which is based on calculating for every request a contribution margin, i.e., the revenue minus the variable costs related to handling, fuel etc., which is then checked against a minimum required margin, the so-called contract entry condition. These conditions are adjusted by route managers on a daily basis using information on historic and current booking profiles, current capacities and the market knowledge of the manager. This paradigm, which is called margin management principle, is very much in the flavor of the bid-pricing policy which is widely used in passenger revenue management.

Kleywegt and Papastavrou (1998, 2001) propose to model the cargo revenue management problem as dynamic stochastic knapsack problem. Yet, the approach allows only one capacity restriction and becomes computational intractable if extended to the two dimensional capacity situation in cargo revenue management. Pak and Dekker (2004) describe a bid-price acceptance policy for cargo revenue management which is based on (approximatively) solving a multidimensional online knapsack problem. This approach respects the multi-dimensionality of cargo capacity and demand, yet, it does assume that for each booking request the itinerary, i.e. the flights its uses is uniquely defined. With this assumption the flexibility (and also some of the complexity) of cargo routing versus passenger routing is neglected.

Our approach is different from all the above mentioned in so far, as it does not treat revenue management as an isolated problem, but respects that the decisions or alternatives at this operational level are constrained through structural decisions and capacity allocations made already on a higher and earlier management level: tactical planning, and, that the decisions made on these two levels should be driven through the same goal and coordinated.

In any airline, market-oriented planning on the tactival level requires the design of a profitable schedule and the allocation of resources like airplanes, staff, ground capacities etc. Here the objective is to maximize contribution to profit which is achieved by generating schedules and assigning airplanes allowing high load factors. When solving this complex design and decision problem highly sophisticated market-models are used to estimate market demand. The results of such forecasts are then used as data for schedule design and resource allocation.

Etschmeier and Matheisel (1985) describe the concept of an iterative schedule planning process based on two phases: schedule construction and schedule evaluation. In schedule construction, a central scheduling department develops a draft schedule, and in schedule evaluation, this schedule is evaluated by various operating departments in terms of feasibility, cost, and economic value. Based on these evaluations, the draft schedule is revised and a modified schedule is generated for a new evaluation.

While this generic process is similar for all types of airlines, the evaluation of schedules is quite different for cargo airlines and requires a special approach. In Antes et al. (1998) we have developed a mathematical model for evaluating cargo schedules. The focus of our model is the determination of the contribution to profit, that will accrue from a given schedule. For that purpose the model calculates the so-called *optimal freight flow* considering yield and operating cost. Here a freight flow is the assignment of demand to the capacities (flights) of a given schedule. Thus, the planning problem is to "determine the freight flow with maximum contribution to profit with respect to fixed market demand".

For passenger airlines, Berge and Hopperstad (1993) have proposed the "Demand Driven Dispatch"-concept. Here airplanes with different capacities but similar crew-requirements are assigned to flights, depending on the actual demand/booking situation to increase the load factor and contribution to profit. For this purpose so-called swap-potentials for resources have to be evaluated in a pro-active manner by (re-)solving fleet assignment problems and aircraft routing problems. This concept has been generalized by Gershkoff (1998). Yet, the problem with these just-in-time scheduling concepts is the complexity of identifying feasible changes/feasible schedules, for instance with respect to maintenance constraints etc., in nearly "real time" and the actual lead times to implement proposed changes in day-to-day operations.

Hence in operational control the schedule and the capacity allocation is assumed to be fixed, and the problem is to adequately respond to bookings, e.g. the actual market demand such that the yield obtained from the schedule is maximized. At this point changes in the schedule and/or capacities are possible only at a marginal level, and the dominant instruments for controlling efficiency are prizing (and over-) booking strategies. Thus, conceptually thinking, the operational problem in cargo revenue management is to "re-optimize freight-flows due to bookings and price-changes". And as we will show, this problem can be tackled by the same model and method which has been developed for schedule evaluation on the tactical planning level.

Thus, in this paper we propose the application of the model and methods which we have developed for medium term tactical planning to the short term operational problem of revenue management. The motivation of this proposal and study is twofold: First, we think that operational control and tactical planning should be governed and guided by the same concepts and factors: objectives, constraints, and data to minimize loss of profit potential during implementation and increase the efficiency of the planning process. Secondly, as will be demonstrated in this paper, the planning methods are functional on the operational level, too.

2 A conceptual model of air cargo transportation

A cargo airline offers the conceptually simple service to transport a certain amount of goods from an origin to a destination within a certain time interval at a given price. For this purpose the airline keeps and/or leases capacities for transporting goods to, from, and between airports. The tactical planning problem of such airlines is to design a (weekly) flight schedule of so-called legs which allows the most profitable service of the estimated market demand for the next period (half a year, say). Then, on the operational level booking requests have to be assigned to specific flights of the fixed flight schedule.

There are several differences between the situation in passenger transportation and cargo transportation which make the development of concepts, models and systems for planning as well as revenue management more complex in the cargo environment:

a) While the demand in the passenger sector is one dimensional (seat) and kind of smooth, the demand in the air cargo sector is heterogenous and lumpy with completely different capacity requirements for single booking requests (compare for instance volume requirements for documents versus industrial aggregates). This leads to a multi-dimensional characterization (weight, volume, classification, handling needs etc.) of cargo units.
b) While passengers usually book return-flights, cargo demand is one-way in general which leads to geographically unbalanced transportation patterns with respect to structure and volume of cargo.
c) Yet, the most subtle difference is that passengers are booking so-called *itineraries*, e.g. specific sequences of flight connections, so-called legs leading the passenger from the origin to the destination of their journey. Thus, a passenger booking a flight from Cologne (CGN) to Chicago (ORD) via Frankfurt (FRA) and New York (JFK) on a given date, has to be given a seat reservation on three specific flights (CGN, FRA), (FRA, JFK) and (JFK, ORD). In the cargo business customers book a transportation capacity from an origin to a destination, with a certain service level, e.g. within a certain time-window. In general, the cargo customer is not interested in the specific routing of his package. Hence the airline has some degree of freedom how and when to assign requests to specific flights and this difference leads to different planning models.

For schedule evaluation as well as revenue management the key problem is to "assign market demand to the schedule". Market demand can be described as a set of estimated and/or booked service requests. And such a request has several attributes with the origin and destination airport being the dominant characteristic. Therefore, such transportation requests are commonly referred to as *O&D's or O&D pairs* in the cargo airlines terminology. Conceptually, the object class OD of *O&D's* can be modelled as a complex 5-ary recursive relationship type between the entity type AIRPORT, playing the role of an origin and destination respectively, TIME, playing the role of availability and due time respectively, and PRODUCT with attributes specifying the quantity (kg), the volume (m^3) and the yield/freight rate ($/kg) etc. The schedule on the other hand can be described as a set of so-called *legs*. The concept of a leg stems again from the airlines terminology, meaning a direct flight between two airports offered at the same time every week. Conceptually, we model the object class LEG as a 4-ary recursive relationship type between the entity type AIRPORT playing the role of an origin (from) and a destination (to) and TIME, playing the role of a departure and an arrival time, respectively, with attributes specifying the capacity of the leg (kg), the operating cost ($/kg) etc. The relevant information reflecting ground handling, that is the time and cost for import, export, and transfer of goods at the airports is modelled via a relationship type HANDLING

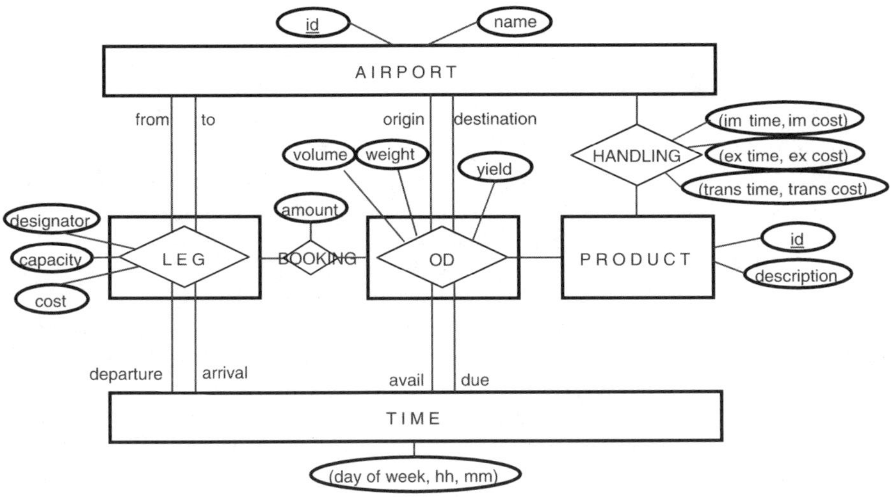

Fig. 1 Conceptual schema for air cargo transportation

associating the entity types AIRPORT and PRODUCT. Figure 1 gives the graphical representation of this conceptual schema, using the common symbols of the entity relationship model (cf. Elmasri and Navathe (2004)).

Note that all attributes mentioned above are data for the decision model and have to be instantiated. Now, the decision problem in planning as well as revenue management, is captured in the relationship type/the association BOOKING between the entity types OD and LEG. More precisely the problem is to determine the values of its attribute "amount" for every (OD, LEG)-pair. Or, in the terminology of databases, the amount-attribute is a so-called "derived attribute" with the integrity constraints defining feasible values as well as the evaluation of different associations captured in a mathematical optimization model. In tactical planning one can think of initializing this model with no associations at all, i.e. the amount-values are zero and, given the data for the entire (expected) demand, we have to simultaneously "derive" optimal amount-values. When using this conceptual model in revenue management, one can think of initializing the model with an eventually large number of fixed associations, which represent bookings or which are derived from additional (expected) demand, and there is just one booking request which has to be evaluated and associated on top.

In the following two sections we will introduce the mathematical optimization models which represent and formalize this view of the decision problems associated with the two scenarios.

3 The model for O&D schedule planning

Every flight schedule defines a so-called *time-space network* $G=(V, E)$ with V being the set of nodes representing the airports at a certain point of time. The arc set E is composed from two subsets, a set of *flight arcs* which represent the legs and connect the associated airports at departure and arrival time, respectively, and a set of *ground arcs* which connect two nodes representing the same airport at two consecutive points in time. Associated with every leg/arc $e \in E$ is the *weight capacity* u_e measured in *kg*, the *volume capacity* v_e measured in m^3 and the operating cost c_e measured in \$ per *kg*.

Now the different O&D's define different commodities which have to be routed through this network, and the so-called multi-commodity flow problem (see Ahuja et. al. (1993)) is the problem to determine optimal assignments of O&D's/commodities to legs/arcs such that (part of) the demand is routed from origins to destinations. Figure 2 shows the representation of two specific itineraries connecting Cologne (CGN) with Chicago (ORD) via Frankfurt (FRA) and New York (JFK), and via Munich (MUC), respectively, in a time space network.

In our models we use the symbol *od* for representing an O&D-commodity and *OD* as the set of all commodities/O&D's. Associated with a commodity/O&D $od \in OD$ is a specific demand b^{od} measured in kg, a value d^{od} giving the volume per kg and a freight rate or yield value y^{od} measured in \$ per unit of commodity *od*. For every leg $e \in E$ let $o(e)$ be the origin airport and $d(e)$ be the destination airport. Also, let *origin* (*od*) be the origin and *destination* (*od*) be the destination of $od \in OD$, respectively.

The first formulation/mathematical model is the so-called *arc-flow-model*, a linear program which is the immediate implementation of the conceptual model presented in Fig. 1. Here we introduce decision variables $x_{e,\,od}$ giving the amount of commodity $od \in OD$ which is transported over the arc $e \in E$. This leads to the mathematical formulation of the planning problem as an arc-flow multi-commodity flow model.

3.1 Arc-flow-model

$$\max \sum_{od \in OD} \left(\sum_{\substack{e \in E \\ o(e)=o(od)}} y^{od} \cdot x_{e,od} - \sum_{e \in E} c_e \cdot x_{e,od} \right)$$

$$\sum_{\substack{e \in E \\ 0(e)=i}} x_{e,od} - \sum_{\substack{e \in E \\ d(e)=i}} x_{e,od} \begin{cases} \leq b^{od} & \text{if } i = o(od) \\ \geq -b^{od} & \text{if } i = d(od) \\ = 0 & \text{else} \end{cases}$$

$$\text{for } i \in V,\, od \in OD$$

$$\sum_{od \in OD} x_{e,od} \leq u^e \quad \text{for } e \in E$$

$$\sum_{od \in OD} d^{od} x_{e,od} \leq v^e \quad \text{for } e \in E$$

$$x_{e,od} \geq 0$$

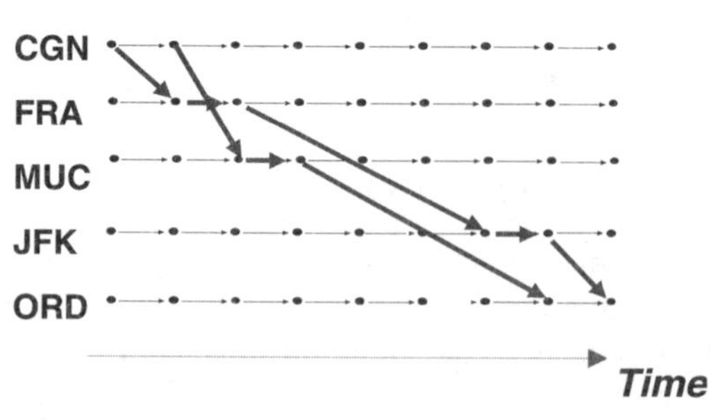

Fig. 2 Itineraries in a time-space network

This model is not very well suited, neither for the planning situation nor for an adaption to revenue management. It does not capture the complex time-window constraints associated with the demand, it does not contain any constraints from ground handling etc. To be able to incorporate these and other constraints into the decision, the so-called path flow formulation should be used which can also be adapted rather nicely for use in revenue management.

The so-called *path-flow model* is based on the obvious fact that any unit transported from an origin to a destination has to be routed over a sequence of legs (possibly only one leg), a so-called path or *itinerary* connecting the origin node with the destination node in the network. A path or itinerary for an O&D/commodity od is a sequence $p=(l_1,\ldots,l_{r(p)})$ of legs $l_i \in E$, $r(p) \geq 1$ with the following properties

$$\begin{aligned} o(l_1) &= origin(od) \\ d(l_i) &= o(l_{i+1}) \quad i = 1, \ldots, r(p) - 1\,. \\ d(r(p)) &= destination(od) \end{aligned}$$

Such a path p is called *od-feasible* if additional requirements are fulfilled which vary with the problem definition. Here we consider several types of constraints which concern due dates, transfer times and product compatibility. For every O&D/commodity od we denote by P^{od} the set of od-feasible itineraries and by S we denote the union of all P^{od} sets. A path may be feasible for many different O&D's. In our model we have to distinguish these roles and consider multiple copies of the same path/the same legs assigned to different commodities/O&D's. Then the relation between arcs/legs and itineraries is represented in a binary indicator δ: $E \times S \rightarrow \{0,1\}$

$$\delta_e(p) := \begin{cases} 1 & \text{if leg } e \in E \text{ is contained in path } p \\ 0 & \text{else} \end{cases}.$$

Note that od-feasibility of paths can be checked algorithmically and, given an itinerary $p \in P^{\mathrm{od}}$, we can calculate $c(p) := \sum_{e \in E} c_e \delta_e(p) = \sum_{e \in p} c_e$ the operating cost as well as $y^{od}(p) := y^{od} - \sum_{e \in p} c_e$ the contribution to profit per kg of commodity od which is transported over p. This calculation as well as the construction of the set P^{od} is done outside our model using a module called "connection builder" and is then given to the model as input data. In the model we introduce a decision-variable $f(p)$ for every $p \in P^{\mathrm{od}}$ giving the amount (in kg) transported via p, and the problem is to select the optimal combination of paths giving maximal contribution to profit which again leads to a linear program.

3.2 Path-flow-model

$$\max \sum_{od \in OD} \sum_{p \in P^{od}} y^{od}(p) f(p) \tag{1}$$

$$\text{s.t.} \sum_{od \in OD} \sum_{p \in P^{od}} \delta_e(p) f(p) \leq u_e \quad \forall e \in E \tag{2}$$

$$\sum_{od \in OD} \sum_{p \in P^{od}} \delta_e(p) d^{od} f(p) \leq v_e \quad \forall e \in E \tag{3}$$

$$\sum_{p \in P^{od}} f(p) \leq b^{od} \quad \forall od \in OD \tag{4}$$

$$f(p) \geq 0 \quad \forall od \in OD \quad \forall p \in P^{od}. \tag{5}$$

The advantage of the itinerary-based model over the leg-based model is the possibility to consider rather general and complicated constraints for feasibility of transportation during the path-construction phase in the connection builder. Keeping this complexity outside the optimization allows to apply the same standard (LP-) solution procedure and standard LP-solver for a wide range of models representing different planning situations.

Moreover, the path-flow-model allows for scaling, i.e. it is not necessary to construct all feasible paths beforehand. Working with a "promising subset" of paths only, reduces the size of the problem instance but may lead to solutions which although not optimal in general, are highly acceptable in quality. Finally, applying the *delayed column generation technique* allows to generate feasible paths on the run during optimization and thus keeps problem size manageable throughout the optimization process. This model and algorithmic approach is the basis for the system SYNOPSE ("System zur Netzoptimierung und Planung Strategischer Entscheidungen", which translates in English as "System for Network Optimization and Planning of Strategic Decisions") (see Antes et al. 1998), an interactive decision support system for analyzing schedules. In Derigs and Zils (2001) we have applied the schedule planning model to the strategic problem of analyzing alliances of cargo airlines. In the following we will demonstrate how this concept and model can be transferred to the operational level, i.e. to the problem of revenue management.

4 The model for O&D revenue management

The model in Section 3 has been developed for the problem to optimally assign an expected demand to a fixed schedule. In revenue management, demand occurs over a longer time period, a year say, and the limited resources have to be "booked" sequentially. When a booking or reservation is made, the airline does not need to assign the O&D to a specific itinerary immediately, it only has to ensure that at operation time capacities are available, i.e. it has to ensure that at any time there exists a feasible routing for all booked O&D's. This situation and requirement is captured in the following optimization model.

In this model we assume a set B of accepted bookings and we distinguish between

- β^k the demand which is already booked and
- b^k the additional forecasted future demand for $k \in OD$ and
- a single booking request r for a commodity $k(r) \in OD$ with demand $\tilde{\beta}^r$,

and we assume an (expected) yield y^k for every commodity $k \in$OD. The model can be used to evaluate the acceptability/profitability of a specific request r at a given price y^r and for "dynamic pricing", i.e. the determination of the minimum acceptable yield for request r.

4.1 Revenue-management-model *RM* (*B*, *r*)

$$\max \sum_{k \in OD} \sum_{p \in P^k} y^k(p) f(p) \tag{6}$$

$$\text{s.t.} \sum_{k \in OD} \sum_{p \in P^k} \delta_e(p) f(p) \leq u_e \quad \forall e \in E \tag{7}$$

$$\sum_{k \in OD} \sum_{p \in P^k} \delta_e(p) d^k f(p) \leq v_e \quad \forall e \in E \tag{8}$$

$$\sum_{p \in P^k} f(p) \leq b^k + \beta^k \quad \forall k \in OD \tag{9}$$

$$\sum_{p \in P^k} f(p) \geq \beta^k \quad \forall k \in OD \backslash \{k(r)\} \tag{10}$$

$$\sum_{p \in P^r} f(p) \geq \beta^r + \tilde{\beta}^r \tag{11}$$

$$f(p) \geq 0 \quad \forall p \in P^k, k \in OD. \tag{12}$$

Note that the application of *RM* (*B*, *r*) is started with no booking, i.e. $\beta^k=0$, $k \in OD$, and no request, which is equivalent to the planning model from Section 3. Then the set *B* is sequentially enlarged introducing accepted booking request. Thus at any time the booking model *RM*(*B*) which is obtained from *RM*(*B*, *r*) by omitting *r* and replacing inequality Eq. 11 by an additional inequality of type Eq. 10 for $k=r$ is feasible and we can start the solution of *RM*(*B*, *r*) from the optimal solution of *RM* (*B*). Now the acceptability/profitability (and the option price) for request *r* depends on the difference between the optimal values of *RM*(*B*, *r*) and *RM*(*B*).

Note that when solving *RM*(*B*, *r*) already booked requests may be rescheduled, i.e. during the entire process booked commodities may be associated to different (sets of) paths. Thus the model and solution process makes use of this flexibility.

We will only outline here the general approach for solving *RM*(*B*, *r*), which is based on applying the column generation technique to the multi-commodity flow problem. Solving *RM*(*B*) by any LP-technique we obtain an optimal dual solution with values

- w_e for every constraint of type Eq. 7,
- z_e for every constraint of type Eq. 8, and
- σ^k for every constraint of type Eqs. 9, 10 and 11.

Based on these values the reduced cost for an itinerary $p \in P^k$, $k \in \mathrm{OD}$, is defined as

$$\bar{c}(p) := y^k - \sum_{e \in p} \left(c_e + w_e + d^k z_e\right) - \sigma^k$$

Thus, if an itinerary $p \in P^k$ has positive reduced cost $= \bar{c}(p)$ then transporting (one unit of) commodity k over this itinerary increases the contribution to profit.

Such a path can be found by determining p^*, the shortest path from *origin*(k) to *destination*(k) with respect to the modified arc cost $c_e' := c_e + w_e + d^k z_e$. If the length of p^* exceeds $y^k - \sigma^k$ then no path of positive reduced cost exists. Otherwise, sending flow over p^* will increase the contribution to profit.

Yet, there is one problem remaining: The capacity δ of the shortest path p^* will in general not be equal to $\tilde{\beta}^r$, the required amount of the current request r. If $\delta \geq \tilde{\beta}^r$, we will increase $f(p^*)$ by $\tilde{\beta}^r$ (and eventually introduce the column associated with p^* into the constraint matrix) and we have solved *RM* (B, r). Otherwise, we can only tentatively increase $f(p^*)$ by δ, update $\tilde{\beta}^r := \tilde{\beta}^r - \delta$ and repeat the process, i.e. the search for another path with unused capacity. More information on the difficulties arising during the process and on several computationally effective strategies can be found in Bartodziej and Derigs (2004). In the following section we will describe the results from a simulation study which we have conducted for evaluating our approach.

5 A simulation study for evaluating the O&D revenue management approach

We have applied the model and method described in Section 4 to a set of benchmark instances. These instances were constructed from data obtained in a marketing study (cf. Zils 1998, 1999). For several airlines O&D-matrices were estimated using a simple gravity model. This study was conducted to support the development and evaluation of models and systems for planning, and its results have been adapted for generating the scenarios of our revenue management simulation. From the different benchmark problems given in Zils (1999), we have extracted three scenarios. The characterization of these three benchmark problem instances P10, P64, and P79 is illustrated in Figs. 3–5.

The data of an instance is maintained in three relations/tables. The schemata of these relations are given in Tables 1, 2, 3.

Note that in the leg table we store information on rotations (i.e. the sequences of legs which are served by the same airplane). This information is necessary to calculate the handling cost. We do not go into more detail on this aspect here.

From this data instances for the schedule evaluation problem were generated and solved to optimality to serve as the starting solution for the revenue management process. During the revenue management process booking requests are generated and evaluated solving problems of type *RM* (B, r), and if accepted they initiate a modification of the production plan and updates of the forecast. This relation between planning and revenue management is reflected in Fig. 6.

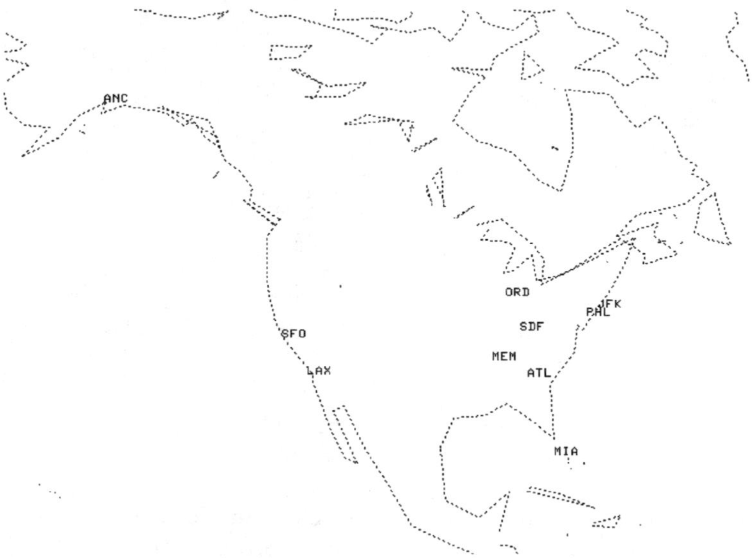

10 airports
624 legs
1338 O&D's

Fig. 3 Characteristics of instance P10

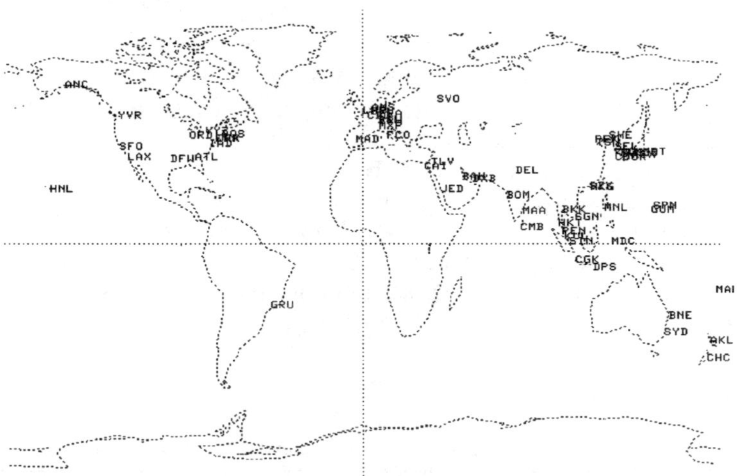

64 airports
1592 legs
3459 O&D's

Fig. 4 Characteristics of instance P64

In the revenue management simulation the (aggregate) O&D-quantities of the planning model have been broken down to the level of od-booking requests in the following way:

Let a^{od} be an estimated total demand for a specific od$\in$OD then we generated a sequence $\alpha_1{}^{od},\ldots,\alpha_{n\,(od)}{}^{od}$ with the following property:

$$\sum_{i=1}^{n(od)} \alpha_i^{od} = a^{od} \text{ and } \frac{a^{od}}{\gamma_1} \leq \alpha_i^{od} \leq \frac{a^{od}}{\gamma_2} \text{ for } i = 1, \ldots, n(od) \text{ with } \gamma_1 > \gamma_2 \geq 1.$$

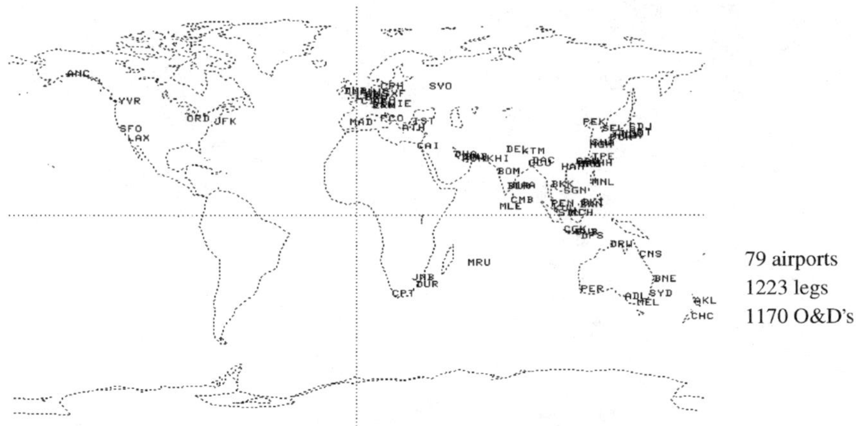

Fig. 5 Characteristics of instance P79

By this construction we ensure that the α_i^{od}'s are within a certain percentage interval with respect to a^{od}. Then we applied a random perturbation to obtain the lumpy booking requests

$$\tilde{\beta}_i^{od} := (1 + \delta) \cdot \alpha_i^{od} \text{ with } \delta \in [-\Delta,\ \Delta]$$

and a random sequence of all generated bookings was generated. Then, this booking pattern of lumpy requests was used to define the $\tilde{\beta}^r$-sequence and the *RM* (*B*, *r*)-problem instances to be solved.

As we have mentioned already, the effectivity of any revenue management method/system is not only driven by the optimization module, but is highly depending on the soundness of the forecasting information. In our computational study we could only simulate the influence of forecasting at a very rudimentary level. Obviously, in a more practical evaluation one would have to integrate the optimization model/system into an integrated environment containing forecasting

Table 1 Leg table (represents the schedule)

Attribute	Type	Default	Example	Comment
leg_id	INTEGER	–	815	Internal identification of leg
departure_station	CHAR3	–	JFK	IATA-Code (mandatory)
departure_day	1..7	–	3	UTC-time (mandatory)
departure_time	TIME	–	15:20	UTC-time (mandatory)
arrival_station	CHAR3	–	ANC	IATA-Code (mandatory)
arrival_day	1..7	–	4	UTC-time (mandatory)
arrival_time	TIME	–	08:15	UTC-time (mandatory)
weight_capacity	INTEGER	–	15,000	Capacity in kg (mandatory)
volume_capacity	INTEGER	–	125	Capacity in m^3 (mandatory)
leg_fix_costs	REAL	–	320,000.00	Fixed costs in US$ (no payload)
leg_var_costs	REAL	–	0.08	Variable costs for one kg in US$
previous_leg	INTEGER	0	814	Pointer to previous leg in rotation
next_leg	INTEGER	0	816	Pointer to next leg in rotation

Table 2 O&D table

Attribute	Type	Default	Example	Comment
od_id	INTEGER	–	3,423	Primary key
origin	CHAR3	–	FRA	IATA-Code
destination	CHAR3	–	JFK	IATA-Code
category	CHAR1	–	S	S=Standard, E=Express, N=No specific time
available_day	1..7	–	2	In UTC-Time
available_time	TIME	–	21:30	In UTC-Time
due_day	1..7	–	3	In UTC-Time
due_time	TIME	–	16:45	In UTC-Time
weight	INTEGER	0	412,34	Demand in kg
yield	REAL	0.0	0.86	Average yield
density	REAL	6.0	5.74	Average density (IATA-Calculation)

and pricing modules. In our simulation study we used the following simple forecast-update: Since in every iteration the actual booking $\tilde{\beta}^r$, $\tilde{\beta}^r = \tilde{\beta}_j^{od}$ say, represents part of the expected/forecasted demand b^{od} of the associated O&D, we have to update the remaining/forecasted demand. Yet, we did not reduce the forecast by $\tilde{\beta}_j^{od}$ but by α_j^{od} to reflect the difference between initial forecast and real booking requests. Simple extensions of our simulation model could correlate the change of the remaining demand with the δ-perturbation of the request, and, also cancellations could easily be incorporated. In Fig. 7 the concept shown in Fig. 6 is presented on a more detailed level.

The efficiency of our approach, i.e. its solution quality and its computational effort is highly dependent on the implementation. And here we have to distinguish between two main aspects which are strongly related and interdependent:

- the implementation of the basic column generation technique, and,
- the integration of the optimization into the revenue management process.

Table 3 Handling table

Attribute	Type	Default	Example	Comment
airport_id	CHAR3	–	FRA	IATA-Code
airline_designator	CHAR2	–	LH	IATA-Code
od_category	CHAR1	–	S	S=Standard, E=Express, N=No special time
import_time	INTEGER	–	140	Time for import in minutes
import_cost	REAL	–	7.06	In US$ per ton
export_time	INTEGER	–	105	Time for export in minutes
export_cost	REAL	–	7.00	In US$ per ton
transfer_time	INTEGER	–	180	Time for transfer in minutes
transfer_cost	REAL	–	15.00	In US$ per ton

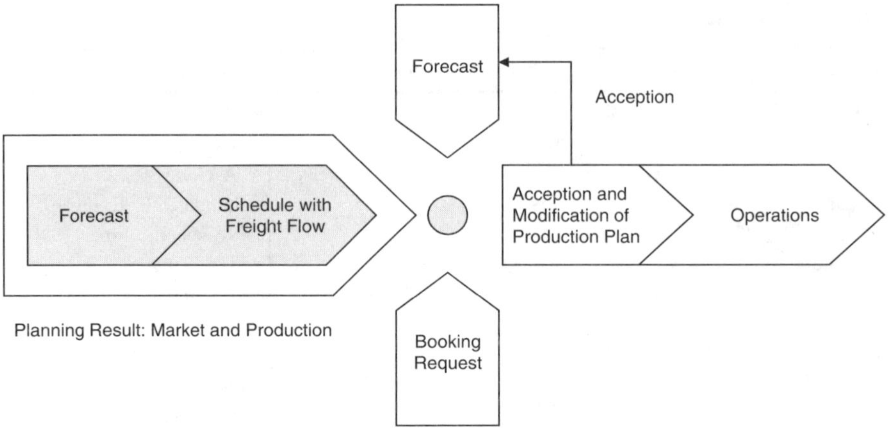

Fig. 6 Revenue management process

We have developed about a dozen different strategies for solving *RM*(B, r) based on different implementations of the column generation technique. In column generation there is a trade-off between quality and speed depending on the frequency of re-optimizing the master problem/calculating the actual dual prices. A detailed description of these computational aspects is beyond the scope of this paper. A computational study on these strategies is given in Bartodziej and Derigs (2004).

Another strategy to reduce the computational effort is to relax the on-line optimization requirement and to accept bookings, which are uncritical with respect to capacity and apparently profitable without evaluation and optimization, and to update the set of booked requests and construct a feasible freight flow using the model in a batch processing kind of mode. Similarly, one could give up the requirement to decide on the acceptance of requests sequentially one by one and to accumulate sequences S=(r_1, r_2,...) of requests to *blocks* or *batches* and then decide on which requests are accepted in one step. And here again one could follow two strategies: to decide on the acceptance of the requests in S based on the optimal dual prices for *RM*(B) or to re-optimize, i.e. to solve a variant of *RM*(B, r) obtained by adapting inequality Eq. 11:

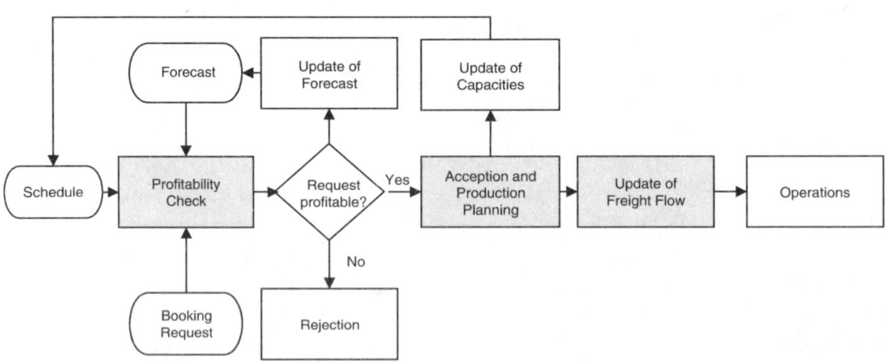

Fig. 7 Concept of revenue management simulation

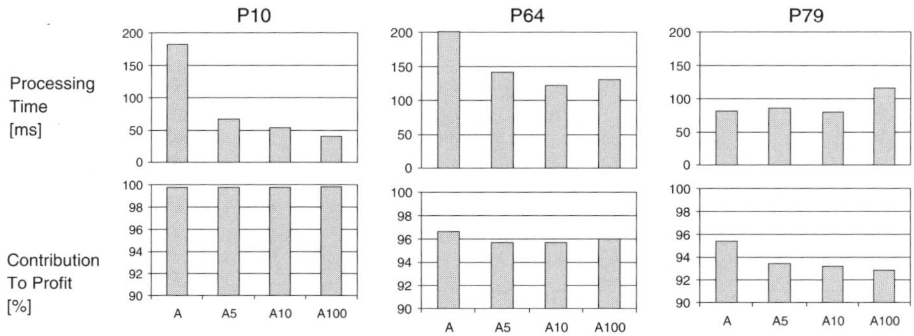

Fig. 8 Comparison of implementation strategies

$$\sum_{p \in P^r} f(P) \geq \beta^r + \tilde{\beta}^r \text{ for } r \in S \qquad (11')$$

Especially, the running time for solving the LP's becomes unacceptable high for problem instances of larger size. Thus in these realistic and time-critical environments additional means to reduce the size of the LP-master problems should be applied. Here one could reduce the number of paths in the search by eliminating the allowable number of legs per path. Another option would be to decompose the network. Then the first step in the revenue management decision would be to design every request to a suitable sub network.

The measurement of the quality of a (final) solution is quite difficult. An obvious measure would be the ratio between the yield potential and the realized yield. Yet, since it is a dynamic process and not a static problem, there is no natural definition for the maximum yield potential. Here, the value of the optimal solution of the planning model can only be an estimation or rational substitute. Using this value as a reference, one can observe rather significant differences among the different strategies with respect to quality.

In Fig. 8 we show the trade-off between quality and computational efficiency with respect to batching. Here, we display results for the following strategies

- sequential evaluation/no batching
- *k*-batching with blocksize "*k*", *k*=5,10,100

The results were obtained for a scenario generated with γ_1=5, $\gamma_2 = 1$ and Δ=0.1 leading to 2,533, 6,521 and 2,206 requests for problems P10, P64 and P79, respectively.

The contribution to profit is given as percentage of the contribution to profit of the optimal solution to the associated planning model. We have divided the total running time for a run by the total number of requests and compare the computational efficiency of the different approaches on the basis of processing time per request (given in CPU-ms). It turns out that the behavior is not monotone over all the examples. Yet, for every example/problem instance one can establish that there exist different "optimal batch sizes" for both measures, and thus an interval of effective batch sizes can be identified.

6 Final remarks

As cargo airlines (and shipping lines for this matter as well) are becoming increasingly under pressure, revenue management could help to improve the performance significantly. One cargo airline was able to increase yield by 5%, while the average industry yield fell by 15% in the same period, only by segmenting their products (express vs. standard) and reserving capacity on fixed routes for higher yield products (see Pompeo and Sapountzis (2002)). Assuming that even more intelligent cargo revenue management systems based on dynamic O&D assignment, as described above, could increase average yield by 1%, the industry could generated additional bottom-line profits of more than US $300 million p. a.

However, to be practical, the necessity to generate answers for booking requests in near real time is a demanding task for its own. For a leading cargo airline the number of O&D's can easily reach more than 50.000 different O&D's (incl. product categories), and over a week hundreds of individual bookings could be requested for a specific O&D. Not only is the use of extremely fast (constrained) shortest path algorithms which first try to use itineraries which are already in the solution an absolute must, moreover, strategies to combine single bookings into blocks which are then evaluated in one iteration seem to be a promising strategy.

After all one can say that using these advanced algorithmic ideas and high speed computers, O&D revenue management based on mathematical planning programs could be(come) feasible in practice and significantly improve the performance of cargo airlines (even if O&D management is only applied to a subset of their overall O&D's and booking requests).

Acknowledgements We want to thank two anonymous referees for their valuable comments on an earlier version of this paper.

References

Ahuja RK, Magnanti TL, Orlin JB (1993) Network flows: theory, algorithms, and applications. Prentice Hall, Englewood Cliffs

Antes J, Campen L, Derigs U, Titze C, Wolle G-D (1998) SYNOPSE: a model-based decision support system for the evaluation of flight schedules for cargo airlines. Decis Support Syst 22:307–323

Bartodziej P, Derigs U (2004) On an experimental algorithm for revenue management for cargo airlines. In: Lecture notes in computer science 3059. Springer, Berlin Heidelberg New York, pp 57–71

Berge ME, Hopperstad CA (1993) Demand driven dispatch: a method for dynamic aircraft capacity assignment, models and algorithms. Oper Res 41(1):153–168

Billings JS, Diener AG, Yuen BB (2003) Cargo revenue optimization. J Revenue Pricing Manage 2:69–79

Cross RG (1997) Revenue management: hard core tactics for market domination. Broadway Books, New York

Derigs U, Zils M (2001) Strategisches Controlling: Strategic Alliance Portfolio Analysis (SAP) - ein modellbasierter Ansatz zur Strategie- und Partnerselektion bei Strategischen Allianzen. In: ZfB-Ergänzungsheft 2/2001, pp 137–159

Etschmaier MM, Mathaisel DF (1985) Arline scheduling: an overview. Transp Sci 19(2):127–138

Elmasri R, Navathe SB (2004) Fundamentals of database systems, 4 edn. Pearson Education Boston

Gershkoff I (1998) An approach for just-in-time airline scheduling, chapter 6. In: Yu G (ed) Operations research in the airline industry, Kluwer, Dordrecht, pp 158–188
Kasilingam RG (1996) Air cargo revenue management: characteristics and complexities. Eur J Oper Res 96:36–44
Kleywegt AJ, Papastavrou JD (1998) The dynamic and stochastic knapsack problem. Oper Res 46:17–35
Kleywegt AJ, Papastavrou JD (2001) The dynamic and stochastic knapsack problem with random sized items. Oper Res 49:26–41
McGill JI, van Ryzin GJ (1999) Revenue management: research overview and prospects. Transp Sci 33(2):233–256
Pak K, Dekker R (2004) Cargo revenue management: Bid-prices for a 0-1 multi knapsack problem, ERIM Report Series Research in Management 55, Erasmus University Rotterdam
Pompeo L, Sapountzis T (2002) Freight expectations. McKinsey Q 2:90–99
Slager B, Kapteijns L (2004) Implementation of cargo revenue management at KLM. J Revenue Pricing Manage 3:80–90
Talluri KT, van Ryzin GJ (2004) The theory and practice of revenue management. Springer, Berlin Heidelberg New York
Zils M (1998) C.A.R.M.A.—Cargo Airline Relative Market Share Analyst: An O&D-based market model for flight network design in the Air Cargo industry, Working Paper, WINFORS, University of Cologne, Germany
Zils M (1999) AirCargo Scheduling Problem Benchmark Instanzen, Working Paper, WINFORS, University of Cologne, Germany

L. H. Lee · E. P. Chew · M. S. Sim

A heuristic to solve a sea cargo revenue management problem

Abstract In this paper, we will introduce a heuristic to solve a single leg revenue management problem with postponement, arising from the sea cargo industry. Based on previous work, it was shown that the optimal policy to allocate the capacity of the ship is a threshold policy. Based on the sample average approximation method, we formulate a mixed integer linear programming problem to determine the stationary threshold policy. A heuristic (known as the perturbation approach) is proposed to solve the problem. From the numerical result, it is shown that our approach performs better than some of the methods used to solve the mixed-integer programming problem.

Keywords Sea cargo · Revenue management

1 Introduction

Recently, many countries are reviewing the regulatory system for liner shipping. These countries include Australia, Canada, the European Union, Japan, South Korea and the United States (The World Shipping Council 2000). One significant event was the amendment of the Ocean Shipping Reform Act (OSRA) by the United States in 1998.

The change gives more legal freedom to negotiation and provision of ocean transportation services in the United States, hence bringing the business relationship between the carriers and the shippers to a new dimension. For example, the United States Department of Agriculture (2001) reported that the contracts for transportation of agricultural product are no longer simply volume discounts, but increasingly contain negotiated and tailored service provisions. This is also observed in other trade areas (see Federal Maritime Commission (2001) for further detail). As a result of the amendment, Federal Maritime Commission (2001) reported that, there is at least a 200 percent increase in the number of service

L. H. Lee (✉) · E. P. Chew · M. S. Sim
Department of Industrial & Systems Engineering, National University of Singapore, Singapore
E-mail: iseleelh@nus.edu.sg

contracts being signed. The number of service contract filed is expected to increase further as Federal Maritime Commission recently agreed to allow Non-vessel-operating common carriers (NVOCCs) to sign confidential contacts with their shipper customers in December 2004.

Motivated by a particular practice in the sea cargo industry, a revenue management model for the carrier is proposed in our working paper (Lee et al. 2005). We have obtained some structural results on the problem and proved that the optimal policy to allocate the capacity of the ship is a threshold policy. To determine the threshold policy, a mathematical model will be presented here in Section 2. In Section 3, an efficient approach to solve the problem will be proposed. Lastly, some numerical results will be presented in Section 4.

2 The sea cargo revenue management model

2.1 Revenue management

The airline industry in the United States started applying revenue management in the 1970s after deregulation of air transportation. With revenue management, the airline carriers have efficiently allocated the airlines seats such that there are enough seats reserved for the full-fare customers arriving at a later time while the remaining available seats are opened to the discount-fare customers. Following the successful stories from the airline industry, revenue management is being applied in other transportation sectors. Kleywegt (2002) presented two models; Contact planning and Booking control, which will help the carriers to determine the optimal strategy to allocate the shipping capacity. The Contact planning model is meant for the carriers to make long-term planning. Given the economical situation and the available capacity on certain voyages, the model seeks to determine a contract that maximizes the carriers' return. With the inputs from the Contract planning model, the Booking control model is used for short-term allocation of capacity in the ship. Pak and Dekker (2004) formulated the cargo revenue management as a multi-dimensional on-line knapsack problem. They showed that a bid-price acceptance policy is asymptotically optimal if demand and capacity increase proportionally and the bid-prices are set correctly.

2.2 Problem description

Often, in service contracts, the carrier and the shipper will sign an agreement stating a specific number of containers for shipment over a period of time. This amount is known as the Minimum Quantity Commitment (MQC). The carrier will reserve the capacity for those shippers who he/she has signed a service contract with. It is assumed that the reserved capacity is constant here. The remaining capacity of the ship will be opened for booking.

Due to the uncertainty in the demand, the number of containers to ship is sometimes more than the MQC, especially during the peak season. We shall call these additional containers from the contractual shippers as contractual containers here. Furthermore, the rate for a contractual container is similar to the rate stated in the service contract.

At the same time, the carrier will also receive orders from shippers whom have not signed any contract with the carrier. These shippers have urgent shipments to make and are generally willing to pay a higher price than the contractual customers for the same amount of capacity. The containers from these shippers will be termed as ad hoc containers here. From the perspective of the carrier, the ad hoc containers will be preferred. However, the carrier must deliver the ad hoc containers in the earliest possible time. As for the contractual containers, the carrier has the flexibility to postpone the delivery to a later date as long as they are delivered to the destination by the agreed date.

Currently, the carrier will accept any incoming containers in the ship arriving at the port. When the ship's capacity runs out, the contractual containers will be allocated to the next ship arriving at the port until its capacity runs out whereas the ad hoc containers will be rejected. This policy will be termed as "zero-threshold" policy here. To better manage the capacity, Lee et al. (2005) introduced a revenue management model.

The revenue management model will be used to decide whether a container (both contractual and ad hoc) should be accepted. If a contractual container is accepted, the model will also determine whether it should be postponed to the next shipment.

In our problem, without loss of generality, we assume that the ship will make a weekly voyage and the carrier has to make a decision to accept or reject the arriving containers (contractual or ad hoc) for that day by the end of each day. Hence, the departure period (i.e. the time interval between the departure of ship from the port) is a week and each departure period will have seven decision periods (i.e. the carrier has to make decision every day). Ship 1 refers to the ship leaving the port at the end of the current departure period while ship 2 refers to the ship leaving the port at the end of the next departure period. Furthermore, we are only considering a single-leg problem here.

We will also like to point out that the term contractual container collectively refers to all orders from contractual shippers requesting for more capacity. Depending on the negotiation between them, each service contract will carry different rate for the same capacity. Although grouping all shippers with service contracts together does not accurately model the problem, it is reasonable to do this based on the following reasons. Firstly, the difference in rate between two different contractual customers is relatively small, compared to the difference between any contractual customer and an ad-hoc customer. Hence, the rate of an ad hoc container will cause a greater impact on the revenue of the carrier. Secondly, the decision made by the carrier when any requests to ship contractual containers arrive, is distinctly different from the decision made when some requests to ship ad hoc containers arrive.

Since the ad hoc container generates higher revenue, it is optimal to accept as many ad hoc containers as possible. As for the contractual containers, they may be accepted as long as the remaining capacity in the ship does not exceed a threshold value. Let the threshold values for allocation of contractual containers in ship 1 and 2 be $\theta_1^{t,k}$ and $\theta_2^{t,k}$ for the tth decision period of kth departure period. The index t is numbered backward so that decision period 1 corresponds to the end of the departure period. The following properties of the threshold values are also obtained in the working paper (Lee et al. 2005):

- For every departure period, $\theta_1^{1,k}=0$.
- For the same departure period, $\theta_1^{t,k} \geq \theta_1^{t-1,k}$.
- For the same departure period, $\theta_2^{t,k}$ is constant for all decision period.

It will not be practical to implement the threshold policy as the number of departure period is large. Hence, a stationary threshold policy, regardless of the departure period:

$$\{\theta_1^N, \theta_1^{N-1}, \ldots, \theta_1^t, \ldots, \theta_1^1, \theta_2\}$$

is desirable for practical application.

Arapostathis et al. (1993) showed that there exists a stationary policy that is discount optimal for all discount factors sufficient close to 1 and is also optimal for average revenue criterion. This is true for problem with a finite state and action space. For our problem, the state space may be large but is bounded by the maximum remaining capacity of the ship. Similarly, the action space cannot be greater than the maximum remaining capacity of the ship. Based on this argument, it is reasonable to use a stationary threshold policy, independent of the departure period when the number of departure period is very large. The determination of the stationary threshold policy is not trivial as there is no closed form solution. We will introduce some heuristic methods to find a good threshold policy.

2.3 The stationary threshold problem

The revenue management problem considered here has many departure periods and each departure period will consist of a number of decision periods. This makes the computation of expected revenue difficult and hence it is not easy to determine the threshold policy for our problem. To solve this, we generate the ad hoc and contractual container demands for $(M+1)$ departure periods. Given these demands, we formulate a deterministic optimization problem to determine the best stationary threshold policy. This method is similar to the SAA method introduced in Shapiro (2001). Denote

S–Maximum capacity of the ship
N–Length of a departure period
t–Index for each decision period
k–Index for each departure period
$x_{t,k}$–The remaining capacity of ship 1 at t^{th} decision period of kth departure period
$y_{t,k}$–The remaining capacity of ship 2 at t^{th} decision period of kth departure period
$A_{t,k}$–The number of ad hoc containers arrived at t^{th} decision period of kth departure period
$C_{t,k}$–The number of contractual containers arrived at t^{th} decision period of kth departure period
$\beta_{AC}^{t,k}$–The number of ad hoc containers accepted at t^{th} decision period of kth departure period when there are A ad hoc and C contractual containers requested to be transported
$\lambda_{AC}^{t,k}$–The number of contractual containers accepted in ship 1 at t^{th} decision period of kth departure period when there are A ad hoc and C contractual containers requested to be transported

$\rho_{AC}^{t,k}$ –The number of contractual containers accepted in ship 2 at t^{th} decision period of kth departure period when there are A ad hoc and C contractual containers requested to be transported

r–The ratio of the revenue of one ad hoc container to one contractual container

The following optimization Problem 1 is obtained:

Problem 1

$$\max_{\theta_1^N \ldots \theta_1^1, \theta_2} \frac{1}{M+1} \sum_{k=0}^{M} R_k$$

$$\text{where } R_k \begin{cases} \sum_{t=1}^{N} \beta_{t,k} \cdot r + \lambda_{t,k} + \rho_{t,k} & k > 0 \\ \sum_{t=1}^{N} \beta_{t,k} \cdot r + \lambda_{t,k} & k = 0 \end{cases}$$

$$\begin{aligned} \beta_{t,k} &\le A_{t,k} \\ \beta_{t,k} &\le x_{t,k} \end{aligned} \tag{1}$$

$$\left.\begin{aligned} &\lambda_{t,k} + \rho_{t,k} \le C_{t,k} \\ &\lambda_{t,k} + \rho_{t,k} \le z_{t,k} + y_{t,k} - \theta_2 \\ &C_{t,k} - \lambda_{t,k} - \rho_{t,k} \le \delta_{t,k}^1 \cdot S' \\ &z_{t,k} + y_{t,k} - \theta_2 - \lambda_{t,k} - \rho_{t,k} \le \left(1 - \delta_{t,k}^1\right) \cdot S' \\ &\lambda_{t,k} \le z_{t,k} \\ &\rho_{t,k} \le y_{t,k} - \theta_2 \end{aligned}\right\} k > 0 \tag{2}$$

$$\left.\begin{aligned} &\lambda_{t,0} \le C_{t,0} \\ &\lambda_{t,0} \le z_{t,0} \\ &C_{t,0} - \lambda_{t,0} \le \delta_{t,0}^1 \cdot S' \\ &z_{t,0} - \lambda_{t,0} \le \left(1 - \delta_{t,0}^1\right) \cdot S' \end{aligned}\right\} k = 0 \tag{3}$$

$$\begin{aligned} &z_{t,k} \ge 0 \\ &z_{t,k} \ge x_{t,k} - \beta_{t,k} - \theta_1^t \\ &z_{t,k} \le \delta_{t,k}^2 \cdot S' \\ &z_{t,k} - x_{t,k} + \beta_{t,k} + \theta_1^t \le \left(1 - \delta_{t,k}^2\right) \cdot S' \end{aligned} \tag{4}$$

$$x_{t,k} = \begin{cases} S & k = M, t = N \\ y_{1,k+1} - \rho_{1,k+1} & k < M, t = N \\ x_{t+1,k} - \beta_{t,k} - \lambda_{t,k} & \text{otherwise} \end{cases} \tag{5}$$
$$y_{t,k} = \begin{cases} S & k > 0, t = N \\ y_{t+1,k} - \rho_{t,k} & \text{otherwise} \end{cases}$$

$$\begin{aligned} &\theta_1^1 = 0 \\ &\theta_2 \geq \theta_1^t \geq \theta_1^{t-1}\ \forall t > 1 \\ &\theta_1^t \in [0,\ S] \\ &\theta_2 \in [0,\ S] \end{aligned} \tag{6}$$

$z_{t,k}$ is equivalent to $(x_{t,k}-\beta_{t,k}-\theta_1{}^t)$
$S' > S$

The allocation of ad hoc and contractual containers by nested threshold policy is described in constraint sets Eq. 1–4. Constraint set Eq. 1 describes the allocation of ad hoc containers while constraint sets Eq. 2–4 consider the allocation of contractual containers. As it is assumed here that the carrier will stop operating after the last departure period (i.e. $k=0$), no contractual containers will be postponed when $k=0$. This explains why a different constraint set Eq. 3 is being used when $k=0$. Constraint set Eq. 4 is added in to ensure that the maximum allocated capacity for contractual containers in ship 1 are non-negative. The binary variables ($\delta_{t,k}^1$ and $\delta_{t,k}^2$) and the constant S' are also introduced at each decision period to model the nested allocation of contractual containers on both ships.

$\delta_{t,k}^1$ determines whether all contractual containers are being accepted at the tth decision period of kth departure period. $\delta_{t,k}^2$ states whether any contractual containers will be allocated in ship 1. When $\delta_{t,k}^1=0$, it means that all contractual containers are being accepted at the tth decision period of kth departure period. If $\delta_{t,k}^2=0$, no contractual containers will be accepted in ship 1 as $z_{t,k}=0$.

Constraint set Eq. 5 shows how the remaining capacity in ship 1 and 2 will be changed as we move from one decision period to the next and constraint set Eq. 6 includes the monotonous property between the threshold values. Although the container allocation in each ship ($\beta_{t,k}$, $\lambda_{t,k}$, $\rho_{t,k}$) and the threshold values (θ_1^t, θ_2) should be discrete for completeness, we relax this restriction here. This is justified by the fact that the capacity of a container ship is usually very large (above 6000 TEU) and hence the effect of rounding off the container allocation to the nearest integer becomes insignificant.

For a stationary threshold problem of k departure periods with t decision periods each, there are $2kt$ binary variables. It is computationally inefficient to solve the problem with the branch-and-bound algorithm when k or t is large. We will introduce a more computationally efficient heuristic to solve the mixed integer programming problem in Section 3.

3 The perturbation approach

3.1 The general idea

3.1.1 Consider

Problem 2

$$\begin{aligned}&\max C_1X_1 + C_2X_2\\&\text{s.t. } A_1X_1 + A_2X_2 + BY \leq b\\&X_1 \geq 0\\&X_2 \geq 0\\&Y_j \in \{0,\ 1\} \quad j = 1, \ldots, p\end{aligned}$$

where A_1 is a $m\times n$ matrix, A_2 is a $m\times n'$ matrix, B is a $m\times p$ matrix, C_1 is a n component row vector, C_2 is a n' component row vector and b is a m component column vector. Due to Y, the optimization problem cannot be solved efficiently. Suppose that each Y_j is arbitrary fixed to $y_j^{\,i}$ at current iteration i, the resulting problem

Problem 3

$$\begin{aligned}&\max C_1X_1 + C_2X_2\\&\text{s.t. } A_1X_1 + A_2X_2 \leq b - BY^i\\&X_1 \geq 0\\&X_2 \geq 0\\&Y_j^i = y_j^i \quad j = 1, \ldots, p\end{aligned}$$

is a linear optimization problem and considered a relatively "easy" problem.

Let z^* and $z^*_{Y^1}$ be the optimal objective value of Problem 2 and 3, it is understood that $z^* \geq z^*_{Y^i}$. Suppose that X_1^i is the optimal solution at i^{th} iteration for Problem (3) with Y^i, if we could obtain Y^{i+1} such that $X_1{}^i$ is also feasible at $(i+1)^{th}$ iteration, then $z^*_{Y^i} \leq z^*_{Y^{i+1}}$. We will explain in the next section on how this idea can work in our problem and show how Y^i can be determined so that $z^*_{Y^i}$ can be improved as the iteration i continues.

3.2 The application of the perturbation approach to our problem

Given the demand data ($A_{t,k}$ and $C_{t,k}$) and the current binary variables ($\delta^1_{t,k}$ and $\delta^2_{t,k}$) at each decision period, it is relatively easy to obtain the corresponding optimal threshold policy (θ_1^N, ..., θ_1^1, θ_2) from a commercial linear programming optimizer such as ILOG CPLEX software. As the optimal solution for a linear programming problem must be an extreme point, at least one of the inequality constraints is binding. In this problem, the binding constraints may help to improve the current solution.

Take for example, if $\delta^1_{t,k}=0$ and $C_{t,k}=z_{t,k}+y_{t,k}-\theta_2$, it means that exactly $C_{t,k}$ contractual containers are accepted and all the allocated capacity for contractual containers in both ships (i.e. $z_{t,k}$ and $y_{t,k}-\theta_2$) is utilized. Suppose that if the value of $\delta^1_{t,k}$ is changed to 1, the current solution is still feasible. Furthermore, a better solution may be obtained when the problem is re-solved by the linear programming optimizer, with $\delta^1_{t,k}=1$.

An intelligent mechanism is next introduced to decide which binary variables should change at each decision period. Given the current threshold policy, if we perturb one of the threshold values, some binary variables may be changed as they restrict the perturbation of the threshold value.

Using the same example above, the capacity allocated for contractual container will be reduced if θ_2 is perturbed up. Hence, it is most likely that not all contractual containers arrived (i.e. $C_{t,k}$) will be accepted. However, $\delta^1_{t,k}=0$ restricts the perturbation of θ_2 as it stipulates that all incoming contractual containers will be accepted. If the value of $\delta^1_{t,k}$ is changed to 1, a better solution may be found. Essentially, by changing $\delta^1_{t,k}$ to one will add S' to the right-hand-side of

$$C_{t,k} - \lambda_{t,k} - \rho_{t,k} \leq 0$$

so that the new constraint is $C_{t,k}-\lambda_{t,k}-\rho_{t,k}\leq S'$. This means that the total number of contractual containers accepted can be less than $C_{t,k}$ and will allow θ_2 to perturb up further. On the other hand, if $\delta^1_{t,k}=1$ and $C_{t,k}=z_{t,k}+y_{t,k}-\theta_2$, it is not required to change any binary variables when θ_2 is perturbed up. This is because no more than $z_{t,k}+y_{t,k}-\theta_2$ contractual containers will be accepted in this case. It is noted that our objective is to use the perturbation of the threshold value to determine which binary variables should change. After the binary variables are changed, the problem will be re-solved by the linear programming optimizer and a new threshold value will be determined.

The special relationship between the threshold values and the binary variables allows us to define a more efficient approach to solve the problem. Perturbation is introduced into the problem via the threshold value. It propagates to other decision period through the change in the total number of container accepted. As a consequence, some of the binary variables in the problem will be updated and a new linear programming problem is obtained and solved. This process will be repeated until no change in binary variables is encountered. It is emphasized that the procedure is used to select which binary variables to change at each iteration. After the binary variables are changed, the threshold Problem 1 will then be re-solved and the threshold value may or may not change.

It is noted that the perturbation approach will obtain a local optimal solution for the stationary threshold problem. This is because, the stationary threshold problem is bounded as the capacity of both ships is not greater than S and the value of S is finite. Furthermore, the value of M (i.e. the number of departure period considered here) is also finite. Since the threshold values are still feasible after the binary variables are changed, the new solution obtained for the next iteration will not be worse than the solution for the current iteration. However, as the stationary threshold problem is not concave, it is not guaranteed that the final solution is globally optimal.

3.3 The general algorithm

Before we describe the algorithm for the perturbation approach, define

δ^{best}—the vector that represent $\delta^1_{t,k}$ and $\delta^2_{t,k}$ at each decision period that give the highest revenue, up to the current iteration

δ^{good}—the vector that represent $\delta^1_{t,k}$ and $\delta^2_{t,k}$ at each decision period that give the highest revenue among all the possible perturbations at current iteration

θ^{best}—the vector that represent θ_1^t at each decision period and θ_2 obtained by solving the stationary threshold Problem 1, given δ^{best}

θ^{good}—the vector that represent θ_1^t at each decision period and θ_2 obtained by solving the stationary threshold Problem 1, given δ^{good}

revenue_{best}—the objective value obtained by solving the stationary threshold Problem 1, given δ^{best}

revenue_{good}—the objective value obtained by solving the stationary threshold Problem 1, given δ^{good}

Algorithm 1 Main perturbation

1. Randomly generate a stationary threshold policy and determine the corresponding vector, δ^{best}.
2. Given δ^{best} at each decision period, solve the stationary threshold Problem 1 and let the objective value obtained be revenue_{best}. Obtain the vector θ^{best}.
3. (a) With the exception of θ_1^1, perturb a threshold value while the other threshold values remain unchanged. Determine the change in $\delta_{t,k}^1$ and $\delta_{t,k}^2$ for every departure period. Solve the resulting stationary threshold Problem 1 with a commercial linear programming optimizer to obtain the new threshold values and objective value.
   Repeat the procedure for other threshold values. It is noted that each threshold value can be perturbed in two directions (i.e. up and down) hence there are $2N$ stationary threshold problems to be solved at this step.
   (b) Among all the perturbations done, select the perturbation that gives the current best improvement in revenue and denote its revenue, binary variables and threshold value obtained as revenue_{good}, δ^{good} and θ^{good}.
4. If there are no improvement ($\text{revenue}_{good} \leq \text{revenue}_{best}$), terminate. Else let

$$\begin{aligned} \delta^{best} &= \delta^{good} \\ \theta^{best} &= \theta^{good} \\ \text{revenue}_{best} &= \text{revenue}_{good} \end{aligned}$$

   and go to step 3.

Algorithm 1 shows the steps taken in the perturbation approach. The reader is referred to our working paper for the detailed procedures to determine the change in $\delta_{t,k}^1$ and $\delta_{t,k}^2$ for every departure period. revenue_{good} will record the highest revenue among all possible perturbations performed in step 3. The algorithm will proceed to the next iteration if revenue_{good} is higher than revenue_{best}, the highest revenue obtained till the current iteration.

4 Numerical result

The following cases in Table 1 are run to illustrate the usefulness of the perturbation approach. The maximum capacity of each ship, S is 200 and the length of a departure period, N is 7. The value of r used is 5. The demand distribution is uniformly distributed with a standard deviation of 2 units. We consider the arrival of containers (ad hoc and contractual) at three levels: low (five containers per

Table 1 The various scenarios tested at N=7 and S=200

Case	1	2	3	4	5	6
Mean demand of ad hoc container (per decision period)	5	15	15	25	25	25
Mean demand of contractual container (per decision period)	25	15	25	5	15	25

decision period), medium (15 containers per decision period) and high (25 containers per decision period). Three cases are omitted as the threshold policy will not cause any impact when the total mean number of containers arrived per departure period is less than S. Hence, only six cases will be studied here.

4.1 How optimal is the perturbation approach?

We will first like to compare the solution of the perturbation approach with that obtained from the CPLEX solver. It is known that it may not be possible to solve a big problem with the branch-and-bound algorithm. Hence, we will only use small problems with five departure periods each. Their results are presented in Table 2. For example, problems SP1_a-SP1_e in Table 2 are instances of demands generated over the five departure periods for case 1.

The starting threshold policy for the perturbation approach is obtained from the random search method. For each problem, the random search method randomly generates 50,000 threshold policies (which obey the monotonous property) and chooses the one that gives the best solution. The result obtained by the starting threshold policy is also given in the tables. For more information on the implementation of the random search method, the reader is referred to our working paper. It is observed that the perturbation approach reaches the optimal solution 27 times out of the 30 problems ran and the range of solution found is at least 98% within the optimal solution.

4.2 Comparison of the methods under time-constraint

In the next experiment, we will evaluate its performance under limited computational time. The algorithm will be terminated when the time limit is reached. Case 1 and 2 will be considered here. For each case, ten test problems with 100 departure periods each will be used. For each test problem, $A_{t,k}$ and $C_{t,k}$ at each decision period will be generated based on the parameters in Table 1. Table 3 shows the results obtained when the time limit is set at 1 hour. For example, problems BP1_A to BP1_J are the ten test problems for case 1. In Table 3, the starting threshold policy for the main perturbation approach is either the zero-threshold policy or obtained from the genetic algorithm.

We use a real-coded steady state genetic algorithm to determine the threshold values here. The typical chromosome $\overline{C}$ for this problem can be represented as:

$$\overline{C} = \{\theta_1^N, \ldots, \theta_1^t, \ldots, \theta_1^1, \theta_2\}$$

Table 2 Average revenue obtained from small problem (case 1–6)

Problem	Optimal solution	Main perturbation approach	Random search
SP1_a	340.8	340.8	337.1
SP1_b	349.6	349.6	340.2
SP1_c	348.0	348.0	345.6
SP1_d	339.2	339.2	338.9
SP1_e	350.4	350.4	346.0
SP2_a	606.4	606.4	597.3
SP2_b	605.6	605.6	601.2
SP2_c	626.4	626.4	619.8
SP2_d	607.8	607.8	605.4
SP2_e	630.4	630.4	626.7
SP3_a	612.0	612.0	611.5
SP3_b	627.2	627.2	627.1
SP3_c	628.0	628.0	620.5
SP3_d	616.0	616.0	613.2
SP3_e	622.4	622.4	618.8
SP4_a	913.6	913.6	908.2
SP4_b	895.2	892.4	850.5
SP4_c	905.6	905.6	905.1
SP4_d	901.6	899.4	888.9
SP4_e	899.2	899.2	890.6
SP5_a	877.6	877.6	865.2
SP5_b	893.6	893.6	887.8
SP5_c	888.8	888.8	879.5
SP5_d	882.4	882.4	880.0
SP5_e	914.4	914.4	909.7
SP6_a	904.0	904.0	799.9
SP6_b	903.2	902.6	891.5
SP6_c	912.8	912.8	908.6
SP6_d	907.2	907.2	900.1
SP6_e	887.2	887.2	885.3

An initial population of chromosomes will be randomly generated and only chromosomes that obey the monotonous property (as described in constraint set Eq. 6 of problem 1) is accepted. For each accepted chromosome, evaluation will be performed according to the primitive threshold policy. The population size used here is 150 and the number of generations run is 15,000.

Under this method, only best two chromosomes will be selected for reproduction. The parent will undergo crossover to produce a child. *BLX-α* (Eshelman and Schaffer 1993) will be used to obtain a crossover child from the two parents. Suppose that $C_1=(c_1^1,\ldots, c_n^1)$ and $C_2=(c_1^2,\ldots, c_n^2)$ are the two chromosomes that have been selected for crossover, the resulting child is $H=(h_1, \ldots, h_n)$ where h_i is a randomly generated number in the interval

$$[c_{\min} - I\cdot\alpha,\ c_{\max} + I\cdot\alpha]\ \text{such that}\ c_{\min} = \min\left\{c_i^1,\ c_i^2\right\}$$

Table 3 Average revenue obtained for various methods

Problem	EMSR heuristic	Branch-and-bound	Genetic Algorithm	Main perturbation approach[a]	Main perturbation approach[b]
BP1_A	316.5	N.s.	339.4	340.1	322.1
BP1_B	316.2	N.s.	339.2	341.6	335.2
BP1_C	319.5	N.s.	344.4	344.6	336.4
BP1_D	317.9	N.s.	341.1	343.7	336.5
BP1_E	313.9	N.s.	342.4	355.3	341.4
BP1_F	315.9	N.s.	305.3	373.1	322.1
BP1_G	318.6	N.s.	318.0	335.2	319.2
BP1_H	321.1	N.s.	327.1	336.1	331.4
BP1_I	318.6	N.s.	318.7	336.5	321.0
BP1_J	319.2	N.s.	320.1	341.4	321.2
BP2_A	594.9	586.5	618.5	620.7	619.2
BP2_B	596.6	580.5	618.5	623.4	620.1
BP2_C	593.5	593.9	618.4	618.6	614.4
BP2_D	597.6	591.3	623.6	626.7	622.0
BP2_E	596.4	592.4	620.4	620.8	612.0
BP2_F	596.4	580.2	603.6	619.2	611.4
BP2_G	594.6	N.s.	597.0	618.6	607.5
BP2_H	596.5	593.9	599.3	614.4	605.6
BP2_I	599.0	N.s.	602.7	619.5	618.9
BP2_J	597.1	N.s.	610.1	612.0	610.0

[a] Starting threshold policy is obtained from the genetic algorithm
[b] Starting threshold policy is zero-threshold policy
N.s. denotes no feasible solution found

$$c_{\max} = \max\{c_i^1,\, c_i^2\}$$
$$I = c_{\max} - c_{\min}$$

A non-uniform mutation (Michalewicz 1992) will be performed on the child after crossover. If the chromosome is $C=(c_1,\ldots c_n)$ and $c_i \in [a_i, b_i]$ is the gene to be mutated, then the new gene c_i', under this operator is:

$$c_i' = \begin{cases} c_1 + \Delta(t,\, b_i - c_i) \text{ if } \tau = 0 \\ c_1 - \Delta(t,\, c_i - a_i) \text{ if } \tau = 1 \end{cases}$$

$$\Delta(x,\, y) = y\left(1 - r^{\left(1-\frac{t}{g_{\max}}\right)^b}\right)$$

where

- t is the current generation
- $g_{\max}$ is the maximum number of generations
- τ is a random number which may have a value of zero or one
- r is a random number from the interval [0, 1]
- b is a parameter that describes the degree of dependency on the number of iterations

Evaluation on the child will next be performed. If the result is better than the worst member in current population, the child will replace the worst member in the next generation.

Herrera et al. (1998) performed experiments on various real-coded genetic algorithms and concluded that *BLX-*α (particularly α=0:5) is among the best crossover operators for real-coded genetic algorithm. They also pointed out that non-uniform mutation is very appropriate for real-coded genetic algorithm. The readers are referred to our working paper for further discussion on the implementation of the genetic algorithm.

It is seen from the table that the CPLEX solver fails to give a feasible solution within the time frame for most problems. When the threshold policy from the genetic algorithm is used as the starting solution, the result obtained is better than the one with the zero-threshold policy as the starting solution. From the result obtained by genetic algorithm, it is observed that the solution improves when the main perturbation approach is used. Apart from this, it is also observed that the perturbation approach obtains better solution than the genetic algorithm for more than 50% of the problems ran even when the zero-threshold policy is used as the starting solution.

Our approach is also compared with a variant of the EMSR heuristic (Belobaba 1987), a popular heuristic used in airline revenue management. The EMSR heuristic determines the threshold value based on the estimated number of full-fare customers arriving in future. The threshold value can be obtained:

$$\theta_1^t = S_A(t)$$
$$\overline{P}_t(S_A(t)) = \tfrac{1}{r}$$

where

$\overline{P}_t(D)$ is the probability of receiving D or more ad hoc containers from t^{th} decision period to departure time and is determined based on $A_{t,k}$ for the 100 departure period

$S_A(t)$ is the capacity reserved for ad hoc containers at t^{th} decision period

To simplify the implementation of the EMSR heuristic in our problem, the threshold value of ship 2 is fixed at θ_1^N. Furthermore, θ_1^1 is fixed at the value of zero as it is proven optimal in our working paper. The result is also shown in Table 3.

We will like to stress that the perturbation approach adopts a different strategy to obtain the threshold policy. Currently, the genetic algorithm and the random search method determine the threshold policy directly while the proposed approach searches for the threshold policy based on the binary variables, $\delta_{t,k}^1$ and $\delta_{t,k}^2$. The perturbation approach uses a selective mechanism to choose the binary variables for change and indirectly obtain a better threshold policy as the iteration proceeds. From the numerical result, it is reasonable to deduce that the strategy taken by our approach is more effective. Due to space constraint, other numerical results are not presented in this paper. The reader is referred to our working paper for details.

References

Arapostathis A, Borkar VS, Fernandez-Gaucherand E, Ghosh MK, Marcus SI (1993) Discrete-time controlled Markov processes with average cost criterion: a survey. SIAM J Control Optim 31:282–344

Belobaba PP (1987) Air Travel Demand And Airline Seat Inventory Manage-ment, Ph.D. thesis, Flight Transportation Laboratory, Massachusetts Institute of Technology, Cambridge, MA

Eshelman LJ, Schaffer JD (1993) Real-coded genetic algorithms and interval-schemata. In: Whitley LD (ed) Foundation of genetic algorithms 2. The Second Workshop on Foundations of Genetic Algorithms, July 26–29. Morgan Kaufmann, San Mateo, CA, pp 187–202

Federal Maritime Commission (2001) The impact of the Ocean Shipping Reform Act of 1998 (Publication No. NR 01-09). Washington, DC

Herrera F, Lozano M, Verdegay JL (1998) Tackling real-coded genetic algorithms: operators and tools for behavioural analysis. Artif Intell Rev 12:265–319

Kleywegt AJ (2002) New approaches for contract planning and booking control for container carriers. Presented in The Logistics Institute Asia-Pacific First TLI-APNTNU Workshop on Multimodal Logistics, Singapore

Lee LH, Chew EP, Sim MS (2005) A revenue management model for the Sea Cargo. ISE Working Paper

Michalewicz Z (1992) Genetic algorithms+data structures=evolution programs. Springer, Berlin Heidelberg New York

Pak K, Dekker R (2004) Cargo revenue management: bid-prices for a 0-1 multi Knapsack problem. Publication No. EI 2004-26, Econometric Institute

Shapiro A (2001) Monte Carlo simulation approach to stochastic programming. Presented in Proceedings of the 2001 Winter Simulation Conference, Virginia, USA

Transportation Services Branch, United States Department of Agriculture (2001) Agricultural ocean transportation trends. Retrieved from http://www.ams.usda.gov/tmd/AgOTT/

World Shipping Council (2000) International liner shipping regulation: its rationale and its benefits. Retrieved from http://www.worldshipping.org

Marta Anna Krajewska · Herbert Kopfer

Collaborating freight forwarding enterprises

Request allocation and profit sharing

Abstract The paper presents a model for the collaboration among independent freight forwarding entities. In the modern highly competitive transportation branch freight forwarders reduce their fulfillment costs by exploiting different execution modes (self-fulfillment and subcontraction). For self-fulfillment they use their own vehicles to execute the requests and for subcontracting they forward the orders to external freight carriers. Further enhancement of competitiveness can be achieved if the freight forwarders cooperate in coalitions in order to balance their request portfolios. Participation in such a coalition gains additional profit for the entire coalition and for each participant, therefore reinforcing the market position of the partners. The integrated operational transport problem as well as existing collaboration approaches are introduced. The presented model for collaboration is based on theoretical foundations in the field of combinatorial auctions and operational research game theory. It is applicable for coalitions of freight forwarders, especially for the collaboration of Profit Centres within large freight forwarding companies. The proposed theoretical approach and the presented collaboration model are suitable for a coalition of freight forwarding companies with nearly similar potential on the market.

Keywords Collaboration · Freight forwarder · Profit sharing · Multi-agent auction

1 Introduction

In the ongoing globalization process large international freight forwarding companies are more competitive than small companies due to their wider portfolio of disposable resources and a higher ranking in the market power structure. The remedy for the medium- and small-sized carrier businesses is to establish coalitions

---

M. A. Krajewska (✉) · H. Kopfer
Department of Economics, University of Bremen, Wilhelm-Herbst-Strasse 5, 28359 Bremen, Germany
E-mail: makr@logistik.uni-bremen.de, kopfer@logistik.uni-bremen.de

in order to extend their resource portfolio and reinforce their market position. Moreover, the structure of large freight forwarding companies frequently assumes autonomously operating subsidiaries, that should, however, cooperate in order to maximize the overall business profit.

The purpose of the cooperation of freight forwarding entities is to find an equilibrium between the demanded and the available transport resources within several carrier entities by interchanging customer requests (Kopfer and Pankratz, 1999).

Section 2 presents the request processing at a single freight forwarding entity. Section 3 introduces the theoretical frame for collaboration modelling. Section 4 investigates the existing models of collaboration in the transportation branch. In Section 5 we present and analyse a model for the collaboration among freight forwarding enterprises.

2 Integrated operational freight carrier planning

A great number of enterprises source transportation tasks out by entrusting independent freight forwarding companies with the execution of the necessary transport activities. For each transportation task the forwarding company is allowed to choose the mode of fulfillment, i.e., own vehicles can be used for the execution of the corresponding entrusted tasks (self-fulfillment) or an external freight carrier (subcontractor) receives a fee for the request fulfillment (subcontraction). Independent shipment contracts of different types and specifications are awarded to the subcontractor for completion. The involvement of the subcontractor can occur due to two incentives (Chu, 2005). In reality, freight forwarders face demand fluctuations. When the total demand is greater than the whole capacity of owned trucks, the logistics managers may consider using outside carriers. Furthermore, integrating the choice of fulfillment-mode into transportation planning may bring significant cost savings to the company, because better solutions can be generated in an extended decision space. This extended problem is known as integrated operational freight carrier planning.

A customer request is assumed to be a pick-up and delivery request describing a single transportation demand, which typically results in a transportation process involving a less-than-truckload packet. The location of the pickup and the location of the delivery are specified as well as the quantities to be moved. Time windows for the loading and unloading operations are also declared. In case of relatively short distances, or in case of a small number of loads per truck, direct transportation is preferred to establishing expensive *hub-spoke* systems, involving inventories or at least reload locations. Therefore, the direct transport from locations of loading to locations of unloading is assumed.

A freight forwarding company generates its profit from the difference between the price that the customer is obliged to pay for the request execution and the costs of request fulfillment. These costs result either from the fulfillment by own transportation capacity, or from the external processing of orders in consequence of involving a subcontractor.

In case of self-fulfillment the execution must be planned and the costs can be optimized by routing and scheduling a fleet of homogenous vehicles with a given capacity in accordance with the general *pick-up-and-delivery-problem with time windows* (PDPTW). The distance and/or time costs are calculated for the round trips of all vehicles. The marginal costs of a single request execution are determined by the additional costs for the used vehicles for the execution of this request.

In contrast to self-fulfillment, the costs of subcontraction cannot be calculated independently, but depend on the shipment contract with the involved subcontractor. The models of integrated operational freight carrier planning proposed so far incorporate different types of subcontraction. Requests can be forwarded to the subcontractors independent of each other, on equal terms (Schmidt, 1994; Greb, 1998). Hence, the freight cost calculation results from isolated price assessment for each request on the basis of a freight cost function. It is also possible to forward complete tours, relative to the tours constructed for self-fulfillment (Savelsbergh and Sol, 1998; Stumpf, 1998). A traditional method of practical relevance for the subcontraction of less-than-truckload packets, called freight flow consolidation (FOP), can be also used (Kopfer, 1990, 1992; Pankratz, 2002). For flow consolidation a least cost flow through a given transportation network under the assumption of request bundling is sought. The costs are calculated in accordance with a tariff which depends on the distance and/or the loading weight. With regard to the last two methods of cost calculation for subcontractor involvement, the marginal costs of a single request execution refer to the additional transportation costs of the corresponding bundle.

3 Preliminaries for collaboration modelling

Collaboration is a powerful measure to improve the integrated operational freight carrier planning of cooperating partners. Bruner (1991) defines collaboration in the following way:

Collaboration is a process of reaching goals that cannot be achieved acting singly (or, at a minimum, cannot be reached efficiently). Collaboration includes all of the following elements: jointly developing and agreeing to a set of common goals and directions; sharing responsibility for obtaining those goals and working together to achieve those goals, using expertise of each collaborator.

For the purpose of formalized collaboration modelling, we introduce some aspects of cooperative game theory. In Operational Research Games, apart from inherent optimisation problems, there arises the natural question of how to allocate the joint cost/benefit among the individual decision-makers (Fernandez et al., 2004). Cooperative games address building coalitions as a crucial aspect. The general problem consists in the analysis of the benefits the players can achieve creating coalitions, in looking for winning coalition and for allocation of benefits which could be accepted by the players (Krus and Brunisz, 2000).

A *cooperative game with transferable utility* (*TU game*) is described (Slikker et al., 2005) by a pair (N,v), where $N=1,2,...,n$ denotes a set of players and $v : 2^N \longrightarrow \Re$ is the characteristic function, assigning to every coalition $S \subset N$ of players a value $v(S)$, representing the maximal total monetary reward the members of this group can obtain when they cooperate. Let v denote the payoff vector $v = (v_i)_{i \in N} \in \Re^n$, specifying for each player $i \in N$ the benefit v_i that this player can expect if he does not cooperate and x the payoff vector $x = (x_i)_{i \in N} \in \Re^n$, specifying for each player the benefit x_i that the player can expect if he cooperates with the other players (Hinojosa et al., 2005). An allocation is called efficient if the payoffs to the various players add up to exactly $v(N)$. $v_0(N)$ denotes the value of the characteristic function if there is no collaboration at all, i.e., $v_0(N) = \sum_{i=1}^{n} v_i$. $I(v)$ is defined as the set of individually rational allocations of the characteristic function v (Borm et al., 2001):

$$I(v) = \{x \in \Re^n \mid \sum_{i \in N} x_i = v(N),\ \forall i \in N : x_i \geq v_i\}.$$

The set $I(v)$ consists of all the payoff vectors with the conditions that the total reward of all players is equal to the monetary reward of the maximal coalition N and that the reward of each player is at least as high as it is without collaboration.

Two desirable properties of a game are superadditivity and monotonicity (Slikker et al., 2005). Superadditivity assures that for any two disjoint coalitions S and T of players $v(S) + v(T) \leq v(S \cup T)$. An important consequence of a superadditive characteristic function is that it is always attractive for two disjoint coalitions to form one big coalition rather than to operate separately. A game is monotonic if the addition of more players will increase the value obtainable, it means $v(S) \leq v(T),\ \forall S \subseteq T$.

As the players are not primarily interested in the benefits of a coalition, but in the individual benefits, the allocation of the additional profit is of main importance. An efficient allocation $x \in \Re^n$ with the property that $x_i \geq v_i$ for all $i \in N$ is individually rational, i.e., $x \in I(v)$. A coalitional game is convex if a player's marginal contribution increases if he joins a larger coalition: $v(S \cup i) - v(S) \leq v(T \cup i) - v(T)$, $\forall S \subseteq T$.

Next, we give a brief introduction to combinatorial auctions. Auctions characterise a general form of multilateral negotiations, where participants interact on the basis of bids (Peters, 2000). Due to complementarities or substitution effects between different assets, the bidders have preferences not just for particular items but for sets or bundles of items. For this reason, economic efficiency is enhanced if participants are allowed to bid on combinations of different assets. The most obvious problem that bids on combinations of items impose consists of selecting the set of winning bids. The problem is called the *Combinatorial Auction Problem* and can be formulated as an Integer Program (de Vries and Vohra, 2003):

Let N be a set of bidders, M a set of m distinct objects. For every subset S of M let $b^j(S)$ be the bid that auction participant $j \in N$ has announced he is willing to pay for S. For all $j \in N$ $b^j(S)$ is superadditive, which corresponds to the idea that the goods complement each other. Let $b(S) = max_{j \in N} b^j(S)$, $x_S : 2^M \rightarrow \{0, 1\}$ and $S_i = \{S \subset M \mid i \in S\}$. $x_S = 1$ is interpreted to mean that the highest bid on the set S is to be accepted, whereas $x_S = 0$ means that no bid on the set S is accepted. In order to determine an optimal set of winning bids we consider the following optimisation model

$$max \sum_{S \subset M} b(S) x_S \tag{1}$$

subject to

$$\sum_{t \in S_i} x_t \leq 1 \, \forall i \in M, \, \forall S \subset M \tag{2}$$

The constraint Eq. 2 ensures that no object in M is assigned to more than one bidder.

The seller is interested in choosing an auction design that will do three things (de Vries and Vohra, 2003):

1. induce bidders to create bids on the basis of their actual evaluations (incentive compatibility)
2. no bidder is worse off (in expectation) by participating in the auction
3. subject to the two above-mentioned conditions the seller maximizes the expected revenue.

Auction designs that satisfy these conditions are called optimal.

Gomber et al. (1997) distinguish between four types of auctions:

1. *English auction*: bids increase until only one bidder is willing to accept the price, he gets the offered good at the price of the last quote.
2. *Dutch auction*: the price decreases until the first bidder accepts the price, he gets the offered good at the current price.
3. *First price sealed bid auction*: the bidders offer their price separately and then the best offer is chosen and the corresponding participant gets the good at the offered price.
4. *Second price sealed bid auction* (*Vickerey auction*): it corresponds to the *First price sealed bid auction*, but the bidder with the best offer gets the good at the price of the second best offer.

4 Existing collaboration models in transport logistics

Kopfer and Pankratz (1999) define a *groupage system* as a logistic interorganisational system which exchanges information and manages capacity balancing by using the cooperation between several independent carriers. Groupage systems enable a request interchange between several forwarding companies to achieve an equilibrium between demanded and available transport resources. The increased number of disposable requests for each individual freight forwarder results in economies of scale. Economies of scope are created due to better capacity utilisation. An additional advantage results from the considerably lower costs of arrangement as in case of external processing of orders. A quasi-merger of freight forwarders to a super-carrier with a central managing entity is not of practical relevance, thus, the decentralization of the collaboration process is recommended (Kopfer and Pankratz, 1999). At the first stage each freight forwarder plans the requests by incorporating self-fulfillment or subcontraction. Only now is the exchange of requests among collaborating forwarders possible.

A model for freight carriers' collaboration was proposed by (Schönberger, 2005). Requests are negotiated among freight forwarding entities. In case of fulfillment of a request by a collaborating forwarder and according to the approach of Schönberger, the entire corresponding revenues are simultaneously shifted to the serving collaboration participant. For the requests that remain unserved an external carrier service is engaged, i.e., the requests are subcontracted at the spot market. The main assumption of Schönberger is that the carrier service incorporation on the spot market is unprofitable, because the charge for each request is higher than the revenues associated with the request. The sum of the external carrier costs is distributed uniformly among the participants of the cooperation. Within the usage of a memetic algorithm, which combines the exploring genetic search and exploiting local search procedures (hill climbers)(Schönberger, 2005), it is proved that the cooperation is able to incorporate significantly more requests, contributing to an increase in the overall profit. The model does not fully support the assumption that each participant should benefit from this cooperation by enlarging its efficiency. Instead, as the preservation of the interests of certain carriers cannot be guaranteed, a 2-step approach is suggested. First, each forwarding entity selects the requests from their own portfolio as well as from portfolios of the other participants, leading to maximal profit contributions. Typically, a single request cannot be served in a profitable way. For this reason, the carrier composes several requests into routes in order to achieve positive profit contributions. The carriers do not only specify single requests but bundles of requests that they can serve in a profitable way. Such a bundle consists of the requests served within one route (Schönberger, 2005). Thus, not the single requests but the subsets resulting from bundling of requests are subject to negotiation. The desired subsets of each forwarder are released. Usually, the most attractive requests are contained in a few subsets. As only one of conflicting subsets can be executed, an independent mediator is introduced. Bundle assignment by the mediator is based on the principles of combinatorial auction. The decision is made with the goal of minimizing the negative sum of avoided carrier costs. The subset of one freight forwarder is accepted and all the other subsets including the request are turned down.

Gomber et al. 1997 present a model of collaboration for transport planning suitable for a freight forwarder agency with several Profit Centres. Profit Centres should be autonomous in request acquisition and negotiations of the price for the request execution with customers. Profit Centres can either fulfill requests with their own vehicle fleet or forward it to the other Profit Centres on the basis of a cooperation structure. The coordination mechanisms for collaboration should meet the following conditions (Gomber et al. 1997):

1. an efficient allocation of requests among Profit Centres
2. no strategic planning, i.e., for each Profit Centre it is profitable to announce the true assessments
3. the requests generating losses should also be dispatched optimally
4. the costs of communication should be acceptable.

In (Gomber et al., 1997) several models for collaboration based on the multi-agent-auction-theory are proposed. The types of cooperation models vary depending on the features of the requests. If the single request forwarding is concerned, the *Vickerey auction* is proposed as the dominant strategy. In order to maximize the probability of getting the request, each participant quotes the maximal price for the request, yet providing profit. In case that a request generates losses, it is assumed that the participants can offer negative bids. *Vickerey auction* functions for negative prices in the same manner as for the positive prices. The bidder is paid for the acceptance of the request the amount of the second "best" bidder price, hence, generating profit. The payment comes from the offering participant who has acquired the request. The mechanism of combinatorial auction, called *Matrix auction*, is proposed for bundles of requests. In principle, it is also based on *Vickerey auction*. Each of the m participants offer the (positive or negative) prices for all $2^n - 1$ combinations of n requests. In order to find the optimal allocation of the requests, a matrix with $2^n - 1$ columns and m rows is constructed. Only one matrix-element can be chosen from each column. Referring to rows, the chosen bundles cannot contain common requests.

5 Proposed model of collaboration

5.1 Description of the collaboration process

Now the profit optimisation and profit sharing of the collaboration among several freight forwarding entities is considered. Each entity operates autonomously. It can quote the price for request execution and decide the method of request fulfillment independently, i.e., each request can be executed by self-fulfillment or by subcontraction. With regard to each request, irrespective of the mode of fulfillment, profit or loss can be generated. It results from the difference between the freight charge received from the customer and the costs of request execution. These costs correspond to the additional travel costs of the vehicle used in case of self-fulfillment, or to the payment for subcontracting. Furthermore, it is assumed that each entity is able to fulfill all the acquired requests within the usage of own disposable resources: own vehicle fleet or subcontractors.

Each freight forwarding entity defines that subset of requests from the self-acquired requests that it does not want to offer to collaborating partners. Those requests are fulfilled within the usage of the own disposable resources: they are planned in the schedule of the own vehicle fleet or forwarded to subcontractors while minimizing the resulting freight costs. All the other requests are included in the collaboration process.

In the collaboration process requests are interchanged among the cooperating freight forwarders. The costs of communication among partners are not considered. Furthermore, it is assumed that each collaboration participant announces their true assessments. There exist incentives for the partners to reveal their true assessments. On one hand, the collaborating entity aims to receive the bundle it is interested in. In order to remain competitive, it quotes the minimal possible costs of bundle execution. On the other hand, it wants to generate profit (or, more precisely, not to generate losses). Thus, the real costs are revealed. In practice, the partners are often interconnected to each other by the formalised market structures, e.g. the partners represent the Profit Centres of one company or holding. In this case, the access to the real costs and profit of the partner is seldom denied.

The collaboration process consists of three phases: preprocessing, profit optimisation and profit sharing.

In the preprocessing phase each partner specifies the lowest costs of fulfillment for each acquired request that they offer to the collaboration partners. These costs are assessed for request execution within the usage of own disposable resources, without participating in the collaboration. It means, the costs of subcontracting and, if it is possible, the costs of self-fulfillment are calculated and the lower amount is chosen. This amount is called potential self-fulfillment costs of the request.

In line with the definition, the main assumption for the collaboration of the freight forwarders is that requests acquired by one partner are allowed to be fulfilled by another cooperating partner if the collective revenues increase. In the profit optimisation phase it is aimed to generate a mapping of requests to collaborating partners. This mapping represents the assignment of requests to the available partners, such that the profit of the entire coalition is maximized. Hence, as the price paid by the customers remains constant, the minimal execution costs for the fulfillment of the offered requests are claimed.

No collaborating participant, except the acquiring enterprise, has to serve requests that it does not want to fulfill. Partners who intend to take over some requests bid on these requests or on a set of requests. Thus, each partner defines bundles of requests it would be able to and wishes to fulfill. For all desired bundles the enterprise evaluates its costs for the fulfillment of the bundle of requests. These costs are called the potential fulfillment costs. Moreover, the potential fulfillment costs have to be specified for each request included in the desired bundles of a particular enterprise as if it were assigned to it separately. Hence, the potential fulfillment costs should be obligatorily specified for all the one-element-bundles of requests for those requests which belong to many-element-bundles considered by a particular collaboration participant. Furthermore, the potential self-fulfillment costs of each request are regarded as a bid on a one-element-bundle that is offered by the acquiring partner itself. The assessments are then revealed and are subject to an optimisation process.

The set of bundles that assures the lowest serving costs for the entire set of requests offered by collaborating partners is determined by solving the Integer Program of the Combinatorial Auction Problem (models 1–2). This set of bundles

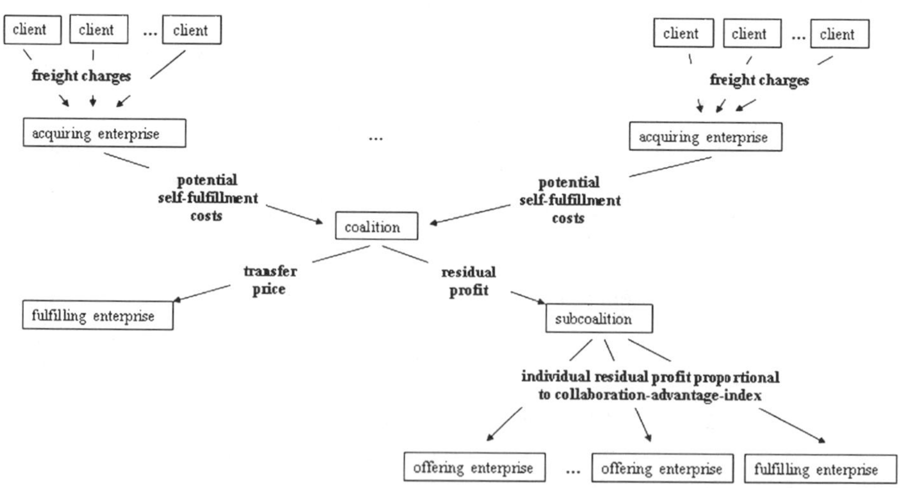

Fig. 1 Payment flows for a single request bundle

assigns all the requests offered by collaborating participants uniquely to one of the bundles. Provided that a one-element-bundle constructed on the basis of potential self-fulfillment costs is included in this set, the request is executed by the offering entity itself. Otherwise, the requests are shifted between partners for execution.

Cooperation is favourable, because the maximal joint profit is always at least as high as the sum of the profits of the players separately. Now the question arises how to allocate the joint benefit among the individual partners in a fair way. The definition of collaboration determines that all freight forwarders should reach at least such a profit as in the case without collaboration, otherwise they should be compensated. Thus, the incentives for each enterprise to participate in collaboration are that they can make additional profit as well as the certainty that their profit in case of operating autonomously is not higher (alternatively loss is not lower) than the one resulting from the collaboration process. In the profit sharing phase the profit resulting from fulfillment of each request is divided among the coalition members. Figure 1 shows the flow of payments for one bundle of requests.

The offering partner holds the payment of the customer freight charge as the reward for request acquisition. Instead, if the request is shifted to another enterprise, the offering partner pays for the request execution the amount of the potential self-fulfillment costs to the coalition. Thus, its financial situation is not worsened in comparison with the situation without collaboration. The amount of profit or loss is maintained.

The transfer price is the payment that the serving enterprise receives from the coalition for bundle fulfillment. In order to set this price, the minimal fulfillment costs for each single request in the bundle are determined. For each request this corresponds to the lowest potential fulfillment costs that have been specified by any partner for the one-element-bundle that contains the considered request. The fulfilling enterprise is awarded the sum of the minimal fulfillment costs for all the requests included in the bundle it should execute. As for the fulfilling entity, the costs for the execution of that bundle can only be equal to or lower than the sum of the minimal fulfillment costs, the participation in the collaboration can exclusively be profitable for the fulfilling entity. The total profit amounts to the difference between the payment the customers offer to the acquiring enterprises and the payment for the fulfilling of bundles by the serving enterprise. The overall residual profit that has not yet been absorbed by the offering and serving partners should be divided among the partners. For one bundle the residual profit consists of the difference between the potential self-fulfillment costs of the requests in the bundle and the transfer price of the bundle. The division corresponds to the benefit that each participant offers to the collaboration and its calculation is based on collaboration advantage indexes. For offering partners the part of the residual profit they receive is calculated for each request they have offered and depends on the benefit of exchanging this request. The collaboration-advantage-index for the offering entity amounts to the difference between the potential self-fulfillment costs and the minimal fulfillment costs. For serving partners their part of the residual profit is determined for the bundle they serve and it depends on the cost reduction that can be achieved by bundling the proper requests. The collaboration-advantage-index for the fulfilling enterprise is equal to the difference between the sum of all potential self-fulfillment costs for the requests in the bundle and the transfer price paid to the serving partner. The residual profit of each bundle is divided among offering and fulfilling coalition members proportional to the collaboration-advantage-indexes.

The formal model of the collaboration process and the proof of satisfying the main assumptions of the collaboration are presented in the subsequent section.

5.2 Formal statement of the collaboration process

Assume a coalition of m independent freight forwarders $P = \{P_1, ..., P_m\}$. Each partner P_k has acquired the set of N_k requests $R^k = \{r_1^k, ..., r_{N_k}^k\}$. First, each participant defines the maximal obtainable profit while only using own disposable resources. Let $F(r_i^k)$ be the freight charge paid by the customer to the acquiring enterprise for the fulfillment of the request r_i^k . The request portfolio of each freight forwarder $R^k = (R_v^{k+}, R_{sc}^{k+}, R^{k-})$ is partitioned into three disjoint sets R_v^{k+} , R_{sc}^{k+} and R^{k-} . Requests from the set R_v^{k+} are executed by the own vehicle fleet. The set R_v^{k+} is dispatched according to the routing plan denoted as $\pi(R_v^{k+})$. Requests from the set R_{sc}^{k+} are forwarded to a subcontractor. The costs of execution of the vehicle scheduling plan refer to $C(\pi(R_v^{k+}))$ and the execution costs of all requests shifted to a subcontractor amount to $C(R_{sc}^{k+})$. R^{k+} denotes $R_v^{k+} \cup R_{sc}^{k+}$ and contains all requests that the freight forwarder does not want to offer to other coalition members. R^{k-} incorporates all the requests that are offered to the collaboration partners.

Preprocessing phase For each request r_i^k from the set R^{k-} the enterprise P_k defines the potential self-fulfillment costs $C(r_i^k)$ as the minimal costs of execution by the usage of own disposable resources (self-fulfillment or subcontraction). The potential profit/loss PR_i^k resulting from the execution of a single request $r_i^k \in R^{k-}$ without collaboration would amount to

$$PR_i^k = F(r_i^k) - C(r_i^k) \tag{3}$$

Hence, the set of requests R^{k-} of the single non-collaborating freight forwarding entity P_k generates the profit of

$$PR^k = F(R^{k-}) - C(R^{k-}) \tag{4}$$

with $F(R^{k-}) = \sum_{r_i^k \in R^{k-}} F(r_i^k)$ and $C(R^{k-}) = \sum_{r_i^k \in R^{k-}} C(r_i^k)$. The overall profit for all members of the coalition P without collaboration refers to the value v_0 of the characteristic function v

$$v_0(P) = \sum_{k=1}^{m} (PR^k) \tag{5}$$

Coalition profit optimisation phase In the profit optimisation phase all the enterprises offer the requests from their sets R^{k-} to the coalition. The requests are then subject to a transfer process between the coalition members, which causes an updating of the request portfolio of each partner in the coalition. Let the set R_i^{kj} , $k \neq j$ denote the transfer of r_i^k from P_k to P_j , i.e., $R_i^{kj} = \{r_i^k\}$ if r_i^k is transferred from P_k to P_j and $R_i^{kj} = \oslash$ else. The updated portfolio of requests $\overline{R}^k$ for P_k should refer to (Schönberger, 2005)

$$\overline{R}^k = R^{k+} \cup (\bigcup_{i=1}^{N_j} \bigcup_{\substack{j=1 \\ j \neq k}}^{m} R_i^{jk}) \setminus (\bigcup_{i=1}^{N_k} \bigcup_{\substack{j=1 \\ j \neq k}}^{m} R_i^{kj}) \tag{6}$$

It is assumed that only disposable requests are transferred: $R_i^{kj} \subseteq R^{k-}$ and that the transfer is unique: $R_i^{kj} \cap R_i^{kl} = \oslash$, $\forall j \neq l$.

Let T be the set of all requests offered to the coalition. The total number of requests involved into the collaboration process amounts to $|T| = \sum_{k=1}^{m} |R^{k-}|$. Let $B_L, L \in \{1, ..., 2^{|T|} - 1\}$, be a bundle of offered requests. The set of all possible bundles is denoted as B . For each bundle B_L the parameter $x_L(r)$ is defined, such that:

$$x_L(r) = \begin{cases} 1 & \textit{if bundle } B_L \textit{ contains request } r \\ 0 & \textit{else} \end{cases} \tag{7}$$

The set of all possible bundles illustrates a pure academic approach. In practice it is impossible to enumerate all bundles, because for a realistic number of offered requests, e.g. 100, there exists an astronomic number of $2^{100} = 1,27 * 10^{30}$ bundles. Therefore, to simplify the combinatorial complexity, only some bundles are specified by the participants.

Each partner P_k defines its potential fulfillment costs $C_k(B_L)$ for each bundle B_L of requests he wants to fulfill and for all one-element-bundles of requests included in the many-element bundles he has defined. For bundles that P_k does not want to fulfill $+\infty$ is assigned to $C_k(B_L)$. All potential self-fulfillment costs $C(r_i^k)$ are regarded as potential fulfillment costs $C_k(\{r_i^k\})$ for one-element-bundles $\{r_i^k\}$ offered by P_k .

A modified *Matrix auction* based on a *first price sealed bid auction* is used to identify the most profitable bundle combination for the coalition and to assign the bundles to coalition partners. Assume the binary variable

$$y_k(B_L) = \begin{cases} 1 & \textit{if bundle } B_L \textit{ is selected to be executed by } P_k \\ 0 & \textit{else} \end{cases} \tag{8}$$

Let $\overline{B}$ be the set of optimal request bundles. Then

$$C(\overline{B}) = \min(\sum_{k=1}^{m} \sum_{B_L \in B} C_k(B_L) * y_k(B_L)) \tag{9}$$

s.t.

$$\sum_{k=1}^{m} \sum_{B_L \in B} x_L(r_i^j) * y_k(B_L) = 1, \quad \forall r_i^j \in T \tag{10}$$

are the minimal total costs the coalition can obtain using the collaboration process. Hence, in accordance with the *Matrix auction*, such a set of request bundles $\overline{B}$ is found that each request is assigned to exactly one partner for execution. This is guaranteed by constraint Eq. 10. As the prequoted payment from the customer for each request is constant and cannot be influenced, the minimization of the potential fulfillment costs for the entire coalition *P*, which is targeted in Eq. 9, concurrently guarantees profit maximization for the coalition. Thus, for the characteristic function *v* of the TU game between the collaborating partners

$$v(P) = \sum_{r_i^k \in T} F(r_i^k) - C(\overline{B}) \tag{11}$$

represents the maximal total monetary reward the members of the coalition can obtain when they cooperate. In particular $v_0(P) \leq v(P)$. $v_0(P) = v(P)$ corresponds to the situation when each coalition member should execute all his acquired requests on his own. A transfer of requests is reasonable only if it improves the total profit of the coalition.

Profit sharing phase In the profit sharing phase it must be assured that the generated solution is acceptable for the partners. Superadditivity is one main prerequisite to guarantee that in the collaboration process no worsening of the financial situation for any participant takes place. The overall new profit NPR^k is the profit that the partner P_k achieves by means of the collaboration. Let NPR_i^{k-} denote the new profit for P_k resulting from offering the request $r_i^k \in T$ to the coalition. NPR_L^{k+} denotes the new profit of P_k for the fulfillment of bundle B_L in result of collaboration. Individually rational allocations of *v*(*P*) are defined as:

$$\begin{aligned} I(v) = \{(NPR^k), k = 1, \ldots, m \mid \\ \sum_{k=1}^{m} NPR^k = v(P), (a) \\ NPR^k \geq PR^k, \forall P_k \in P\} (b) \end{aligned} \tag{12}$$

Assume that $R_i^{kj} \neq \oslash,\ k \neq j$. Each offering enterprise P_k holds the payment from its customer. If it forwards the request to the coalition, it pays the self-defined potential self-fulfillment costs for the request execution and gets additionally some part of the residual profit (RPR_L^k). Hence, the profit increases, respectively, loss decreases for the offering entity, $NPR_i^{k-} = F(r_i^k) - C(r_i^k) + RPR_L^k$, i.e., no worsening of its situation is guaranteed: $NPR_i^{k-} \geq PR_i^k, \forall r_i^k \in R^{k-}$.

Next, the payment received by the enterprise P_k for the fulfillment of bundle B_L, called transfer price TP_L^k , is determined. In order to define the transfer price the Matrix auction based on the *first price sealed bid auction* is performed, but now only one-element-bundles, that include only single requests $B_L^* \in B$ are subject to consideration. The solution of the models 13–14, that assures the minimal fulfillment costs of each request r_i^j, should be found.

$$min \sum_{k=1}^{m} \sum_{B_L^* \in B} C_k(B_L^*) * y_k(B_L^*) \tag{13}$$

s.t.

$$\sum_{k=1}^{m} \sum_{B_L^* \in B} x_L(r_i^j) * y_k(B_L^*) = 1, \ \ \forall r_i^j \in T \tag{14}$$

The minimal fulfillment costs of each one-element-bundle B_L^* can easily be determined by

$$C^*(r_i^j) = \min_{\{k=1,\dots,m\}} C_k(B_L^*) \tag{15}$$

The minimal fulfillment costs $C^*(r_i^j)$ of a particular request correspond to the potential self-fulfillment costs $C(r_i^k)$ of the offering enterprise, if no other coalition member is able to execute this single request at a lower price than the offering partner.

Bundles specified in $\overline{B}$ can include requests offered by different participants. Assume that the bundle B_L consists of L_n requests offered by L_m different participants. One bidder P_k is chosen to serve the bundle. P_k is granted the transfer price of

$$TP_L^k = \sum_{r_i^j \in B_L} C^*(r_i^j) \tag{16}$$

for bundle fulfillment.

The models 13–14 conforms to the models 9–10 with the only exception that models 13–14 are limited to one-element-bundles. In models 9–10 a bundle B_L is assigned to a coalition partner P_k for fulfillment only if its potential fulfillment costs $C_k(B_L)$ are not higher than the sum of minimal fulfillment costs of all one-element-bundles belonging to the assigned bundle B_L. Thus, all the bundles $B_L \in \overline{B}$ satisfy the assumption Eq. 17.

$$C_k(B_L) \leq TP_L^k \tag{17}$$

The new profit for P_k for the fulfillment of B_L amounts to

$$NPR_L^{k+} = TP_L^k - C_k(B_L) \tag{18}$$

which is always positive. Hence, collaboration cannot be unfavourable for any fulfilling enterprise.

The residual overall profit of the entire coalition amounts to

$$RPR = \sum_{B_L \in \overline{B}} \sum_{r_i^j \in B_L} [C(r_i^j) - C^*(r_i^j)] \tag{19}$$

For each bundle $B_L \in \overline{B}$ assume the subcoalition P_L that consists of coalition members offering requests included in the bundle and the coalition member executing this bundle. The residual profit RPR_L, resulting from the collaborative fulfillment of the bundle B_L amounts to

$$RPR_L = \sum_{r_i^j \in B_L} [C(r_i^j) - C^*(r_i^j)] \tag{20}$$

RPR_L is divided among members of the subcoalition P_L. The collaboration-advantage-index CAI_k is calculated for each $P_k \in P_L$ in the following way.

If P_k offers requests to bundle B_L, then its collaboration-advantage-index is defined as the sum of the differences between the potential self-fulfillment costs and minimal execution costs for all requests offered by P_k:

$$CAI_k^- = \sum_{r_i^k \in B_L} [C(r_i^k) - C^*(r_i^k)] \tag{21}$$

The collaboration-advantage-index for the fulfilling entity P_k is defined as the difference between the sum of all potential self-fulfillment costs of the requests in the bundle and the transfer price:

$$CAI_k^+ = \sum_{r_i^j \in B_L} C(r_i^j) - TP_L^k \tag{22}$$

Each subcoalition member $P_k \in P_L$ that participates in the collaborative execution of the bundle B_L holds the individual residual profit that refers to

$$RPR_L^k = \frac{CAI_k * RPR_L}{\sum_{j=1}^{|P_L|} CAI_j} \tag{23}$$

$CAI_k \geq 0,\ \forall P_k \in P_L$. Hence, the individual residual profit $RPR_L^k \geq 0,\ \forall P_k \in P_L$. $NPR^k = \sum_{B_L \in B} RPR_L^k + \sum_{r_i^k \in R^{k-}} NPR_i^{k-} + \sum_{B_L \in B} NPR_L^{k+} \geq PK^k,\quad \forall P_k \in P$ and assumption (12b) is completed for each coalition member. The entire profit of the coalition, $v(P)$, is divided among collaboration partners, satisfying assumption (12a). The assumption (12) is maintained, all the partners have incentives to participate in the coalition.

5.3 Example

Asume a coalition of three freight forwarding entities. In the preprocessing phase the freight forwarders specify the potential self-fulfillment costs. The following requests are offered to the collaboration participants:

P_1 offers portfolio $R^{1-} = \{R_1^1(F = 20, C = 30), R_2^1(F = 30, C = 15)\}$
P_2 offers portfolio $R^{2-} = \{R_1^2(F = 27, C = 22)\}$
P_3 offers portfolio $R^{3-} = \{R_1^3(F = 22, C = 20), R_2^3(F = 17, C = 16)\}$

The request R_1^1 generates losses, while all the other requests are profitable for the acquiring freight forwarders. The overall costs of the coalition partners without collaboration amount to 103 monetary units. The profit from the request execution without collaboration is equal to 13 units.

In the profit optimisation phase the freight forwarders specify the potential fulfillment costs for request execution of the bundles they are interested in. They specify also the potential fulfillment costs for the particular requests from the bundles they would like to serve. The costs of $+\infty$ are assigned to all the other combinations. Table 1 presents the specifications of the example.

Next, the optimal combination for the entire coalition is found on the basis of the *Matrix auction*. Optimal bundles are:

$$\overline{B_1} = \{R_1^1\} \rightarrow P_1$$

$$\overline{B_2} = \{R_2^1, R_1^2, R_1^3\} \rightarrow P_3$$

$$\overline{B_3} = \{R_2^3\} \rightarrow P_1$$

The costs of request execution in case of collaboration amount to 99 monetary units. The total additional profit from the cooperation is then equal to four monetary units.

In the profit sharing phase this profit should be divided among the cooperating freight forwarders. First, the minimal fulfillment costs of each request from one-element-bundles are specified:

$C_1^*(R_1^1) = 30$, $C_1^*(R_2^1) = 15$, $C_1^*(R_1^2) = 20$, $C_2^*(R_1^3) = 20$, $C_1^*(R_2^3) = 15$

The transfer prices for such bundle execution are as follows:

$TP_1^1 = 30$, $TP_2^3 = 15 + 20 + 20 = 55$, $TP_3^1 = 15$

The profit for the fulfilling freight forwarder amounts to:

$NPR_1^{1+} = 0$, $NPR_2^{3+} = 55 - 54 = 1$, $NPR_3^{1+} = 0$

Table 1 Potential (self-)fulfillment costs

bundle	P_1	P_2	P_3
$\{R_1^1\}$	30	33	$+\infty$
$\{R_2^1\}$	15	$+\infty$	25
$\{R_1^2\}$	20	22	21
$\{R_1^3\}$	20	20	20
$\{R_2^3\}$	15	$+\infty$	16
$\{R_1^1,R_1^3\}$	$+\infty$	52	$+\infty$
$\{R_1^2,R_2^3\}$	35	$+\infty$	$+\infty$
$\{R_1^3,R_2^3\}$	$+\infty$	48	$+\infty$
$\{R_2^1,R_1^2,R_1^3\}$	$+\infty$	$+\infty$	54
$\{R_1^2,R_1^3,R_2^3\}$	58	$+\infty$	$+\infty$
$\{R_2^1,R_1^2,R_1^3,R_2^3\}$	$+\infty$	$+\infty$	70

The overall residual profit of the coalition is equal to three monetary units. It is split among the bundles as follows:

$RPR_1 = 0$, $RPR_2 = 2$, $RPR_3 = 1$

Next, the specification how it is divided among the freight forwarders takes place.

The collaboration-advantage-index for $\overline{B_2}$ is equal to:

$CAI_1^- = 0$, $CAI_2^- = 15 + 22 + 20 - 15 - 20 - 20 = 2$, $CAI_3^- = 0$, $CAI_3^+ = 2$

Then, each participating coalition member holds the singular residual profit that refers to:

$RPR_2^1 = 0$, $RPR_2^2 = \frac{2*2}{2+2} = 1$, $RPR_2^3 = 1$

In case of $\overline{B_3}$ the residual profit is shared as follows:

$CAI_1^- = 1$, $CAI_3^+ = 1$, $RPR_3^1 = \frac{1}{2}$, $RPR_3^3 = \frac{1}{2}$

Concluding, the overall profit from the collaboration is shared among the participants: P_1 , P_2 and P_3 are awarded $\frac{1}{2}$ monetary unit, one monetary unit and $2\frac{1}{2}$ monetary units.

The profit from the request execution has risen to 17 monetary units. No freight forwarder has generated a loss in result of collaboration, the sum of generated profit/loss is either maintained, or the financial situation improves.

6 Conclusions and future work

The collaborative freight carrier planning is of high practical importance in the modern transportation branch. However, there hardly exist any theoretical frames for the market actors in the literature. As far as we are aware there is no approach in the literature for the collaboration of freight forwarders including the choice of fulfillment mode for each forwarder and the exchange of orders among independent cooperating partners. The model we propose is based on the combinatorial auction theory as well as on the operations research game theory. Its main strength is that each participant generates no losses in consequence of the collaboration and has a realistic chance to increase its profit by participating in the coalition. The collaboration-advantage-indexes have been chosen in a way that all participating coalition members can expect positive payoff-vectors. Therefore each partner has strong incentives to join and to maintain the coalition they belong to.

The presented collaboration model forms the theoretical frame for request exchange, profit optimisation and profit sharing for a coalition of freight forwarding entities. It is assumed that the market forces of all the coalition members are equal or strongly similar. Therefore, in order to receive empirical results, it would be recommendable to apply the cooperation mechanism to a forwarding company with several autonomous nearly similar Profit Centers. In a practical case study of cooperating Profit Centres we will analyse and investigate whether the collaboration profit resulting from such a mechanism is high enough to create an incentive for establishing a coalition. Secondly, the question arises, whether the potential self-fulfillment costs are easy to assess for the offering partners and whether the other Profit Centres are willing to execute the requests at lower costs than the subcontractors from the spot market.

In additional future work the model could be adapted for collaboration scenarios where not all partners have similar potential on the market. In general, the residual profit can be divided among the partners on the basis of different mechanisms, the proposed collaboration-advantage-indexes can be adapted to different situations. Especially in the case that the requests offered to the coalition are most unfavourable for all the partners, it could be possible to increase the reward for the fulfilling partner while decreasing that of the offering partners. If transaction costs should be taken into account, some part of the reward should be transferred to the coalition itself. Anyhow, the proposed model is a useful basis for developing application-specific profit-sharing mechanisms.

Acknowledgments This work was supported by the Deutsche Forschungsgemeinschaft (DFG) within the SFB637 and by Bremen Innovationsagentur (BIA) sponsored from the EU-fond (project number 2FUE0342B).

References

Borm P, Hamers H, Hendrickx R (2001) Operations research games: a survey. Sociedad de Estadistica e Investigacion Operativa TOP 9(2):139–216

Bruner C (1991) Thinking collaboratively: ten questions and answers to help policy makers improve children's services, education and human services. Consortium, Washington District of Columbia

Chu C-W (2005) A heuristic algorithm for the truckload and less-than-truckload problem. Eur J Oper Res 165(3):657–667

Fernandez F, Hinojosa M, Puerto J (2004) Multi-criteria minimum cost spanning tree games. Eur J Oper Res 158(2):399–408

Gomber P, Schmidt C, Weinhardt C (1997) Elektronische Märkte für die dezentrale Transportplanung. Wirtschaftsinformatik 39(2):137–145

Greb T (1998) Interaktive Tourenplanung mit tabu search. Dissertation, Universität Bremen

Hinojosa MA, Marmol AM, Thomas LC, (2005) Core, least core and nucleolus for multiple scenario cooperative games. Eur J Oper Res 164(1)225–238

Kopfer H (1990) Der Entwurf und die Realisierung eines A*-Verfahrens zur Lösung des Frachtoptimierungsproblems. OR Spektrum 12(4):207–218

Kopfer H (1992) Konzepte genetischer Algorithmen und ihre Anwendung auf das Frachtoptimierungsproblem im gewerblichen Güterfernverkehr. OR Spectrum 14(3):137–147

Kopfer H, Pankratz G (1999) Das Groupage-Problem kooperierender Verkehrsträger. Oper Res Proc 1998:453–462, Springer, Berlin Heidelberg New York

Krus L, Bronisz P (2000) Cooperative game solution concepts to a cost allocation problem. Eur J Oper Res 122(2):258–271

Pankratz G (2002) Speditionelle Transportdisposition. Deutscher Universitätsverlag, Wiesbaden

Peters R (2000) Elektronische Märkte und automatisierte Verhandlungen. Wirtschaftsinformatik 42(5):413–421

Savelsbergh M, Sol M (1998) DRIVE: dynamic routing of independent vehicles. Oper Res 46 (4):474–490

Schmidt J (1994) Die Fahrzeugeinsatzplanung im gewerblichen Güterfernverkehr—ein graphentheoretischer Ansatz zur Planung von Fahrzeugumläufen. Verlag Peter Lang, Frankfurt am Main

Schönberger J (2005) Operational Freight Carrier Planning, Springer, Berlin Heidelberg New York

Slikker M, Fransoo J, Wouters M (2005) Cooperation between multiple news-vendors with transshipments. Eur J Oper Res 167(2):370–380

Stumpf P (1998) Tourenplanung im speditionellen Güterfremdverkehr, Schriftstelle der Gesellschaft für Verkehrsbetriebswirtschaft und Logistik (GVB) e.V. Band 39, Nürnberg

de Vries S, Vohra R (2003) Combinatorial auctions: a survey. INFORMS Journal of Computing 15(3):284–309

Printing: Krips bv, Meppel
Binding: Stürtz, Würzburg